Surface Treatment c
for Adhesi

Surface Treatment of Materials for Adhesive Bonding

Second Edition

Sina Ebnesajjad, PhD
President, FluoroConsultants Group, LLC

assisted by

Cyrus F. Ebnesajjad

ELSEVIER

AMSTERDAM • BOSTON • HEIDELBERG • LONDON
NEW YORK • OXFORD • PARIS • SAN DIEGO
SAN FRANCISCO • SINGAPORE • SYDNEY • TOKYO
William Andrew is an imprint of Elsevier

William Andrew is an imprint of Elsevier
The Boulevard, Langford Lane, Kidlington, Oxford, OX5 1GB
225 Wyman Street, Waltham, MA 02451, USA

First published 2006
Second edition 2014

British Library Cataloguing-in-Publication Data
A catalogue record for this book is available from the British Library

Library of Congress Cataloging-in-Publication Data
A catalog record for this book is available from the Library of Congress

ISBN: 978-0-323-26435-8

For information on all William Andrew publications
visit our website at **store.elsevier.com**

Printed and bound by CPI Group (UK) Ltd, Croydon, CR0 4YY

Transferred to digital print 2012

Contents

Preface

The first edition (2006) of this book was received well by the practitioners of adhesion, thus prompting a second edition. The second edition is an expansion of the first book. Three new chapters have been added to describe the effect of additives on surface properties and adhesion, surface treatment of ceramics, and adhesion promoters.

Chapter 6 of the first edition describes, dealing with surface treatment methods and techniques. It has been divided into two chapters to focus and devote more space to plasma treatment methods. Both plasma treatment techniques under vacuum and at atmospheric pressure have been covered. These techniques are gaining popularity because they are "dry." The term dry means plasma treatment does not require wet chemicals, which often generate toxic chemical waste that needs disposal. The necessity of reducing the environmental impact of surface treatment operations has generated a great deal of research and development in the area of plasma technologies. Even though plasma treatment techniques require small amounts of gases, they are considered environmentally safe and friendly.

Chapter 10, dealing with the effect of additives on surface treatment of plastics, delves into the impact of additives on surface tension, and the fact that adherability of a material can be positive, neutral, or negative. Some additives actually enhance the bondability of a surface, some have little or no effect, and finally some including silicones and fluoropolymers reduce the ability of a surface to form a bond using an adhesive.

Chapter 11, dealing with surface treatment of ceramics, focuses on the surface treatment of this important group of materials. Ceramics have innumerable applications in industry, dental restorations, consumer products, and art. In addition to ceramics which occur naturally, many new ceramics have been (and continue to be) developed for special applications. By and large, developments in dental restoration have led to the creation of the largest number of ceramics.

Chapter 12, dealing with adhesion promoters, authored by Dr. Peter Pape, sheds light on the use of silanes and other adhesion promoters. Adhesion promoters, or coupling agents, are chemicals that act at the interface between an organic polymer and an inorganic substrate to enhance adhesion between the two materials.

Some readers have commented about the space devoted to the discussion of fluoropolymers. The answer is that fluoropolymers are simply the most difficult plastics to bond. Traditionally, sodium etching has been used to render the surface of prefluoropolymers such as polytetrafluoroethylene bondable. These plastics represent one extreme type of material because of their unique low surface energy (<40 dynes/cm) and nonwetting characteristics. The other end of the spectrum of materials include metal oxides and metals which have high surface energy (>100 dynes/cm). In spite of the relatively small volume of fluoropolymer consumption, as compared to say polyolefins, they play extremely important roles in human life. Applications include medical devices such as pacemakers, automotive applications, fuel and hydraulic hoses in aerospace, chemical processing, semiconductor fabrication, pharmaceuticals, biopharmaceuticals, and food and beverages.

The authors hope that the second edition of the book is helpful to those who deal with surface treatment of materials and adhesion bonding.

None of the views or information presented in this book reflect the opinion of any of the companies or individuals that have contributed to the book. If there are errors, they are oversights on the part of the authors. A note indicating the specific error to the publisher, for the purpose of correction of future editions, would be most appreciated.

Sina Ebnesajjad
Cyrus F. Ebnesajjad
2013

Preface to First Edition

A basic industrial process is the bonding of alike and different materials to each other. Adhesion using an *adhesive* has proven to be an effective means of material bonding. This book describes treatment methods that must be applied to a material surface before successful adhesive bonding is possible. There are numerous examples ranging from wallpaper in a house and paint on surfaces, to parts used in the construction of aircraft.

The aim of this text is to explain, in a simple yet complete manner, all that is required to achieve successful adhesion bonding of different materials. Fundamental material considerations have been given priority to enable the use of the contents of this book in different industries. The book is both a reference and a source for learning the basics, and is useful for all involved in the product value chains. The book also offers information helpful to engineers, chemists, students, and all others involved in material adhesion and processing.

This book is intended to be a handbook for reference of processes and techniques, but it is written with a technical tone. This approach increases the availability of the whole text to readers with a college level background in mathematics and chemistry, but an in-depth knowledge of adhesion technology is not necessary. Chapters 6 through 8 can be used by almost anyone interested in surface treatment of materials and do not require college level background.

The book is organized in two parts. The first five chapters discuss definitions, surface characterization and analysis, surface energy measurement methods, adhesion mechanisms, and failure modes. Chapters 6 through 8 provide extensive information about surface treatment of materials. Chapter 6 is focused on surface preparation and treatment of metals and plastics to render surfaces bondable and to enhance the strength and durability of the bond. This chapter also includes the surface treatment techniques for fluorinated polymers, which are resistant to most techniques. Chapter 7 reviews the techniques applied to specific metals to make them bondable. Chapter 8 describes the surface preparation methods for plastic, thermosetting, and elastomeric materials. Some of the materials in Chapters 7 and 8 are updated from the excellent reference and classic Noyes publication, *Adhesives Technology Handbook*, by Arthur H. Landrock (1985), now out of print.

Attempt has been made throughout the text to balance the practical with the theoretical. Every chapter has been arranged so that it can be studied independently. For those who are interested in in-depth information, content and references have been provided for surface, adhesion, and polymer science in Chapters 1 through 5. The reader interested only in the practical information can directly consult Chapters 6 through 8. The references listed at the end of each chapter serve as both a bibliography and source for additional reading. Review papers are helpful as a starting point for finding additional sources for concentrated reading in a selected area.

The sole purpose of surface treatment in the present context is to render surfaces adherable. It is virtually impossible to understand or achieve adequate surface preparation and treatment without an intimate knowledge of the science and practice of adhesion.

None of the views or information presented in this book reflects the opinion of any of the companies or individuals that have contributed to the book. If there are errors, they are oversights on the part of the authors. A note indicating the specific error to the publisher, for the purpose of correction of future editions, would be much appreciated.

Sina Ebnesajjad
Cyrus F. Ebnesajjad
2006

Acknowledgments

No massive undertaking can be completed without the contributions of many people and organizations. Many companies and individuals have contributed information and data to this book. The authors are grateful to each company and individual for their contributions. Diligent attempts have been made to acknowledge every contributor. Any oversight errors brought to the attention of the publisher or the authors will be corrected in future editions.

The authors wish to offer their sincere appreciation to companies whose information and data appear in this book. It is thanks to the contributions of these companies that the contents of this book are connected to the real world of practice and applications.

We would like to especially thank our friend, Dr. Laurence W. McKeen, for his support with the development of the first edition. He made an in-depth review of the manuscript, suggestions, and corrections, and contributed the section on the grit blasting of metals. We are indebted to Larry for his friendship and care.

Credit goes to later Arthur H. Landrock for the materials updated (found in Chapters 7 and 8 of this book) from his excellent *Adhesives Technology Handbook* (Noyes Publications, 1985).

We honor the memory of our deceased friend Bill Woishnis, the cofounder of William Andrew Publishing (acquired by Elsevier in 2009), for his support of the authors in the development of the first edition of the book.

We would like to thank Matthew Deans, the Senior Publisher of William Andrew, for his leadership and invaluable support. Thanks to Matthew's wisdom and guidance, William Andrew continues to grow in both the number of titles and the breadth of subject matters it offers. David Jackson, Associate Editor in William Andrew, is not only brilliant, but a most pleasant and hardworking young man. We appreciated his support throughout the revision of this book. Ms. Lisa Jones' support during the typesetting and production of this book was invaluable.

Our special thanks go to our assistant Brenna Kirby for her editorial support and assistance with manuscript preparation.

We dedicate this book to our family members, who patiently supported us throughout the entire revision process. Without their generosity and devotion this revision would have not been possible.

Sina Ebnesajjad
Cyrus F. Ebnesajjad

Part I

Background and Theory

Chapter 1

Introduction to Surface Preparation

1.1 DEFINITION OF SURFACE PREPARATION, ADHESIVES, AND ADHESIVE BONDING

Surface preparation or *treatment* for adhesion is defined as one or a series of operations including cleaning, removal of loose material, and physical and/or chemical modification of a surface to which an adhesive is applied for the purpose of bonding. In plastics bonding, surface preparation is aimed at increasing the surface polarity, improving surface wettability, and creating sites for adhesive. Metal surfaces are typically covered with oxide layers, rendering the surface highly polar and fit for good adhesive bonding. Surface preparation is intended to enhance the bonding strength to metal surfaces and, significantly, to improve the durability of the bond, especially when exposed to humid environments.

An adhesive is a material applied to surfaces of articles to permanently join them by a bonding process. It is a substance capable of forming bonds to each of the two or more part interfaces comprising the final object [1]. A feature of adhesives is the relatively small quantities required relative to the weight of the final object.

An entirely satisfactory definition for *adhesion* has not been found. One of the better definitions has been proposed by Wu [2]. "Adhesion refers to the state in which two dissimilar bodies are held together by intimate interfacial contact such that mechanical force or work can be transferred across the interface. The interfacial forces holding the two phases together may arise from van der Waals forces, chemical bonding, or electrostatic attraction. The mechanical strength of the system is determined not only by the interfacial forces, but also by the mechanical properties of the interfacial zone and the two bulk phases."

The primary function of adhesives is to join parts together. Adhesives accomplish this goal by transmitting stresses from one member to another in a manner that distributes the stresses much more uniformly than mechanical fasteners. Adhesive bonding often provides structures that are mechanically equivalent to or stronger than conventional assemblies at lower cost and weight.

Surface Treatment of Materials for Adhesive Bonding. DOI: http://dx.doi.org/10.1016/B978-0-323-26435-8.00001-0

In mechanical fastening, the strength of the structure is limited to that of the areas in contact with the fasteners [3]. It is not unusual to obtain adhesive bonds that are stronger than the strength of the adherends (surfaces that are to be bonded together).

1.2 INTRODUCTION TO SURFACE TREATMENT

The main reasons for applying surface treatments prior to bonding are:

1. to remove or prevent the later formation of a weak layer on the surface of the substrate;
2. to maximize the degree of molecular interaction between the adhesive or primer and the substrate surface;
3. to optimize the adhesion forces which develop across the interfaces and therefore ensure sufficient joint strength, initially and during the service life of the bond;
4. to create specific surface microstructure on the substrate.

Normally, optimum surface energy and structure is achieved by chemical surface treatment of plastic, rubber, and metal surfaces. The chemical composition and the morphology of the surfaces are altered so that the surface energy of the substrate is elevated sufficiently for adhesion. Chemical treatments also increase the chances that hydrogen, dipole, van der Waals, ionic, and/or covalent bonding can form at the substrate/adhesive interface. A "quick and dirty" test for adequate chemical treatment is to place a drop (bead) of water on the surface of the part and note whether the water spreads. If so, the contact angle it forms with the surface is very small. There are special fluids called *dyne* liquids that can fairly precisely bracket the numeric value of the surface energy of a material surface, e.g., between 40 and 42 dynes/cm.

Alternative methods to chemical treatment (with similar surface changes) include plasma, corona, and flame treatment methods. These methods act in ways similar to chemical treatments, usually with less hazardous waste generation. Mechanical abrasion is another means of surface preparation. This method is superior to other methods due to the simplicity of the process and significantly less waste generation. Mechanical abrasion works by providing a clean surface and increasing the contact area between the substrate and adhesive. In ordinary gluing of objects, where performance requirements are not stringent, these methods are not needed since maximum adhesion bond strength is not critical. When a strong adhesion bond is required, the minimum surface preparation consists of cleaning and removal of dirt and grease from the contact surfaces.

To give the reader an idea of the basic methods of surface treatment for adhesive bonding, short descriptions of degreasing, abrasion, and chemical treatments are provided in this chapter. These and the more complex methods have been described in great detail in Chapters 6, 7, and 8.

1.2.1 Degreasing

Removal of all traces of oily contamination and grease from substrate surfaces is vital to the formation of strong adhesive bonds. Different methods are available for degreasing, which should be carried out even if the contact surfaces appear clean to the eye.

A part may be submerged in a pure, uncontaminated solvent such as trichloroethylene, perchloroethylene vapors, or hydrocarbons. As the vapors condense on the part, contaminants are dissolved and drip off with the condensed solvent. When a vapor-degreasing unit is not available, the joint surfaces can be wiped with a cloth soaked with still, use acetone solvent, followed by complete evaporation from the joint surfaces. These solvents are toxic both in liquid and vapor form, and under all circumstances require a well-ventilated work environment. Banning of chlorinated solvents has virtually eliminated them as options for degreasing.

Another technique that is increasingly used consists of scrubbing the joint surfaces in a solution of detergent in water. Metals can be immersed in or sprayed with an aqueous alkaline degreasing agent followed by rinsing with clean, hot water, and thorough drying by hot air, steam, or ambient air.

Ultrasonic degreasing produces excellent results for small components. Acetone, heavy alcohols and hydrocarbons are among good solvents for ultrasonic degreasing. An effective solvent is 1,1,1-trichloroethane because of its low toxicity and flammability, though it is now banned by the Montreal Protocol. Ethyl alcohol, gasoline, or paint thinners are not recommended. To verify the cleanliness of a surface after degreasing, water or dyne liquids are used. If a drop of water forms a film on the surface, the surface is free from contamination, and if the drop retains its shape, degreasing of the surface must be repeated.

1.2.2 Abrasion

Surfaces roughened by abrasion usually form stronger adhesive joints than do highly polished surfaces, primarily due to larger contact surface areas. A properly abraded surface should not contain any smooth or polished areas. Abrasion treatment should be followed by a degreasing treatment to ensure the removal of loose particles. Grit-blasting removes surface deposits such as tarnish, rust, or mill scale from metal surfaces. If grit-blasting equipment is not available or the metal is too thin to withstand blast treatment, clean the joint surfaces with a wire-brush, emery cloth, or sandpaper. Painted surfaces should be stripped down to the substrate with a stripper prior to preparation to insure the strength of the adhesive joint.

1.2.3 Chemical Treatment

Degreasing alone or followed by abrasion and further degreasing is sufficient to meet the requirements of many adhesive bonds. To obtain maximum strength, reproducibility, and resistance to deterioration, however, a chemical or electrolytic pretreatment is required. Careful attention should be paid in the preparation of chemical solutions to assure correct proportioning of components required for formation of adequate bond strengths. Exposure time in the solution application is critical. If the application is too short, it does not activate the surfaces sufficiently, while overexposure to the solution builds up a layer of chemical reaction products that may interfere with the adhesion bond formation.

Some of the surface preparation treatments use chemicals which may have serious health effects unless used with extreme caution. As with all chemicals, consult the Material Safety Data Sheet (MSDS) before using. Organic solvents must be handled with care since their use may cause fire, human exposure and/or toxic hazard. Read the material safety data sheet before handling. Acids and sodium hydroxide are aggressive (corrosive) chemicals. Always wear protective clothing and a face shield when using them.

In summary, the objective of any surface treatment method is to enhance the adhesive bond strength and durability. The basic requirements for a good adhesive bond are the following [4]:

- Proper choice of adhesive.
- Good joint design.
- Cleanliness of surfaces.
- Wetting of *adherends* (surfaces that are to be bonded together) by the adhesive.
- Proper adhesive bonding process (solidification and cure).

REFERENCES

[1] Modified from ASTM D907-12a, Standard Terminology of Adhesives, published in Volume 15.06: Adhesives, 2012 Annual Book of ASTM Standards.

[2] Wu S. Polymer interface and adhesion. 1st ed. New York: CRC Press; 1982.

[3] Staff written, Joining Techniques, Section 4, Machine Design, Fastening and Joining Reference Issue, 48(26):155–162. (November 18, 1976).

[4] Petrie EM. Plastics and adhesives as adhesives. In: Harper CA, editor. Handbook of plastics and elastomers. New York: McGraw-Hill; 1975.

Chapter 2

Surface Tension and Its Measurement

2.1 INTRODUCTION

Surface science is a branch of physical organic chemistry that studies the behavior and characteristics of molecules at or near a surface or interface. The interface can be between solids, liquids, gases, or combinations of these states. Complex apparatuses have been developed to identify and quantify surfaces and interfaces. Polymer surfaces are of special interest in industrial and biological applications; examples of the latter include dental implants and body-part prosthetic devices. Surface *modification of* these devices allows formation of controlled interfaces to achieve characteristics such as bondability and compatibility.

Adhesion is a phenomenon that takes place at the interfaces of adherends and adhesives. An understanding of the forces that develop the interfaces is essential in the selection of the right adhesive, proper surface treatment of adherends, and effective and economical processes to form bonds. This chapter is devoted to the discussion of the thermodynamic principles and the work of adhesion that quantitatively characterize surfaces of materials. In Chapter 3, laboratory techniques for surface characterization are described, allowing an understanding of the chemical and physical properties of material surfaces.

2.2 WHAT IS AN INTERFACE?

Two solid or liquid phases in contact have atoms/molecules on both sides of an imaginary plane called the *interface*. The interfacial particles differ energetically from those in the bulk of each phase because of being on the phase boundary and interactions with the particles of the other phases. The composition and energy vary continuously from one phase to the other throughout the interface. This region has a finite thickness, usually less than 0.1 μm [1].

Surface Treatment of Materials for Adhesive Bonding. DOI: http://dx.doi.org/10.1016/B978-0-323-26435-8.00002-2

2.3 SURFACE TENSION

The molecules of a liquid are held together by attractive forces. The sum of all attractive forces on any molecule present in the bulk of a liquid averages zero. The net force (also known as *cohesion* force) on a surface molecule is a non-zero quantity in the direction towards the bulk (Fig. 2.1). This is the force that must be counteracted to increase the surface area; the energy consumed by this process is called *surface energy*. The unbalanced forces on the interface cause it to contract to a minimum surface area value. Therefore, water droplets are spherical because a sphere has the minimum surface area of all shapes for a given volume. Surface tension and *surface free energy* of a liquid are equal, while the same is not true for a solid surface.

Surface tension is defined as the work required to increase the area of a surface isothermally and reversibly by a unit amount. Surface tension (γ) is expressed as surface energy per unit area and alternatively as force per unit length. Surface tension of liquids can be measured directly and expressed in the units of work or energy per unit area (erg/cm^2), which is then simplified to dyne/cm (erg/cm^2 = dyne; cm/cm^2 = dyne/cm). Chapters 2 and 3 review methods for the measurement of surface tension. The challenge has been to find methods to determine the surface tensions of solid surfaces (Ch. 3).

Surface tension of polymers can be divided into two components, *polar* (γ^p) and *dispersion* (γ^d), to account for the type of attraction forces at the interfaces [2]. The chemical composition of the surface determines the relative contribution of each component to surface tension. The polar component is comprised of various polar molecular interactions including hydrogen bonding, dipole energy, and induction energy, while the dispersion component arises from London dispersion attractions. The attractive forces (van der

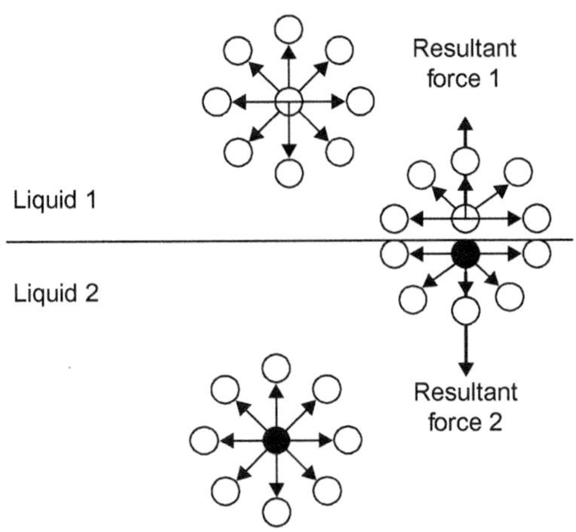

FIGURE 2.1 Liquid/liquid interface and balance of forces on molecules of liquids.

Waals and London dispersion) are additive, which results in the surface tension components being additive: $\gamma = \gamma^p + \gamma^d$.

2.4 SURFACE FREE ENERGY

A hypothetical example is used to describe the concept of surface free energy. Suppose a box with a sliding cover is filled with a liquid (Fig. 2.2). The sliding cover is assumed to have no interfacial tension with the liquid. If the cover is slid back to uncover a surface area of dA, the necessary reversible work will be (γdA). For a pure substance, the increase in the free energy of the system at constant temperature and pressure is defined by Eq. (2.1).

$$dG = \gamma dA \tag{2.1}$$

The total free energy of the system is comprised of the energy of the bulk liquid and the surface liquid. The latter is equal to the surface free energy per unit area (G_S) multiplied by the surface area as shown in Eq. (2.2). Combining Eqs. (2.1) and (2.2) results in Eq. (2.3), which illustrates that the free surface energy of a pure substance is equal to its surface tension.

$$dG = G_S\, dA \tag{2.2}$$

$$G_S = \left(\frac{dG}{dA}\right)_{T,P} = \gamma \tag{2.3}$$

In a reversible system, the heat (q) associated with the elimination of the new surface can be related to entropy (S) or *surface entropy* (Eq. 2.4), where S_S represents surface entropy per unit area. Equation (2.5) is a thermodynamic relationship applied to the liquid surface in which T represents absolute temperature. Equation (2.6) is obtained by substituting for G_S from Eq. (2.3).

$$dq = T\, dS = T\, S_S\, dA \tag{2.4}$$

$$\left(\frac{dG_S}{dT}\right)_P = -S_S \tag{2.5}$$

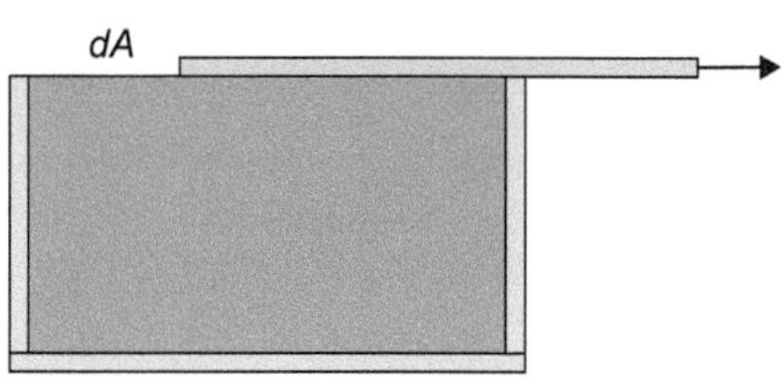

FIGURE 2.2 An "ideal liquid" box.

$$\frac{d\gamma}{dT} = -S_S \tag{2.6}$$

The total surface energy (E_S in Eq. 2.7) can be calculated by applying the enthalpy relationship (H_S) with Gibbs free energy and entropy to the liquid surface.

$$H_S = E_S = G_S + T\,S_S \tag{2.7}$$

Equation (2.8) is the result of substitution from Eq. (2.6) into Eq. (2.7).

$$E_S = G_S - T\frac{d\gamma}{dT} \tag{2.8}$$

Surface tension of most liquids decreases with increasing temperature in a linear manner. A well known equation (Eq. 2.9) defining the relationship between temperature and surface tension has been attributed to EÖTVÖS [3].

$$\gamma V^{2/3} = k(T_c - T) \tag{2.9}$$

V is the molar volume, k has the same value for most liquids (2.1 erg/K), T_c is the critical temperature of the liquid and T is the liquid temperature. The expectation is that surface tension of a liquid approaches zero at its critical temperature. There are other equations that express the behavior of liquids as a function of temperature [1].

2.4.1 Surface Energy of Solids

A solid is defined as a material that is rigid and resists stress. A solid surface may be characterized by its surface free energy and surface energy. Surface energy (tension) of a solid cannot be measured in a similar manner to that of a liquid, due to the difficulty caused by the reversible formation of its surface. The methods for determining the surface energy of solids are described in this chapter.

Solid material surfaces can be divided into one of two categories: high or low surface energy [4]. High surface energy materials include metals and inorganic compounds such as oxides, silicates, silica, diamond, and nitrides. Surface tension of high-energy materials is 200–500 dynes/cm. Low-energy materials are mainly comprised of organic compounds including polymers with critical surface tension less than 100 dynes. Polymer surfaces have been classified [5] as low, medium, or high surface energy. Low surface energy solids have critical surface tension in the range of 10–30 dynes/cm, medium energy solids, from 30 to 40 dynes/cm, and high energy solids, greater than 40 dynes/cm.

Low-energy materials such as oils are spontaneously absorbed by the high-energy surfaces, due to the reduction in the free surface energy of the system. This means that a clean, high-energy surface exposed to the normal ambient environment will not remain clean for long because the absorption of water and especially organic particulate matter from the environment will contaminate it. This is the reason a surface cleaning operation is included in many processes directly before the actual application of the adhesive or coating to prevent prolonged exposure of the cleaned substrate (adherend) to the factory environment. Some applications require coating the surface with a protective primer or corrosion inhibiting primer. Another approach is to apply a protective film to the clean surface, which is removed immediately prior to adhesive coating.

2.4.2 Work of Adhesion

The work of adhesion is defined as the reversible thermodynamic work required to separate the interface from the equilibrium state of two phases to a separation distance of infinity. Equation (2.10) shows the work of adhesion for a liquid–solid combination. This definition is attributed to the French scientist A. Dupre.

$$W_a = \gamma_L + \gamma_S - \gamma_{SL} \tag{2.10}$$

γ_L is the surface energy (tension) of the liquid phase, γ_S is the surface energy of the solid phase, γ_{SL} is the interfacialsurface tension, and W_a is the work of adhesion. A rise in the interfacial attraction results in an increase in the work of adhesion. Equation (2.10) can be rewritten to determine the work of cohesion (W_c) when the two phases are identical ($\gamma_L = \gamma_S$ and $\gamma_{SL} = 0$) and no interface is present as shown in Eq. (2.11) for a solid phase.

$$W_a = 2\gamma_S \tag{2.11}$$

2.5 CONTACT ANGLE (YOUNG'S EQUATION)

Most liquids wet solid surfaces and exhibit a contact angle. A contact angle in a static system can be measured at equilibrium. Figure 2.3 illustrates the contact angle in an ideal system where the solid surface is homogeneous, smooth, planar, and rigid. The interfacial tensions designated as γ represent equilibrium values at the point that three phases intersect. The subscripts L, S, V denote liquid, solid, and vapor phases. The γ° designation is used to indicate that the solid surface must be in equilibrium with the liquid's saturated vapor; that is, a film of the liquid is absorbed on the solid surface. Young [6] described Eq. (2.12) without presenting a proof; it has been since proven by different researchers [7–9].

$$\gamma_{LV}\ \cos\theta = \gamma^o_{SV} - \gamma_{SL} \quad \text{Young's Equation} \tag{2.12}$$

One route to proving Young's equation is using the Gibbs free energy of the wetting, proposed by Poynting and Thompson [10]. After the liquid droplet

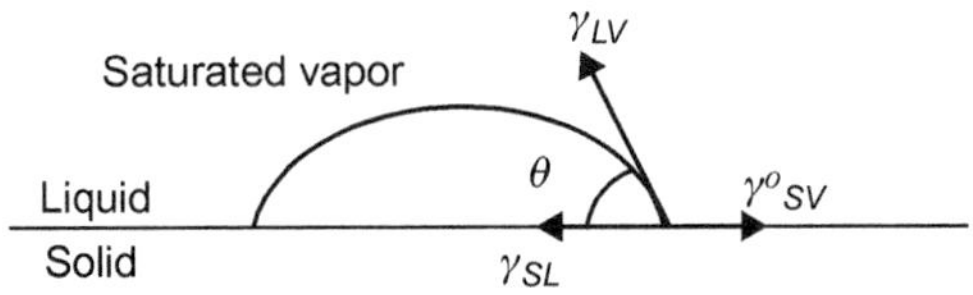

FIGURE 2.3 Equilibrium contact angle on an ideal surface.

forms the meniscus and reaches equilibrium, the variation in Gibbs free energy is zero. An assumption in Eq. (2.13) is the neglect of the gravitational force.

$$dG = 0 \tag{2.13}$$

$$dG = \gamma_{LV}\, dA_{LV} + \gamma^o_{SV}\, dA_{SV} + \gamma_{SL}\, dA_{SL} = 0 \tag{2.14}$$

In Eq. (2.14), *dA* represents small incremental increases or decreases in the surface or interfacial area. Changes in the interfacial areas are given by Eqs. (2.15) and (2.16). Any increase in the solid–liquid interface is countered by a decrease in the solid–vapor interface. Substitution from these two equations into Eq. (2.14) will yield Young's equation (Eq. 2.12).

$$dA_{SL} = -dA_{SV} \tag{2.15}$$

$$\begin{aligned} dA_{LV} &= \cos\theta \;\, dA_{SL} \\ \gamma_{LV}(\cos\theta\, dA_{SL}) + \gamma^o_{SV}(-dA_{SL}) + \gamma_{SL} dA_{SL} &= 0 \end{aligned} \tag{2.16}$$

$$\gamma_{LV} \cos\theta = \gamma^o_{SV} - \gamma_{SL} \tag{2.12}$$

The difference between the equilibrium surface energy of solid–vapor and solid–liquid is sometimes called adhesion [11]. Note that the work of adhesion and adhesion tension involves the solid–vapor equilibrium rather than that of the solid–liquid:

$$A_{SLV} = \gamma^o_{SV} - \gamma_{SL} = \gamma_{LV} \;\cos\theta$$

Most surfaces have heterogeneous composition and are not perfectly smooth. Wetting of such a surface may reach equilibrium or remain in a metastable state. In the case of an ideal surface, the addition or removal of a small volume of liquid from the drop will result in the advancement or recession of the drop. The contact angle will return to its equilibrium value. In the case of a surface that may contain roughness and heterogeneity, there is a delay in the movement of the liquid drop in response to the addition or removal of liquid. This phenomenon, called *hysteresis*, requires a revision of the definition of contact angle.

The contact angle, formed as a result of the addition of liquid, is referred to as the *advancing angle*. The angle formed by the removal of liquid is called the *receding angle*. The contact angle of a liquid on a real surface is measured in both advancing and receding modes. Typically, after the addition or removal of the liquid, there is a delay followed by a sudden motion in the drop of the liquid. The maximum angle for the advancing mode and the minimum angle for the receding mode are defined, respectively, as *advancing* and *receding contact angles*.

Harkins and Livingston [12] have proposed a correction to Young's equation when the surface of the solid carries a film of the liquid's vapor. The surface energy of a solid surface that contains an adsorbed vapor layer (γ_{SA}) is less than that of a "clean" surface. This concept has practical significance because clean surfaces tend to adsorb the ambient vapors and oils and must therefore be protected prior to the application of adhesive. Harkins and Livingston's correction, known as *spreading coefficient* (π_E), is shown in Eq. (2.17), thus resulting in Eq. (2.18) after substitution into Eq. (2.12).

$$\gamma_{SV}^{o} = \gamma_{SA} - \pi_E$$
$$\gamma_{LV}\ \cos\theta = \gamma_{SA} - \pi_E - \gamma_{SL} \tag{2.17}$$

$$\gamma_{SA} = \gamma_{LV}\ \cos\theta + \gamma_{SL} + \pi_E \tag{2.18}$$

The spreading coefficient can be measured by a technique developed by Padday[13]. This method assumes that a sessile drop, when successively increased in volume, reaches some constant maximum height (h) for a given solid–liquid system, provided the system is aged to reach equilibrium. This maximum height is related to the spreading coefficient by Eq. (2.19), in which ρ is the density of the liquid and g is the gravitational acceleration.

$$\pi_E = -\frac{\rho g h^2}{2} \tag{2.19}$$

Finally, by substituting for the interfacial tension from the modified Young's equation (Eq. 2.12) into the work of adhesion equation (Eq. 2.20) for a solid–liquid system, the equation for the work of adhesion can be simplified to Eq. (2.21), also known as Young–Dupre's equation.

$$W_a = \gamma_{LV} + \gamma_{SV}^{o} - \gamma_{SL} \tag{2.20}$$

$$W_a = \gamma_{LV}(1 + \cos\theta) \tag{2.21}$$

This means that the work of adhesion can be calculated by measuring the contact angle and the surface tension of the liquid.

2.6 LAPLACE'S EQUATION

This equation is the governing relationship for the shapes of all bubbles and drops of liquids. It is also the basis for measuring the static surface/interface tensions of fluids. Laplace's equation states that the pressure drop across a curved surface is proportional to the capillary forces as shown in Eq. (2.22) [6]

$$\Delta P = P_a - P_b$$
$$P_a - P_b = \gamma\left(\frac{1}{R_1} + \frac{1}{R_2}\right) \tag{2.22}$$

P_a and P_b are pressures in the two phases and R_1 and R_2 are the main radii of curvature. Dimensionless forms of Laplace's equation have been derived that are more convenient to apply [14]. For a spherical surface, $R_1 = R_2$, simplifying Eq. (2.22) to the following form:

$$\Delta P = P_a - P_b = \frac{2\gamma}{R} \tag{2.23}$$

2.7 EFFECT OF TEMPERATURE ON SURFACE TENSION

An important variable affecting surface tension is temperature, which has practical value during the adhesion bonding of plastics. Surface tension of both adhesive and polymer are influenced by temperature. Guggenheim's equation (Eq. 2.24) is applicable to liquids comprised of small molecules [15] and polymers. In this equation, γ_0 is surface tension at $T = 0$ K and T_c is the critical temperature (K) of the liquid. The values of γ_0 and T_c can be determined by fitting a line to the surface tension data as a function of temperature. According to the Guggenheim equation, surface tension decreases as temperature increases [16]. The rate of decrease of surface tension as a function of temperature is 0.1 dynes/°C cm for liquids with small molecules [15,17].

$$\gamma = \gamma_0\left(1 - \frac{T}{T_c}\right)^{11/9} \tag{2.24}$$

2.8 SURFACE TENSION MEASUREMENT

Surface tension measurement techniques are divided into methods for solids and liquids. There are two modes for measuring surface tension of liquids: static and dynamic. Values reported in the literature are often static surface tensions of liquids. Tables 2.1–2.3 present a brief description of the common techniques for surface tension measurement of liquids and solids. Some of these methods are described in further detail.

Several standards specify methods for measuring contact angles for different applications (Table 2.4). These methods provide procedures for the comparison of surface energy of different industrial materials.

TABLE 2.1 Static Surface Tension Measurement Methods for Liquids

Technique	Brief Description
Du Nouy Ring	Traditional method for the measurement of surface and interfacial tension measurement. Maximum pulling force on a ring by the surface is measured. Wetting properties of liquids have no influence on this technique.
Wilhelmy Plate	This technique is broadly applicable to liquids, especially when surface tension must be measured over a long time period. A vertical plate with known perimeter is attached to a balance and submerged in the liquid. The force due to wetting is measured by the balance.
Spinning Drop	Suitable for low interfacial tensions. The diameter of a drop within a heavy phase is measured while both are being rotated.
Pendant Drop	Surface and interfacial tension are measured by this method.

TABLE 2.2 Dynamic Surface Tension Measurement Methods for Liquids

Technique	Brief Description
Bubble Pressure	This method is used for determining surface tension at short surface ages. The actual measured variable is the maximum pressure of each bubble.
Drop Volume	A method employed for determining interfacial tension as a function of time (interface age). The time between two consecutive drops is measured when liquid of one density is pumped into a second liquid of a different density.

TABLE 2.3 Surface Tension Measurement Methods for Solids

Technique	Brief Description
Sessile Drop	This method involves measurement of the contact angle optically and is used to estimate wetting properties of a localized region on a solid surface. The angle between the baseline of the drop and the tangent at the drop boundary is determined. This technique is ideal for curved samples or when one side of the sample surface has different properties than the other side.
Dynamic Wilhelmy	This method requires uniform geometry for calculating average advancing and receding contact angles on solids. Both sides of the solid must have the same properties. Wetting force on the solid material is measured as the solid

(*Continued*)

TABLE 2.3 (Continued)

Technique	Brief Description
	is submerged in, or withdrawn from, a liquid of given surface tension.
Single Fiber Wilhelmy	This is a dynamic Wilhelmy method, applied to single fibers to measure advancing and receding contact angles.
Powder Contact Angle	This procedure allows the measurement of average contact angle and adsorption speeds of powders and other porous species. The change in the weight as a function of time is measured.
Liquid Homolog Method (Zissman)	Called *critical surface tension*, it is based on measuring the contact angle of several liquids and plotting the cosine of the angle vs. surface tension of the liquids. The surface tension at which cosine of the contact angle is equal to one, obtained by extrapolation, is the critical surface tension of the solid.

TABLE 2.4 Standard Contact Angle Measurement Methods for Various Materials

Technique	*Title* – Scope
ASTM D724-99 (2003)	*Standard Test Method for Surface Wettability of Paper.* Method covers the quantitative determination of the resistance of paper surfaces to wetting by measuring the behavior of a drop of liquid applied directly to the paper surface.
ASTM D5725-99 (2008)	*Standard Test Method for Surface Wettability and Absorbency of Sheeted Materials using an Automated Contact Angle Tester.* Method measures the contact angle of a test liquid on a flat sample of a film or a paper substrate under specified conditions.
ASTM C813-90 (2009)	*Standard Test Method for Hydrophobic Contamination on Glass by Contact Angle Measurement.* Method covers the detection of hydrophobic contamination on glass surfaces by means of contact angle measurement.
ASTM D5946-09	*Standard Test Method for Corona-treated Polymer Films using Water Contact Angle Measurement.* Method covers the measurement of contact angle of water droplets on corona-treated polymer film surfaces; the results are used to estimate the film's wetting tension.
TAPPI T458*	*Surface Wettability of Paper.* In this method, the contact angle between air and liquid on a paper surface is taken as a measure of the resistance of the paper surface to wetting by the liquid. The initial angle of contact or initial wettability is considered to be a measure of the ruling quality of the paper. The rate of change in the wettability is considered to be a measure of the writing quality.

**See TAPPI (Technical Association of the Pulp and Paper Industry) web site: <www.tappi.org>.*

2.8.1 Measurement for Liquids: Du Nouy Ring and Wilhelmy Plate Methods

The du Nouy ring and Wilhelmy plate methods (Fig. 2.4) are the two most frequently used techniques for measuring surface tension at a liquid air interface or interfacial tension at a liquid–liquid interface. Only the du Nouy method can be applied to measure interfacial tension. Both of these techniques are based on pulling an object with a well-defined geometry of the surface of liquids and measuring the pull force. These techniques are also known as *pull-force* methods. In the Wilhelmy method, a plate is the pull object, while in the du Nouy technique, a ring is used. These techniques are named after the two scientists who conducted some of the early research in the field of surface tension measurement. In 1863, Wilhelmy [18] described the measurement of capillary constants without a detailed calculation of surface tension. Lecomte du Nouy illustrated the shortcomings of the past surface tension determination methods in a paper published in 1919 [19].

Equation (2.25) shows the elements required for the calculation of surface tension by the du Nouy ring method. In this equation, P_T is the total force on the ring, P_R is the weight of the ring, R is the radius of the ring, and γ_{ideal} is the ideal surface tension. In practice, a meniscus correction factor is required because the size and shape of the surface inside and outside the ring are not the same. Surface tension must, therefore, be corrected for the shape of the ring by a factor (f) as shown in Eq. (2.26). The correction factors have been determined and tabulated [17,20,21].

$$P_T = P_R + 4\pi R \cdot \gamma_{\text{ideal}} \tag{2.25}$$

$$\gamma = f\ \gamma_{\text{ideal}} \tag{2.26}$$

The Wilhelmy method does not require a correction factor for the meniscus shape, although it does require correction if the plate is partially or completely submerged in the liquid. In Eq. (2.27), P_T is the total force on the plate being measured, P_P is the weight of the plate, p is the perimeter of the plate, and γ_{ideal} is the ideal surface tension. A buoyant force term must be

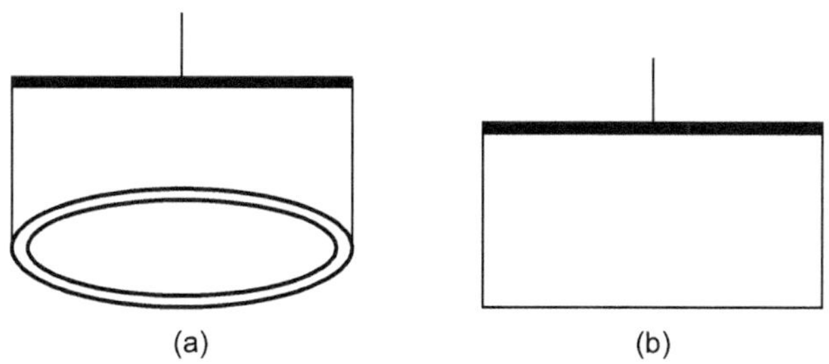

FIGURE 2.4 Schematic of (a) du Nouy Ring and (b) Wilhelmy Plate.

added to or subtracted from the second part of the equation, depending on whether the plate is above or below the level of the free liquid. In Eq. (2.28), h is the height above or below the free liquid level, A is the cross-sectional area of the plate, and γ is the surface tension.

$$P_T = P_P + p\gamma_{\text{ideal}} \tag{2.27}$$

$$P_T = P_P + p\gamma + \rho ghA \tag{2.28}$$

The total force (P_T) acting on the ring (du Nouy) or the plate (Wilhelmy) can be measured by a balance connected to either device. Substituting for the total force and the other parameters in Eq. (2.28) allows the value of surface tension (γ) to be calculated.

2.8.2 Measurement for Solids: Liquid Homolog Series

Surface tension of solid plastics is determined indirectly, usually by contact angle methods. A problem with the direct measurement of surface tension arises from the difficulty in the reversible formation of a solid surface. Table 2.3 shows a list of methods that can be applied to measure the surface energy of solids.

An alternative method uses a concept called *critical surface tension*, proposed by Fox and Zissman [4,22,23], to characterize the surface energy of solids. A cosine plot of the contact angle (cos θ) and liquid–vapor surface tension (γ_{LV}) yields a straight line for a homologous series of liquids (Fig. 2.5). Non-homologous liquids yield a curved line that may not be easily extrapolated. The intercept of the line at cos θ equal to one is defined as the *critical surface tension* of the polymer (γ_c). Values of 18 dynes/cm for polytetrafluoroethylene and 30 dynes/cm for polyethylene are obtained according to this procedure (Figs. 2.5 and 2.6). Tables 2.5 and 2.6 present surface free energies of solids and surface tension of liquids.

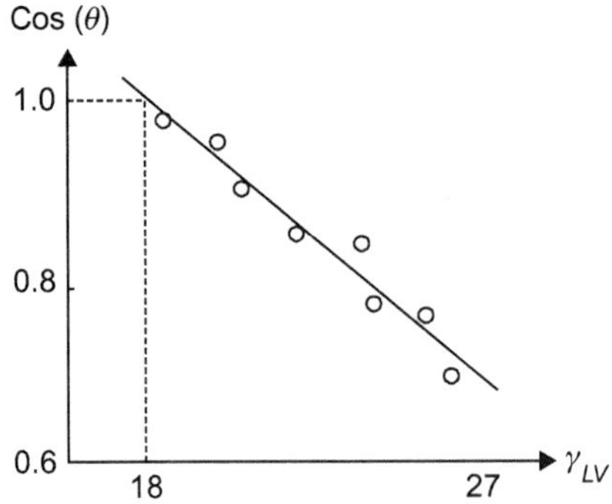

FIGURE 2.5 Zissman plot for polytetrafluoroethylene using *n*-alkanes as the testing liquids [4,23].

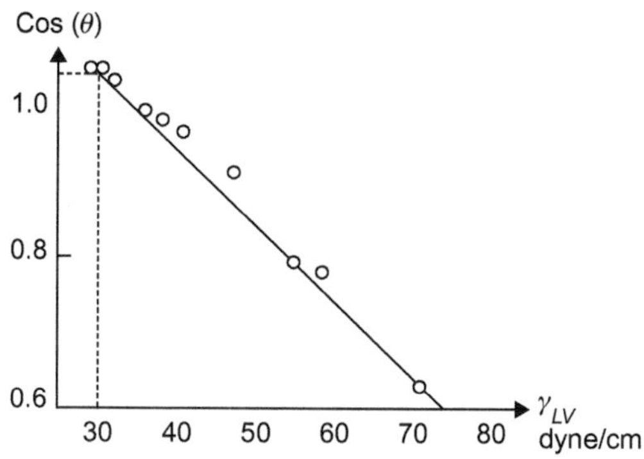

FIGURE 2.6 Zissman plot for polytethylene using *n*-alkanes as the testing liquids [4,23].

TABLE 2.5 Surface Free Energy of Select Plastics

Plastic Material	Surface Free Energy, dynes/cm
Polytetrafluoroethylene	18–19
Polytrifluoroethylene	22
Polyvinylidene Fluoride	25
Polyvinyl Fluoride	28
Polypropylene	29
Polyethylene	30–31
Ionomer (low) Polystyrene	33
Ionomer (high) Polystyrene	37
Polymethylmethacrylate	38
Polyvinyl Chloride	39
Cellulosics	42
Polyester	43
Nylon	46

TABLE 2.6 Surface Tension of Select Liquids

Liquid	Surface Free Energy, dynes/cm
n-Hexane	18
Alcohols	22
Cyclohexane	25
Toluene, Xylene	29
Phenol	41
Aniline	43
Glycol	47
Formamide	58
Glycerol	63
Water	72

One can obtain a relationship (Eq. 2.29) between the critical surface tension and the solid–vapor surface tension by setting the contact angle to zero in Young's equation (Eq. 2.12). Critical surface tension is, therefore, smaller than solid–vapor surface tension. Figure 2.7 shows the effect of temperature on critical surface tensions of two plastics [24]. The Zissman plot for poly (methyl methacrylate) is given in Figure 2.8.

Surface energy of plastics decreases with increasing temperature as seen in Figure 2.9. Consequently a molten polymer has a lower surface than its

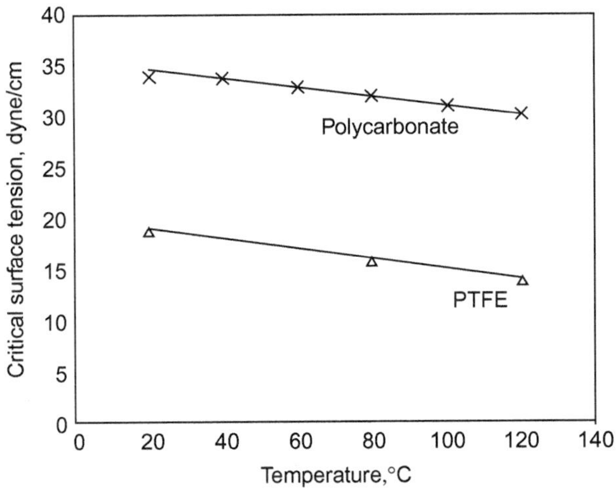

FIGURE 2.7 Effect of temperature on critical surface tension of two plastics [24].

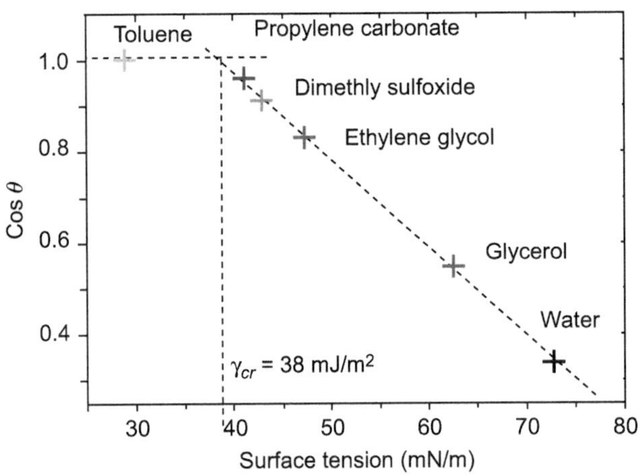

FIGURE 2.8 Zissman Plot for poly(methyl methacrylate) [25].

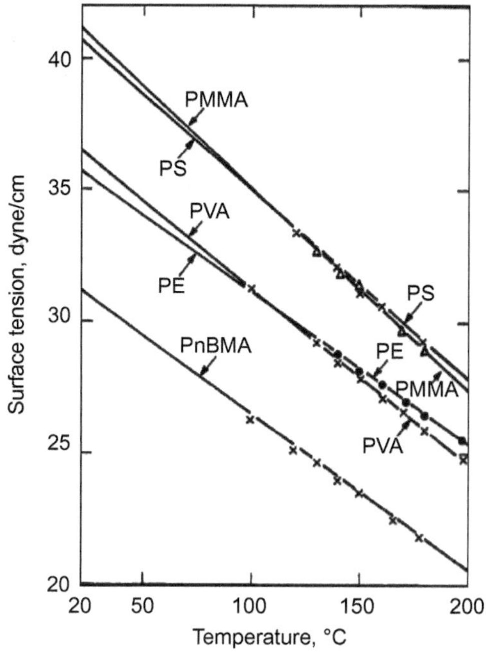

FIGURE 2.9 Surface tensions of some polymer melts. PMMA; poly(methyl methacrylate); PS, polystyrene; PnBMA, poly(n-butyl methacrylate); PE, polyethylene; PVA, poly(vinyl acetate) [26].

TABLE 2.7 Solid Surface Energy Data for Common Polymers

Name	CAS Ref. No.	Surface Free Energy (SFE) at 20°C in mN/m	Temp. Coefficient SFE in mN/(m K)	Dispersive Contrib. of SFE in mN/m	Polar Contrib. of SFE in mN/m
Polyethylene-linear PE	9002-88-4	35.7	−0.057	35.7	0
Polyethylene-branched PE	9002-88-4	35.3	−0.067	35.3	0
Polypropylene-isotactic PP	25085-53-4	30.1	−0.058	30.1	0
Polyisobutylene PIB	9003-27-4	33.6	−0.064	33.6	0
Polystyrene PS	9003-53-6	40.7	−0.072	(34.5)	(6.1)

(*Continued*)

TABLE 2.7 (Continued)

Name	CAS Ref. No.	Surface Free Energy (SFE) at 20°C in mN/m	Temp. Coefficient SFE in mN/(m K)	Dispersive Contrib. of SFE in mN/m	Polar Contrib. of SFE in mN/m
Poly-α-methyl styrene PMS (polyvinyltoluene PVT)	9017-21-4	39.0	−0.058	(35)	(4)
Polyvinyl fluoride PVF	24981-14-4	36.7	–	(31.2)	(5.5)
Polyvinylidene fluoride PVDF	24937-79-9	30.3	–	(23.3)	(7)
Polytrifluoroethylene P3FEt/PTrFE	24980-67-4	23.9	–	19.8	4.1
Polytetrafluoroethylene PTFE (Teflon™)	9002-84-0	20	−0.058	18.4	1.6
Polyvinylchloride PVC	9002-86-2	41.5	–	(39.5)	(2)
Polyvinylidene chloride PVDC	9002-85-1	45.0	–	(40.5)	(4.5)
Polychlorotrifluoroethylene PCTrFE	25101-45-5	30.9	−0.067	22.3	8.6
Polyvinylacetate PVA	9003-20-7	36.5	−0.066	25.1	11.4
Polymethylacrylate (polymethacrylic acid) PMAA	25087-26-7	41.0	−0.077	29.7	10.3
Polyethylacrylate PEA	9003-32-1	37.0	−0.077	30.7	6.3
Polymethylmethacrylate PMMA	87210-32-0	41.1	−0.076	29.6	11.5
Polyethylmethacrylate PEMA	9003-42-3	35.9	−0.070	26.9	9.0
Polybutylmethacrylate PBMA	25608-33-7	31.2	−0.059	26.2	5.0
Polyisobutylmethacrylate PIBMA	9011-15-8	30.9	−0.060	26.6	4.3
Poly(t-butylmethacrylate) PtBMA	–	30.4	−0.059	26.7	3.7
Polyhexylmethacrylate PHMA	25087-17-6	30.0	−0.062	(27.0)	(3)
Polyethyleneoxide PEO	25322-68-3	42.9	−0.076	30.9	12.0
Polytetramethylene oxide PTME (polytetrahydrofurane PTHF)	25190-06-1	31.9	−0.061	27.4	4.5

(Continued)

TABLE 2.7 (Continued)

Name	CAS Ref. No.	Surface Free Energy (SFE) at 20°C in mN/m	Temp. Coefficient SFE in mN/(m K)	Dispersive Contrib. of SFE in mN/m	Polar Contrib. of SFE in mN/m
Polyethyleneterephthalate PET	25038-59-9	44.6	−0.065	(35.6)	(9)
Polyamide-6,6 PA-66	32131-17-2	46.5	−0.065	(32.5)	(14)
Polyamide-12 PA-12	24937-16-4	40.7	–	35.9	4.9
Polydimethylsiloxane PDMS	9016-00-6	19.8	−0.048	19.0	0.8
Polycarbonate PC	24936-68-3	34.2	−0.04	27.7	6.5
Polyetheretherketone PEEK	31694-16-13	42.1	–	36.2	5.9

Ref: <www.surface-tension.de/solid-surface-energy.htm>, February 2013.

solid form. Surface energies of a number of common polymers are presented in Table 2.7.

$$\gamma_c = \lim_{\theta \to 0}(\gamma_{LV}\ \cos\theta) = \gamma_{SV} - \gamma_{SL} \tag{2.29}$$

In summary, the experimental and analytical methods described in this chapter enable the reader to both measure and calculate surface energy of liquids and solids. Surface preparation techniques often change the surface energy of materials, which can be determined by the methods provided in this chapter.

REFERENCES

[1] Wu S. Polymer interface and adhesion. 1st ed. New York: Marcel Dekker, Inc., 1982.
[2] Wu S. J Adhesion 1973;5:39–55.
[3] Adam NK. The Physics and Chemistry of Surfaces. 3rd ed. Oxford University Press; 1941.
[4] Fox HW, Zissman WA. J Colloid Interface Sci 1950;5:514.
[5] Lee LH. Fundamentals of adhesion. New York: Plenum Press; 1991.
[6] Young T. Phil Trans Roy Soc (London) 1805;95:65.
[7] Gibbs JW. The collected works of J. W. Gibbs. New York: Longmans, Green; 1931.
[8] The scientific papers of J. Willard Gibbs, vol. 1. Thermodynamics. New York: Dover; 1961.

[9] Johnson Jr RE. J Phys Chem 1959;63:1655.

[10] Poynting JH, Thompson JJ., pub. by A textbook of physics: properties of matter. 8th ed. London: Charles Griffin; 1920

[11] Bartell FE, Bartell LS. J Am Chem Soc 1934;56:2205.

[12] Harkins WD, Livingston HK. J Chem Phys 1942;10:342.

[13] (a) Padday JF. Apparatus for measuring the spreading coefficient of a liquid, on a solid surface, J Sci Instr, vol. 36 (June 1959); (b) Padday, JF, Proceedings of the Second International Conference of Surface Activity, III:136, London: Butterworths Scientific Publications 1958.

[14] Bashforth S, Adams JC. An attempt to test the theory of capillary action, Cambridge Univ. Press and Deighton Pub, Bell and Co., London, UK, 1892.

[15] Guggenheim EA. J Chem Phys 1945;13:253.

[16] Paul DR, Newman S. Polymer blends. New York: Academic Press; 1978.

[17] Padday JF. In: Matijevic E, editor. Surface and Colloid Science, 1. New York: Wiley; 1969.

[18] Wilhelmy L. Ueber die Abhangigkeit der Capillaritats – Constanten des Alkohol con Substanz und Gestalt des benetzten festen Korpers. Ann Physik 1863;119:177–217.

[19] Lecomte du Nouy P. A new apparatus for measuring surface tension. J Gen Physiol 1919;1:521–4.

[20] Harkins WD, Jordan HF. J Am Chem Soc 1930;52:1756.

[21] Freud BB, Freud HZ. J Am Chem Soc 1930;52:1772.

[22] Fox HW, Zissman WA. J Colloid Sci 1952;7:109.

[23] Fox HW, Zissman WA. J Colloid Sci 1952;7:428.

[24] Petke FD, Jay BR. J Colloid Interface Sci 1969;31:216.

[25] Abhinandan Agrawal, Surface Tension of Polymers, Hatsopoulos Microfluids Laboratory, Department of Mechanical Engineering, Massachusetts Institute of Technology, <http://web.mit.edu/nnf/education/wettability/summerreading-2005short.pdf>; June 29, 2005.

[26] Wu S. Surface and interfacial tensions of polymer melts. II, poly(methyl methacrylate), poly(n-butyl methacrylate) and polystyrene. J Phys Chem 1970;74(3).

Chapter 3

Surface Energy of Solids and Its Measurement

3.1 INTRODUCTION

In Chapter 2, we reviewed surface tension measurement techniques, including the *Zisman* method for solid surfaces. There are a number of empirical equations proposed for relating the surface tension (energy) of a solid surface with a contact angle and other parameters. This chapter is devoted to the review of more significant empirical equations, with a focus on polymeric materials, and provides a fairly detailed description of each equation.

Metals and metal oxides generally have significantly higher surface energy than polymeric materials, which simplifies the surface preparation methods from a technological standpoint. In contrast, polymeric surfaces have low energy that requires a variety of treatment techniques, some of which are quite complex. The discussions in this chapter are concentrated on polymeric materials. The reader is encouraged to consult the references for a deeper and broader understanding of the topic.

3.2 DETERMINING THE SURFACE ENERGY OF SOLIDS

Let us consider Young's equation (Eq. 3.1) once again. The right side of this equation can be calculated by measuring the surface tension of the liquid and the contact angle. There are still two unknowns (γ_{SV}^o and γ_{SL}) left in Young's equation, and to calculate γ_{SV}^o, the value of γ_{SL} must be known. A number of models have been developed to determine the surface energy of solids from contact angle measurements. The significant models include the equation of state, polymer melt, surface tension components, and the critical surface tension.

$$\gamma_{LV} \cos\theta = \gamma_{SV}^o - \gamma_{SL} \quad \text{Young's Equation} \tag{3.1}$$

First, we will discuss an important concept called *fractional polarity*, which is helpful in understanding the different methods for determining the surface energy of solids.

Surface Treatment of Materials for Adhesive Bonding. DOI: http://dx.doi.org/10.1016/B978-0-323-26435-8.00003-4

3.2.1 Fractional Polarity

Wu [1] has proposed separating the surface energy into nonpolar (dispersion) and polar components (Eq. 3.2). The superscripts d and p designate nonpolar and polar components. The concept of the additive nature of surface energy components has been accepted by researchers such as Fowkes [2] and Meyer *et al.* [3,4] The *polar* component of surface tension includes various dipole interactions and hydrogen bonding. The components are lumped together to simplify the discussions. The *dispersion* component includes the nonpolar fraction of surface energy. Fractional polarity is defined by Eq. (3.3) and nonpolarity by Eq. (3.4).

$$\gamma = \gamma^d + \gamma^p \tag{3.2}$$

$$X^p = \frac{\gamma^p}{\gamma} \tag{3.3}$$

$$X^d = \frac{\gamma^d}{\gamma} \tag{3.4}$$

Fractional polarity can be defined for both solid and liquid phases based on Eqs. (3.3) and (3.4).

3.3 EQUATION OF STATE

Here, *equation of state* is defined as a proposed relationship between contact angle and critical surface tension. A series of liquids are tested to measure their contact angles against a solid surface. The critical surface tension values (γ_c) are plotted against the liquid–vapor surface tension (γ_{LV}); this plot is called an *equation of state plot.* The maximum value of the critical surface tension curve is called the *surface tension of the solid.* The equations which mathematically illustrate this description are derived in this section.

We begin with the theory put forward by Good and Girifalco [5–7]. Equation (3.5) relates work of adhesion (W_a) to the interfacial surface tension for two phases, denoted by 1 and 2. Good and Girifalco introduced an interaction parameter (ϕ) defined by Eq. (3.6), in which W_c is the work of cohesion. A combination of Eqs. (3.5) and (3.6) yields the equation of Good and Girifalco (Eq. 3.7).

$$\gamma_{12} = \gamma_1 + \gamma_2 - W_a \tag{3.5}$$

$$\phi = \frac{W_a}{(W_{c1} W_{c2})^{1/2}} \tag{3.6}$$

$$W_{c1} = 2\gamma_1 \quad \text{and} \quad W_{c2} = 2\gamma_2$$

$$\gamma_{12} = \gamma_1 + \gamma_2 - 2\phi(\gamma_1\gamma_2)^{1/2} \tag{3.7}$$

Young's equation, in its general form, Eq. (2.18), i.e., containing the Harkins and Livingston's correction term (π_E), is combined with Eq. (3.7). Critical surface tension is defined as the value of γ_{LV} when the contact angle approaches zero, Eq. (3.9).

$$1 + \cos\theta = 2\phi\left(\frac{\gamma_{SV}}{\gamma_{LV}}\right)^{1/2} - \frac{\pi_E}{\gamma_{LV}} \tag{3.8}$$

$$\gamma_c = \lim_{\theta \to 0} \gamma_{LV} \tag{3.9}$$

Next, Eqs. (3.8) and (3.9) are combined and rearranged, the square of which results in Eq. (3.10). Expansion of Eq. (3.10) yields Eq. (3.11), which can be simplified to Eq. (3.12). When Eq. (3.12) is used to substitute for γ_c, it yields Eq. (3.13), which rapidly converges to Eq. (3.14).

$$\begin{aligned} 1 + 1 &= 2\phi\left(\frac{\gamma_{SV}}{\gamma_c}\right)^{1/2} - \frac{\pi_E}{\gamma_c} \\ \left(\frac{\gamma_E}{\gamma_c} + 2\right)^2 &= 4\phi^2\left(\frac{\gamma_{SV}}{\gamma_c}\right) \end{aligned} \tag{3.10}$$

$$\pi_E^2 + 4\pi_E\gamma_c + 4\gamma_c^2 = 4\phi^2\gamma_c\gamma_{SV} \tag{3.11}$$

$$\gamma_c = \phi^2\gamma_{SV} - \pi_E - \frac{\pi_E^2}{4\gamma_c} \tag{3.12}$$

$$\gamma_c = \phi^2\gamma_{SV} - \pi_E - \frac{\pi_E^2}{4\left(\phi^2\gamma_{SV} - \pi_E - \frac{\pi_E^2}{4\gamma_c}\right)} \tag{3.13}$$

$$\gamma_c = \phi^2\gamma_{SV} - \pi_E \tag{3.14}$$

γ_c is a function of the interaction parameter (ϕ). By substituting for γ_{SV} in Eq. (3.8), one obtains Eq. (3.15). Equation (3.16), called the *equation of state*, is the product of rearranging Eq. (3.15). In this equation, the surface tension of the liquid (γ_{LV}) is known, and by measuring the contact angle, γ_{SV} can be calculated.

$$1 + \cos\theta = 2\left(\frac{\gamma_{SV}}{\gamma_{LV}}\right)^{1/2} - \frac{\pi_E}{\gamma_{LV}} \tag{3.15}$$

At equilibrium, $\pi_E = 0$ when the liquid has wet the solid surface.

$$\gamma_c = \frac{1}{4}(1 + \cos\theta)^2\gamma_{LV} \tag{3.16}$$

A series of liquids can be used to measure contact angle and surface tension of the solid surface. The maximum value of the interaction factor is reached when the polarities of both the liquid and solid are the same ($\phi = 1$). In that case, Eq. (3.14) can be written as Eq. (3.17). When the contact angle is larger than zero, Eq. (3.18) applies.

$$\gamma_c = \gamma_{SV} - \pi_E \tag{3.17}$$

In Eq. (3.14), π_E is sufficiently smaller than γ_{SV} (i.e., spreading pressure is small) that it can be ignored. In conclusion, Eq. (3.18) states that critical surface tension of a solid is equal to the surface energy of the solid when it is measured using specific liquids. The liquids have matching polarity with the solid and do not form a low contact angle.

$$\begin{aligned} \gamma_c &= \phi^2 \gamma_{SV} \\ \gamma_c &= \gamma_{SV} \end{aligned} \tag{3.18}$$

A more recent equation of state (Eq. 3.19) was proposed by Kwok and Neumann [8] in 1999, based on a rather extensive phenomenological basis. Equation (3.19) is similar to Good and Girifalco's (Eq. 3.7), in which the interaction parameter (ϕ) is defined by the exponential term. The parameter (β) is a constant with a value of 0.0001247 $(\mathrm{m^2/mJ})^2$. The combination of Eq. (3.19) and Young's equation yields an equation of state (Eq. 3.20). In this equation, the surface tension of the liquid (γ_{LV}) is known, and by measuring the contact angle, γ_{SV} can be calculated.

$$\gamma_{SL} = \gamma_{LV} + \gamma_{SV} - 2(\gamma_{LV}\gamma_{SV})^{1/2} e^{-\beta(\gamma_{LV} - \gamma_{SV})^2} \tag{3.19}$$

$$1 + \cos\theta = 2(\gamma_{LV}\gamma_{SV})^{1/2} e^{-\beta(\gamma_{LV} - \gamma_{SV})^2} \tag{3.20}$$

3.4 SURFACE TENSION COMPONENTS

The approach of measuring the components of surface tension is attributed to Frank Fowkes [9]. The components of surface tension include those generated by London forces, permanent dipole forces, hydrogen bonds, induction forces, and metallic interactions. These forces have been divided into dispersive and polar, in which the polar term is a sum of all forces.

Figure 3.1 shows the energies associated with typical interaction forces and bonds. Covalent and ionic bonds (sometimes called *primary valency forces*) have the highest energy, with decreasing energy values for the other types of interactions (sometimes called *secondary valency forces*).

In general, this concept may be applied to the interface of any two phases. Let us assume that a solid surface only embodies dispersive forces, thus limiting the interaction between this solid and a liquid to the dispersion type. Linear polyethylene has been a favorite example of such a surface due

Type	Example	E (kJ/mol)
Covalent	C–C	350
Ion–Ion	Na^+ Cr	450
Ion–Dipole	Na^+ CF_3H	33
Dipole–Dipole	CF_3H CF_3H	2
London Dispersion	CF_4 CF_4	2
Hydrogen Bonding	H_2O H_2O	24

FIGURE 3.1 Examples of energies of Lifshitz–van der Waals interactions and chemical bonds.

to the virtual absence of polar interactions. On the liquid side, a paraffin such as hexane or heptane is envisioned as a liquid that only interacts with the surface by dispersion forces. Good and Girifalco's [5–7] equation can be written (Eq. 3.21) for a two-phase, solid–liquid system where the interaction parameter (ϕ) is equal to one.

$$\gamma_{SL}^{d} = \gamma_{SV}^{d} + \gamma_{LV}^{d} - 2(\gamma_{SL}^{d}\gamma_{LV}^{d})^{1/2} \tag{3.21}$$

Owens and Wendt [10] and Kaelble [11] extended Fowkes' proposed equation by considering polar forces in addition to dispersive forces. They proposed a *geometric mean* equation that, in combination with Young's equation, yields Eq. (3.22). Equation (3.23) also applies and can be used to substitute for γ_{LV}.

$$\gamma_{LV}(1 + \cos\theta) = 2(\gamma_{SV}^{d}\gamma_{LV}^{d})^{1/2} + 2(\gamma_{SV}^{p}\gamma_{LV}^{p})^{1/2} \tag{3.22}$$

$$\gamma_{LV} = \gamma_{LV}^{d} + \gamma_{LV}^{p} \tag{3.23}$$

$$(\gamma_{LV}^{d}\gamma_{LV}^{p})(1 + \cos\theta) = 2(\gamma_{SV}^{d}\gamma_{LV}^{d})^{1/2} + 2(\gamma_{SV}^{p}\gamma_{LV}^{p})^{1/2} \tag{3.24}$$

The surface tension of the solid (γ_{SV}) is calculated by measuring the contact angle (θ) of two liquids with known dispersion and polar surface tension components, leaving two simultaneous equations and two unknown parameters (γ_{SV}^{d} and γ_{SV}^{p}). Equation (3.25) is used to calculate the surface energy of the solid.

$$\gamma_{SV} = \gamma_{SV}^{d} + \gamma_{SV}^{p} \tag{3.25}$$

Wu [12] proposed another empirical equation in which the *harmonic mean* is used to calculate the interfacial tension. A combination of Young's equation and the harmonic mean produces Eq. (3.26). After substituting for γ_{LV} from Eq. (3.23), the contact angle is measured using two liquids with known dispersion and polar surface tension components. This leaves two

TABLE 3.1 Surface Tension Components of Water and Methylene Iodide [12]

Liquid	Surface Tension, dyne/cm (20°C)		
	γ^d	γ^p	$\gamma = \gamma^d + \gamma^p$
Harmonic mean			
Water	22.1	50.7	72.8
CH_2I_2	44.1	6.7	50.8
Geometric mean			
Water	21.8	51.0	72.8
CH_2I_2	49.5	1.3	50.8
	48.5	2.3	50.8

equations with two unknowns (γ_{SV}^d and γ_{SV}^p), thus allowing the calculation of a solid's surface energy from Eq. (3.25).

$$\gamma_{LV}(1 + \cos\theta) = 4\left(\frac{\gamma_{SV}^d \gamma_{LV}^d}{\gamma_{SV}^d + \gamma_{LV}^d} + \frac{\gamma_{SV}^p \gamma_{LV}^p}{\gamma_{SV}^p + \gamma_{LV}^p}\right) \tag{3.26}$$

Water and methylene iodide are two liquids used for the determination of the components of surface tension of solid polymers by measuring contact angles. Table 3.1 gives the values of the components of their surface tensions.

In another approach, proposed by Fowkes *et al.* [13–15], the surface tension of a solid is divided into two components: one resulting from non-covalent, long-range interactions and the second one arising from short-range, Lewis acid–base interactions. (The reader can find a description of Lewis acid–base interactions in Chapter 5.) Equation (3.27) thus applies to the components of the surface tension. The first component consists of long-range interactions such as dispersion forces, polar forces, and others called *van der Waals* forces or, more accurately, *Lifshitz–van der Waals* interactions. The short-range, acid–base component is a function of the geometric mean of two parts called *electron-acceptor* (γ^+) and *electron-donor* (γ^-), shown in Eq. (3.28).

$$\gamma = \gamma^{LW} + \gamma^{AB} \tag{3.27}$$

Here, *LW* stands for Lifshitz–van der Waals and *AB* for Lewis acid–base.

$$\gamma^{AB} = 2\sqrt{\gamma^+ \gamma^-} \tag{3.28}$$

Equation (3.29) is obtained by combining Eqs. (3.27) and (3.28) with Young's equation, and can be used to determine the solid surface energy.

The three unknowns for the solid surface ($\gamma_{SV}^{LW}, \gamma_{SV}^{+}$, and γ_{SV}^{-}) can be determined by solving three simultaneous equations. These equations can be obtained by measuring the contact angle for three liquids that have known long-range and acid–base surface tension components ($\gamma_{LV}^{LW}, \gamma_{LV}^{+}$, and γ_{LV}^{-}). Two of these liquids must be polar.

$$\gamma_{LV}(1 + \cos\theta) = 2(\gamma_{LV}^{LW}\gamma_{SV}^{LW})^{1/2} + 2(\gamma_{SV}^{+}\gamma_{LV}^{-})^{1/2} + (\gamma_{LV}^{+}\gamma_{SV}^{-})^{1/2} \tag{3.29}$$

3.5 POLYMER MELT METHOD

In the case of polymers, one can extrapolate the surface tension of melts to lower temperatures at which the polymer is a solid. Surface tension versus temperature can be fitted with the Guggenheim [16] equation, which was originally developed for liquids (Eq. 3.30). This equation has been found to be applicable to polymers. Typically, the relationship between surface tension and temperature is linear in the 0°C–200°C range.

$$\gamma = \gamma_0\left(1 - \frac{T}{T_c}\right)^{11/9} \tag{3.30}$$

T_c is the critical temperature of the polymer and γ_0 is the surface tension at 0 K. It must be noted that the critical temperature of polymers is very high, around 700°C, according to Wu [12].

3.6 CRITICAL SURFACE TENSION

Zisman and Fox [17–19] first proposed the concept of *critical surface tension* in the early 1950s. Zisman's method investigated the relationship between the contact angle of various liquids on a solid and the surface tension of the liquids. Specifically, cos θ (θ is the contact angle) is plotted against γ_{LV} (known as the *Zisman plot*) in which a straight line is often obtained when a homologous series of liquids are used to wet the solid's surface. The same relationship between cos θ and γ_{LV} for non-homologous liquids yields a curved line, as shown in Figure 3.2.

One can obtain a relationship (Eq. 3.31) between the critical surface tension and the solid–vapor surface tension by setting the contact angle to zero. Critical surface tension is, therefore, smaller than solid–vapor surface tension. The interfacial tension γ_{SL} varies with the nature of the liquid and is very large if the polarities of the solid and the liquid are very different from each other.

$$\gamma_c = \lim_{\theta \to 0}(\gamma_{LV}\cos\theta) = \gamma_{SV} - \gamma_{SL} \tag{3.31}$$

The linear relationship between cos θ and γ_{LV} can be written in the form of Eq. (3.32), where K is the slope of the line in the Zisman plot. The value

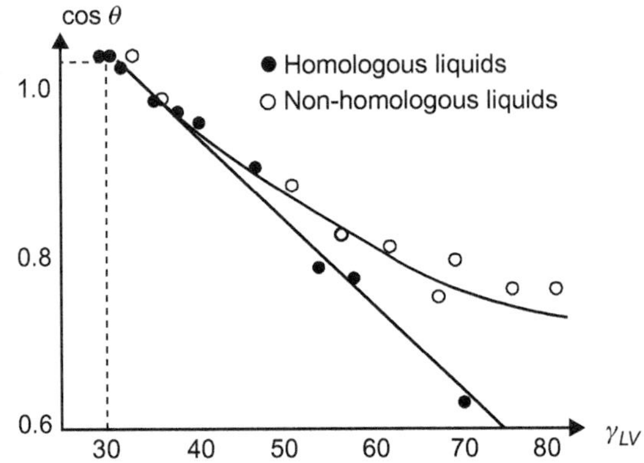

FIGURE 3.2 Zisman plot for homologous and non-homologous liquid series.

of K varies based on the testing liquid, but is usually in the range of 0.03–0.04 cm/dyne [20]. A constant slope would yield a linear Zisman plot, while a variable slope results in curvature. Equation (3.32) is an empirical relationship due to the failure by many workers to derive it theoretically [2,21,22]. An explanation for the curvature in the Zisman plot has been attributed to interactions such as hydrogen bonding between the liquid and solid phases. This explanation does not apply to an essentially non-interacting (neutral) surface such as polytetrafluoroethylene, which also exhibits a curved Zisman plot. An explanation can be found in the variability of the value of slope, K, illustrated here:

$$\cos\theta = 1 + K(\gamma_c - \gamma_{LV}) \tag{3.32}$$

The result of differentiation in Eqs. (3.32) and (3.8) is shown in Eqs. (3.33) and (3.34), respectively. Substitution from Eq. (3.14) into Eq. (3.34), assuming π_E is sufficiently smaller than γ_{SV}, yields Eq. (3.35). Clearly, K is a function of the liquid surface tension, thus explaining the curvature of the Zisman plot.

$$\frac{d(\cos\theta)}{d\gamma_{LV}} = -K \tag{3.33}$$

$$\frac{d(\cos\theta)}{d\gamma_{LV}} = \phi\left(\frac{\gamma_{SV}}{\gamma_{LV}^{3}}\right)^{1/2} \tag{3.34}$$

$$K = \left(\frac{\gamma_c}{\gamma_{LV}^{3}}\right) \tag{3.35}$$

Finally, Table 3.2 shows a comparison of the surface energy of a few polymers determined by the methods discussed in this chapter. There is general agreement among the results obtained using harmonic mean, geometric

TABLE 3.2 A Comparison of Solid Surface Tension (at 20°C) of Polymers by Different Methods [1]

Polymer	Harmonic Mean	Geometric Mean	Molten Polymer	Zisman Critical Surface Tension
Polyethylene	36.1	33.2	35.7	31
Polytetrafluoroethylene	22.5	19.1	26.5	18
Polymethyl methacrylate	41.2	40.2	41.1	39
Polyethylene terephthalate	42.1	41.3	44.6	43
Polyvinyl chloride	41.9	41.5	43.8	39
Polystyrene	42.6	42.0	40.7	33
Polyhexamethylene adipamide	44.7	43.2	46.5	46
Paraffin wax	31.0	25.4	35	23

TABLE 3.3 A Comparison of Fractional Polarity of Polymers by Different Methods [1]

Polymer	Harmonic Mean	Geometric Mean	Molten Polymer
Polyethylene	0.022	0	0
Polytetrafluoroethylene	0.089	0.026	–
Polymethyl methacrylate	0.245	0.107	0.281
Polyethylene terephthalate	0.221	0.085	–
Polyvinyl chloride	0.146	0.036	0.11
Polystyrene	0.099	0.014	0.168
Polyhexamethylene adipamide	0.344	0.211	–
Paraffin wax	0	0	0

mean, and polymer melt methods, with few exceptions. Table 3.3 gives the fractional polarity of the polymers listed in Table 3.2 using the equations in Section 3.2.1.

In summary, this chapter describes significant empirical equations for calculating surface tension (energy) of solid polymeric surfaces using contact angles and other parameters. A fairly detailed description of each equation has been presented to enable the reader a gain a deeper understanding of each technique.

3.7 OTHER METHODS FOR ESTIMATION OF SURFACE TENSION OF MOLTEN POLYMER

Since the surface tension of polymers is a manifestation of intermolecular forces, it may be expected to be related to other properties derived from intermolecular forces, such as work of cohesion internal pressure, and compressibility. Some of the relationships for calculation of surface tension of solids are listed [23].

3.7.1 Grunberg and Nissan's Relationship

The relationship between the work of cohesion, molecular volume, and surface tension was derived by Grunberg and Nissan in 1949 [24].

$V =$ molar volume
$N_A =$ *Avogadro number*
$V/N_A =$ molecular volume
$(N_A/V)^{2/3} =$ molecules/unit surface area
$\gamma =$ surface energy

$$\frac{\gamma V^{2/3}}{N_A^{2/3}} = \textit{surface energy per molecule}$$

$$\gamma N_A^{1/3} V^{2/3} = \textit{molar surface energy}$$

$$2\gamma N_A^{1/3} V^{2/3} = \textit{work of cohesion}$$

$$W_{coh} = 2\gamma N_A^{1/3} V^{2/3} \tag{3.36}$$

3.7.2 Hildebrand and Scott's Equation

An interesting empirical relationship between surface tension and solubility parameter was found by Hildebrand and Scott in 1950) [25].

$$\frac{\gamma}{V^{1/3}} \sim \frac{\textit{molar surface energy}}{V} = \frac{\Delta E_V}{V} = \delta^2$$

$\delta =$ Solubility parameter
$\Delta E_V =$ the energy of vaporization

$$\delta = 4.1\left(\frac{\gamma}{V^{1/3}}\right)^{0.43} \quad \text{(cgs units)} \tag{3.37}$$

3.7.3 McGowan's Equation

The surface tension is a manifestation of intermolecular forces; it may, therefore, be expected to be related to other properties derived from intermolecular forces, such as internal pressure, compressibility, and cohesion energy density [23]. This is indeed found to be so. In the first place there exists a relationship between compressibility and surface tension. According to McGowan [26] the correlation is seen in Eq. (3.38):

$$\begin{aligned} &K = compressibility \\ &K\uparrow \rightarrow \gamma\downarrow \\ &K\cdot\gamma^{2/3} = 1.33\times10^{-8}(\text{cgs units}) \end{aligned} \tag{3.38}$$

3.8 SHUTTLEWORTH'S EQUATION

This section is not entirely relevant to the subject matter of this chapter. It does, however, discuss an equation which purports to relate the surface tension of an unstrained solid with Helmholtz free energy of the solid's surface. Roughly speaking the approach was somewhat analogous to that used for liquids.

In 1950 Shuttleworth (*Proc Phys Soc London*, A 63, p. 444, 1950) proposed a relation between surface tension and surface energy on an unstrained solid, as given by the Shuttleworth equation in Eq. (3.39). This equation suggests surface tension of a solid can be derived from mechanics, although it provides no additional relation to the physics of surfaces.

$$\gamma = F + A\frac{\partial F}{\partial A} \tag{3.39}$$

A is surface area, γ is surface tension and F is surface free energy per unit area. γ was defined as the tangential stress (force per unit length) in the surface layer and F as the total Helmholtz free energy (H) per unit area of a surface per Eq. (3.39).

$$F = \frac{H}{A} \tag{3.40}$$

Shuttleworth's equation has been even considered as the second most important equation in surface physics after Young's equation. Shuttleworth's equation is viewed to be a thermodynamic equation that provides the excess surface stress at a solid surface [27]. In particular, F has been widely interpreted as the thermodynamic energy required for forming a unit area of new unstrained surface by cleaving, and the second term on the right-hand side in Eq. (3.39) as the term which separates the surface tension on a liquid from that on a solid.

There have been a number of claims that the derivations of the Shuttleworth equation are flawed [28–35]. The issue is conceptually complex and has resulted in a great deal of controversy [36–38]. Several published papers have criticized the application of the Shuttleworth equation. Other papers have rebuffed those criticisms as incorrect (e.g. Refs [36,39]) and made statements supporting its use. In 2012 Makkonen from VTT Technical Research Centre of Finland [27] published a paper in which he stated the interpretation of the Shuttleworth equation as a relation between surface tension and surface energy on an unstrained crystal was inappropriate. He also demonstrated, via mathematical derivation, how the misunderstanding has arisen. The author concluded the Shuttleworth equation should not be interpreted. Because it reduces to a basic definition, it does not provide an additional law of surface physics.

Makkonen [27] stated the evaluation of the surface stress should be based on its mechanical definition, as outlined by Gurtin and Murdoch [40] and Wolfer [41].

REFERENCES

[1] Wu S. J Polym Sci 1971;C34:19.
[2] Fowkes FM. Chemistry and physics of interfaces. Washington, DC: American Chemical Society; 1965, pp. 1–2.
[3] Meyer EF, Wagner RE. J Phys Chem 1966;70:3162.
[4] Meyer EF, Renner TA, Stec KS. J Phys Chem 1971;75:642.
[5] Good RJ, Girifalco LA. J Phys Chem 1957;61:904.
[6] Good RJ. Advan Chem Ser 1964;43:74.
[7] Good RJ. In: Patrick RL, editor. Treatise on adhesion and adhesives, vol. 1. New York: Marcel Dekker; 1967.
[8] Kwok DY, Neumann AW. Contact angle measurement and contact angle interpretation. Adv Colloid Interf Sci 1999;81:167–249.
[9] Fowkes FM. J Phys Chem 1962;66:382.
[10] Owens DK, Wendt RC. J Appl Polym Sci 1969;13:1741.
[11] Kaelble DH. J Adhesion 1970;2:50.
[12] Wu S. Polymer interface and adhesion. 1st ed. New York: Marcel Dekker, Inc.; 1982.
[13] Fowkes FM. Acid–base interactions in polymer adhesion. In: Mittal KL, editor. Physicochemical aspects of polymers surfaces, vol. 2. New York: Plenum Press; 1983.
[14] Fowkes FM, Maruchi S. Org Coat Plast Chem Prep 1977;37:605.
[15] Fowkes FM, Mostafa MA. Ind Eng Chem Prod Res Dev 1978;17:3.
[16] Guggenheim EA. J Chem Phys 1945;13:253.
[17] Fox HW, Zisman WA. J Colloid Sci 1950;5:514.
[18] Fox HW, Zisman WA. J Colloid Sci 1952;7:109.
[19] Fox HW, Zisman WA. J Colloid Sci 1952;7:428.
[20] Adamson AW. Physical chemistry of surfaces. 2nd ed. New York: Wiley-Interscience; 1967.
[21] Fowkes FM. Chemistry and physics of interfaces, vol. 1. Washington, DC: American Chemical Society; 1965, pp. 1–12.
[22] Fowkes FM. Chemistry and physics of interfaces, vol. 2. Washington, DC: American Chemical Society; 1971, pp. 153–168.

[23] van Krevelen DW, te Nijenhuis K. Properties of polymers. Oxford, UK: Elsevier; 2009.
[24] Grunberg L, Nissan AH. Trans Faraday Soc 1949;45:125.
[25] Hildebrand JH, Scott RL. The solubility of non-electrolytes. 3rd ed. New York: Reinhold Publishing Corp; 1950.
[26] McGowan JC. Polymer 1967;8:57.
[27] Makkonen L. Misinterpretation of the Shuttleworth equation. Scr Mater 2012;66:627–9.
[28] Gutman EM. J Phys: Condens Matter 1995;7:L663.
[29] Lang G, Heusler KE. J Electroanal Chem 1999;472:168.
[30] Bottomley DJ, Ogino T. Phys Rev B 2001;63:165412.
[31] Makkonen L. Langmuir 2002;18:1445.
[32] Bottomley DJ, Ogino T. Phys Rev B 2004;70:159903 (E)
[33] Bottomley DJ, Makkonen L, Kolari K. Surf Sci 2008;603:97.
[34] Gutman EM. J Phys: Condens Matter 2010;22:428001.
[35] Marichev VA. Prot Met Phys Chem Surf 2011;47:25.
[36] Kramer D, Weissmuller J. Surf Sci 2007;601:3042.
[37] Vermaak JS, Maus CW, Kuhlmann-Wilsdorf D. Surf Sci 1968;12:128.
[38] Marichev VA. Philos Mag 2009;89:3037.
[39] Rusanov AI. Surf Sci Rep 2005;58:111.
[40] Gurtin ME, Murdoch AI. Int J Solids Struct 1978;14:431.
[41] Wolfer WG. Acta Mater 2011;59:7736.

Chapter 4

Surface and Material Characterization Techniques

4.1 INTRODUCTION

In the analysis of adhesion bonding or troubleshooting, it is of great importance to have a working knowledge of the composition and structure of the adherends' surfaces. Characterization of bonding surfaces can aid in both the design of a bond and the failure analysis if an adhesion bond fails.

Surface analysis requires the use of a number of analytical techniques including microscopic, spectroscopic, chemical, and physical methods that provide different types information about the surface of a sample. The term "sample" refers to any piece of material, structure, device, or substance being studied. The probed region may be the outermost layer of atoms (the only true surface, for purists), or it may extend to several microns (millionths of a meter) beneath the sample surface, depending on the technique used. The analysis is done to provide information about such characteristics as the chemical composition, the level of trace impurities, or the physical structure or appearance of the sampled region. Such information is of importance to researchers or manufacturers, who must understand the materials in order to verify a theory or make a better product [1]. In addition to surface-specific methods, this chapter describes other analytical techniques which can be applied to characterize the bulk properties of materials.

4.2 SURFACE ANALYSIS TECHNIQUES

Many of the techniques used to probe surfaces utilize a beam of ions [e.g., secondary ion mass spectroscopy (SIMS)] to strike the surface and knock off atoms of the sample material. These atoms are ionized, identified, and measured using a technique known as mass spectrometry. Another technique is ion scattering spectroscopy (ISS), which probes the outermost layer of the surface. If a beam of ions is directed at a sample surface, then a certain

Surface Treatment of Materials for Adhesive Bonding. DOI: http://dx.doi.org/10.1016/B978-0-323-26435-8.00004-6

number of beam particles will be elastically reflected. Measurement of the energy of the backscattered particles can be used to identify the mass of these atoms. The intensity of the scattered ions, as a function of the angle of emission, provides information regarding the surface crystallographic structure. The variation in the intensity of the scattered beam is partly due to the shadowing of substrate atoms by adsorbed atoms. By use of scattering theory, knowledge of the sites that the surface atoms occupy can be derived [2,3].

In other techniques, surface is struck with electrons [Auger spectrometry, energy dispersive X-ray spectroscopy (EDS), or EDX] or X-rays [electron spectroscopy for chemical analysis (ESCA), also called X-ray photoelectron spectroscopy (XPS) and total reflection X-ray fluorescence (TXRF)], to measure the resulting electron or photon emissions to probe the sample. Extremely high energy helium nuclei bounce off a sample and are measured as a sensitive indicator of surface layer composition and thickness [Rutherford backscattering spectrometry (RBS)].

Surface structure, or topography, on a microscopic scale, is observed by using optical microscopy (OM), scanning electron microscopy (SEM), environmental SEM, and atomic force or scanning probe microscopy (AFM/SPM). Note that most of these techniques must be performed in a high-vacuum environment. What is characterized is the substrate under those conditions, not at the atmospheric conditions under which most bonding processes are operated.

Surface characterization techniques are often constrained by the information obtained from the measurements, but also with the conditions required for the analysis. For instance, techniques that require vacuum cannot be used for analyzing liquid or volatile chemicals. This limitation means that a given technique may have a limited range of application. It is thus beneficial to employ more than one technique and combine results from several methods to get as much information as possible. Among the criteria for selection of one or more of the techniques are:

1. The required depth of the surface analysis;
2. The required chemical information – elemental or functional group;
3. Quantification – quantitative, semi-quantitative or qualitative requirements;
4. Lateral resolution – whether very small features or spots need to be detected; and
5. Is other information required, such as topographical information, about the sample surface.

Table 4.1 and Figure 4.1 show a comparison of the sampling depth for a select number of common methods and relatively new techniques which are capable of measurements in the nano-scale. These analyses can focus on a much shallower thickness of the surface and virtually yield analyses of the outermost layers of a polymer article.

TABLE 4.1 Comparison of Average Sample Depth for Various Surface Analysis Techniques [4]

Analysis Method	Sampling Depth
Fourier transform infrared spectroscopy (FTIR), Raman spectroscopy, conventional secondary ion mass spectroscopy (conventional SIMS)	<2 μm
Energy dispersive X-ray (EDS or EDAX)	<5,000 Å
Rutherford backscattering (RBS)	<400 Å
Surface secondary ion mass spectroscopy (surface SIMS)	<300 Å
Electron spectroscopy for chemical analysis (ESCA), also called X-ray photoelectron microscopy (XPS)	<100 Å
Auger electron spectroscopy (AES)	<100 Å
Ion scattering spectroscopy (ISS), also called low-energy ion scattering (LEIS)	<2 Å
Time-of-flight secondary ion mass spectroscopy (TOF-SIMS)	1–3 monolayers
Atomic force microscopy (AFM) or scanning probe microscopy (SPM)	0.1 Å

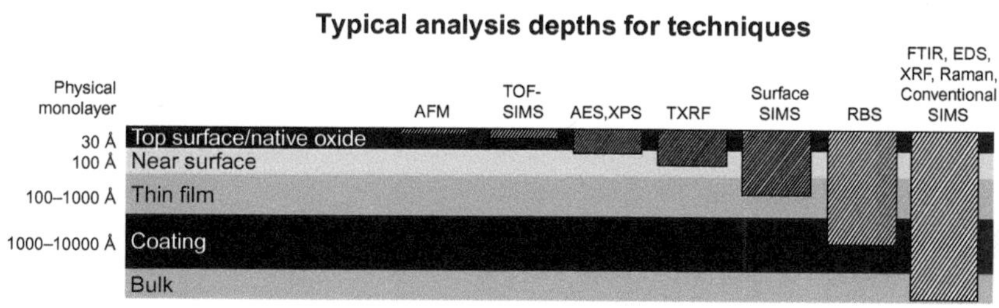

FIGURE 4.1 Sampling depth for surface analysis techniques [1].

4.3 OPTICAL (LIGHT) MICROSCOPY

One of the first techniques used to study the topography of a surface is optical microscopy, also called light microscopy. An optical microscope (Fig. 4.2) usually has a single eyepiece which can often be fitted with a camera for photography. In contrast another device called a stereomicroscope (Fig. 4.3) uses a light microscope with dual eyepieces. The same depth of field versus magnification relationships are present. It differs from the stereomicroscope examination in that the specimen examined is sometimes embedded in a holding material (Bakelite or epoxy, usually) called a metallurgical

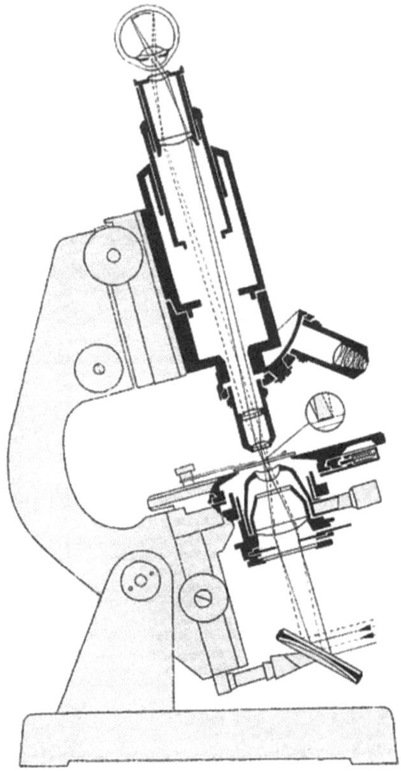

FIGURE 4.2 Basic design of a transmission optical microscope.

mount, and the microscope possesses an adjustable stage. This type of specimen configuration provides a planar cross-sectional view of a fracture surface along with adjacent material for evaluation.

Conventional optical microscopes have a resolution limited by the size of submicron particles approaching the wavelength of visible light (400–700 nm). The two types of available optical microscope, which depend on the type of light exposure, include:

1. Transmission: beam of light passes through the sample; and
2. Reflection: beam of light reflected off the sample surface. An example is the polarizing or petrographic microscope for which the samples are usually fine powder or thin slices (transparent). Another example is the metallurgical or reflected light microscope which is used for the surfaces of materials, especially opaque ones.

For metals, examination is often conducted when it is in an as-polished, or un-etched, condition, which reveals features such as corrosion, erosion, cracks,

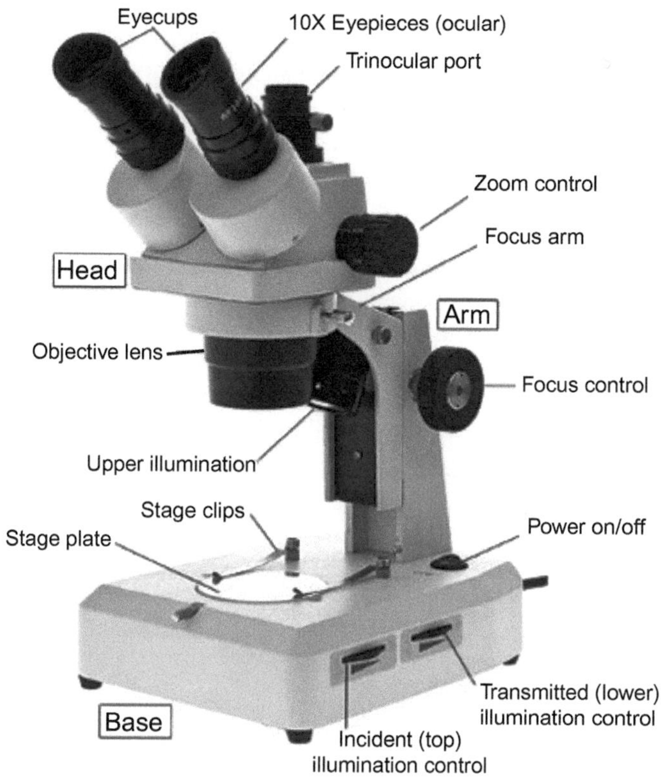

FIGURE 4.3 An example of a stereomicroscope.

laps/folds, gas and/or shrinkage porosity, cold shuts, inclusions, and graphite forms for cast iron. Once etched with an appropriately selected etchant, optical microscopy of metals reveals microstructural phases such as ferrite, martensite, and austenite (for steel), and grain size and shape, any multi-phase structures, flow lines, alloy depletion, decarburization, and remnants of a dendritic structure if the part was cast.

Similarly, the optical microscope can be used to evaluate the surfaces of non-metallic parts such as ceramics and plastics or elastomers. These are similar to metal samples, but the softness of polymers can present difficulties. Experience is important in obtaining good results in the analysis of polymeric surfaces for optical microscopy.

Knowledge of the above assists the failure analyst in determining part history and conformance to specifications. Optical microscopy also allows the analyst to observe the general structure and configuration of the part in the region of fracture origin. This can be critical to a failure analysis if detrimental microstructural features are present (decarburization, alpha case, inclusions, etc.)

in the vicinity of a fracture origin. Changes in fractographic features observed during scanning electron microscopy might be explained when optical microscopy reveals changes in metallurgical structure throughout a part's thickness.

Once a properly prepared mount is available for examination of a polymer or other soft material, this technique can be as revealing as it is for metals. Knit lines, porosity, and other molding issues are revealed. The size and shape of reinforcing fibers or particles can be observed and measured and any changes throughout the part in their distribution can be observed and documented. As with metals, crack profiles can be studied and microcracking revealed. Occasionally, in strained regions of ductile polymers that are not yet fully formed, cracks can be observed, providing information about how a part might be absorbing service stresses.

Figures 4.4 and 4.5 show examples of a normal photograph and one taken under an optical microscope. The optical micrograph provides useful quantitative data about the surface in addition to the more specific shape of surface

FIGURE 4.4 Photograph of overbraid of a flexible hose.

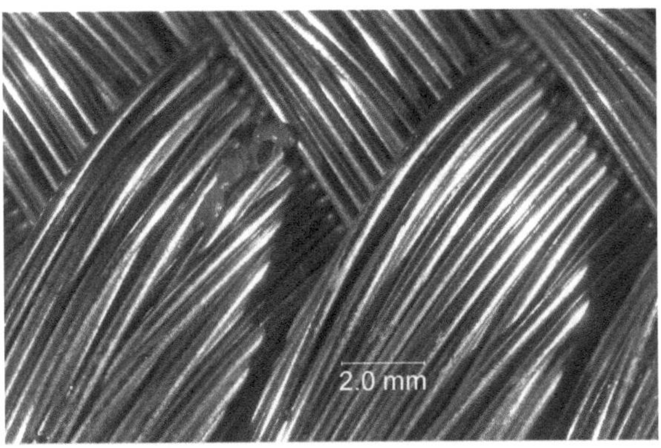

FIGURE 4.5 Optical micrograph of overbraid of a flexible hose.

topography. Missing are fine structures and chemical composition of surface species, which are obtained by scanning electron microscopy techniques and energy dispersive X-ray spectroscopy.

4.4 INFRARED SPECTROSCOPY

During analysis by infrared spectroscopy (IR), a sample is subjected to electromagnetic radiation in the infrared region of the spectrum. The wavelengths absorbed by the sample depend on the nature of the chemical groups present. These wavelengths are defined by a wave number (cm^{-1}) obtained by dividing the number 10,000 by the wavelength in microns. The ranges of various wave numbers are given in Table 4.2. Absorbance is defined in Eq. (4.1) and varies from 0 (100% transmission) to infinity (0% transmission). Absorbance is related to concentration of the absorbing species by Beer's law [Eq. (4.1)].

$$\text{Absorbance} = \text{Log}\frac{100}{\%\text{Transmission}} \tag{4.1}$$

$$\text{Absorbance} = K.c.l \tag{4.2}$$

K = a constant that is occasionally called the *extinction coefficient*
c = concentration of the absorbing species
l = sample thickness

Equation (4.2) can be used to determine the concentration of a compound in a solution when the value of K is known for that compound. Chemical bonds, C–O, O–H, etc., absorb different amounts of infrared energy over various wavelengths. Absorption patterns vary from sharp to broad for different bonds. Peak IR absorption wavelength (wave number) is characteristic of a chemical bond. Absorption over the infrared spectrum is a fingerprint characteristic of an organic material. Qualitative identification can be achieved by obtaining and analyzing the IR spectrum of a material.

TABLE 4.2 Definition of Electromagnetic Wave Ranges

Definition of Range	Wave Number (10^4 cm^{-1})	Wave length (μm)
Ultraviolet	5–2.5	0.2–0.4
Visible	2.5–1.42	0.4–0.7
Near IR	1.42–0.4	0.7–2.5
IR	0.4–0.02	2.5–50
Far IR	0.02–0.0012	50–830

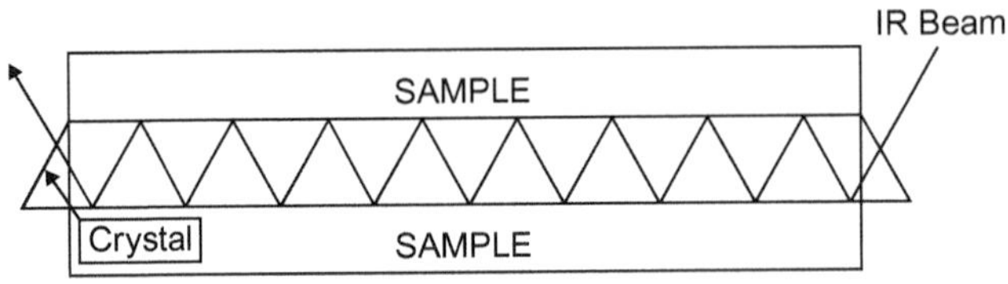

FIGURE 4.6 Schematic of the sample configuration relative to the reflective crystal in attenuated total reflectance IR.

Infrared spectra can also be obtained by reflecting the IR beam on the surface of a sample. This technique is applied when it is not possible to obtain an IR spectrum by a *transmission* technique. Attenuated total reflectance (ATR), also known as ATIR (attenuated total internal reflectance), is based on multiple internal reflectance of the IR beam on the sample surface using a high refractive index crystal (e.g., thallium bromo-iodide). The IR beam is entered into the crystal at an angle, and after about 25 internal reflections (5 cm), we may obtain a spectrum similar to that of the transmission. The sample is in tight contact with both surfaces on the crystal, as seen in Fig. 4.6.

Modern infrared spectrometers use Fourier transform calculations. The method is called Fourier transform infrared spectroscopy (FTIR).

4.5 RAMAN SPECTROSCOPY

When incident light strikes a sample, part of the light is scattered. Most of the scattered light is of the same wavelength as the incident light; this is called Rayleigh scattering. Some of the light is scattered at a different wavelength; this is called Raman scattering. The energy difference between the incident light and the Raman scattered light is called the Raman shift. This difference is equal to the energy required to vibrate or rotate the molecule. Several different Raman shifted signals can often be observed in a single sample; each is associated with different vibrational or rotational motions of molecules in the sample. The particular molecule and its environment will determine what Raman signals will be observed. In practice, because the Raman effect is so small, a laser is used as the source of the incident light. A plot of Raman intensity versus the frequency of the Raman shift is known as a Raman spectrum. It usually contains sharp bands characteristic of the functional groups of the compounds or materials. This information can be used to determine chemical structure and identify the compounds present. It is complementary to FTIR because it uses a different method to measure molecular vibrations.

Raman spectroscopy is a functional technique for qualitative analysis as well as for discrimination of organic and/or inorganic compounds in mixed materials. A Raman spectrum can be obtained from samples that are as small as 1 μm [1] (see Fig. 4.7). The intensities of bands in a Raman spectrum depend on the sensitivity of the specific vibrations to the Raman effect and are

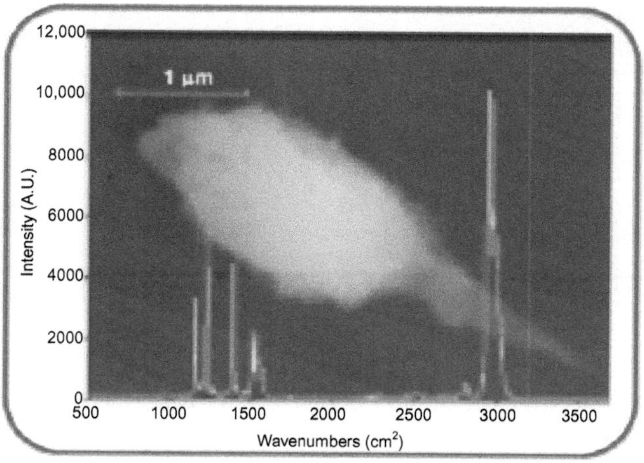

FIGURE 4.7 Raman spectroscopy identified this micron-sized particle as polyethylene.

proportional to the concentration. Thus, Raman spectra can be used for semi-quantitative and quantitative analysis. The technique is used for identification of organic molecules, polymers, biomolecules, and inorganic compounds both in bulk and as individual particles. Raman spectroscopy is particularly useful in determining the structure of different types of carbon (diamond, graphitic, diamond-like carbon, etc.) and their relative concentrations.

4.6 SCANNING ELECTRON MICROSCOPY (SEM)

Scanning electron microscopy (SEM) is a useful technique for the analysis of plastic surfaces as well as any surface that survives in a vacuum. Almost all SEMs start by sputtering the surface with a thin layer of gold metal. This ensures that the surface is conductive, which is a requirement. SEM involves a finely collimated beam of electrons that is focused into a small probe that scans across the surface of a specimen. The interactions between the beam and the material result in the emission of electrons and photons as the electrons penetrate the surface. The emitted particles are collected with the appropriate detector to yield information about the surface. The final product of the electron beam collision with the surface topology of the sample is an image of the surface (see Fig. 4.8).

Figure 4.8 shows the result of topical SEM analysis of a drug-infused polymer coating on a stainless steel device. Scanning electron microscopy of the surface at 750 times magnification is shown in Fig. 4.8(a) and at 35,000 times magnification in Fig. 4.8(b). The higher magnification revealed the presence of crystals on the surface of the coating.

A helpful attachment to the SEM is the *electron microprobe*. An electron beam is focused on a sample surface causing ionization to a depth of a few

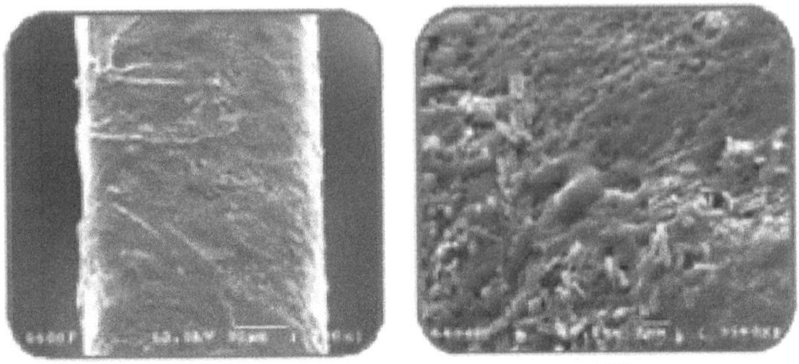

(a) (b)

FIGURE 4.8 SEM images of a polymer-coated device: (a) 750 times magnification; and (b) 35,000 times magnification [1].

micrometers. Energies and wavelengths of the emitted X-ray during the de-excitation cycle are characteristic of the elements present in the top layers of the sample. The result is not a true surface analysis, although the electron microprobe allows analysis of various spots of the sample surface [5].

4.7 RUTHERFORD BACKSCATTERING THEORY (RBS)

Rutherford backscattering (RBS) is based on collisions between atomic nuclei, and derives its name from Lord Ernest Rutherford, who in 1911 was the first to present the concept of nuclei in atoms. It involves measuring the number and energy of those ions in a beam that backscatter after colliding with atoms in the near-surface region of a sample. With this information, it is possible to determine atomic mass and elemental concentrations versus depth below the surface. RBS is ideally suited for determining the concentration of trace elements heavier than the major constituents of the substrate. It has, however, poor sensitivity for light masses, and for the makeup of samples well below the surface.

When a sample is bombarded with a beam of high-energy particles, the vast majority of particles are implanted into the material and do not escape. This is because the diameter of an atomic nucleus is approximately 1×10^{-15} meters, while the spacing between nuclei is closer to 2×10^{-10} meters. A small fraction of the incident particles undergo a direct collision with a nucleus of one of the atoms in the upper few micrometers of the sample. This "collision" does not actually involve direct contact between the projectile ion and target atom. Rather, energy exchange occurs because of Coulombic forces between nuclei in close proximity to one another. However, the interaction can be modeled accurately as an elastic collision using classical physics.

The energy measured for a particle backscattering at a given angle depends upon two processes. Particles lose energy while they pass through the sample, both before and after a collision. The amount of energy lost is dependent on that material's stopping power. A particle will also lose energy as a result of the collision. The ratio of the energy of the projectile before and after collision is called the kinematic factor.

The numbers of backscattering events that occur from a given element depend upon two factors: the concentration of the element and the effective size of its nucleus. The probability that a material will cause a collision is called its *scattering cross section.*

4.8 ENERGY DISPERSIVE X-RAY SPECTROSCOPY (EDS)

Energy dispersive X-ray spectroscopy (EDS) is a standard method for identifying and quantifying elemental compositions in a very small sample of material (even a few cubic micrometers). In a properly equipped SEM, the atoms on the surface are excited by the electron beam, emitting specific wavelengths of X-rays that are characteristic of the atomic structure of the elements. An energy dispersive detector (a solid-state device that discriminates among X-ray energies) can analyze these X-ray emissions. Appropriate elements are assigned, yielding the composition of the atoms on the specimen surface. This procedure is called energy dispersive X-ray spectroscopy (EDS) and is useful for analyzing the composition of the surface of a specimen (Figs. 4.9 and 4.10).

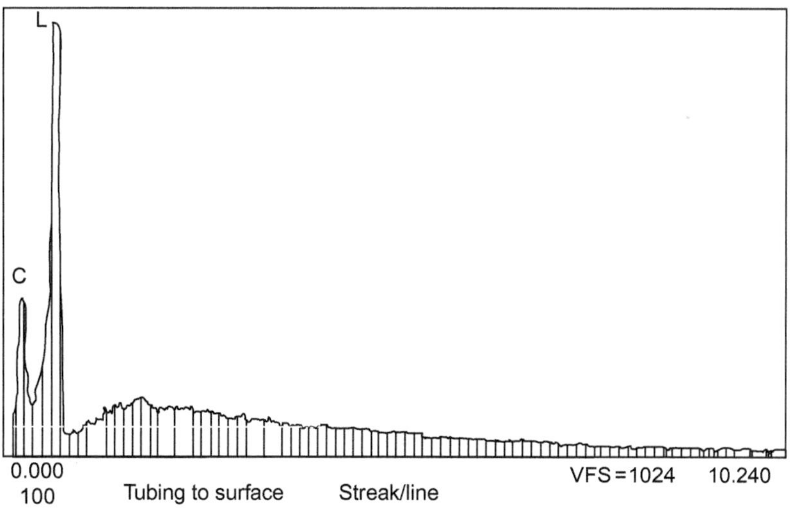

FIGURE 4.9 EDX elemental analysis of a clean PTFE surface. (*Courtesy of DuPont Fluoroproducts.*)

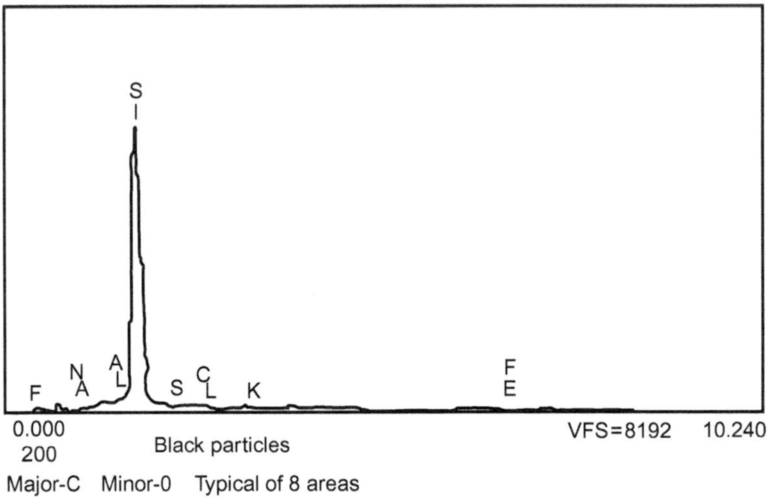

FIGURE 4.10 EDX elemental analysis of a contamination spot on the PTFE surface. (*Courtesy of DuPont Fluoroproducts.*)

4.9 TRANSMISSION ELECTRON MICROSCOPY (TEM)

Transmission electron microscopy (TEM) is similar to SEM with the exception that the beam passes through the sample. A high voltage (80–200 keV) highly focused electron beam is passed through a thin, solid sample, typically 100–200 nm in thickness. Electrons undergo coherent scattering or diffraction from lattice planes in the crystalline phase of materials, yielding phase identification. Characteristic X-rays that are generated can be seen in a separate detector permitting qualitative elemental analysis.

Figure 4.11 shows a TEM micrograph of particles of polyvinylidene fluoride produced by emulsion-polymerization in an aqueous phase. Figure 4.12 shows the capability of TEM in revealing the structure of a material, as illustrated in the graphite sample. Lattice thickness and interplanar spacing of graphite can be measured from the micrograph due to its excellent resolution.

4.10 ELECTRON SPECTROSCOPY FOR CHEMICAL ANALYSIS (ESCA)

Electron spectroscopy for chemical analysis (ESCA) is a widely used analytic technique for characterizing polymer surfaces. ESCA, also called X-ray photoelectron spectroscopy (XPS), is able to detect all elements except hydrogen. A sample is irradiated by X-ray beams, which interact with the inner electron shell of atoms. Photon energy of the X-rays is transferred to an electron in the inner shell, enabling it (the photoelectron) to escape from the sample surface.

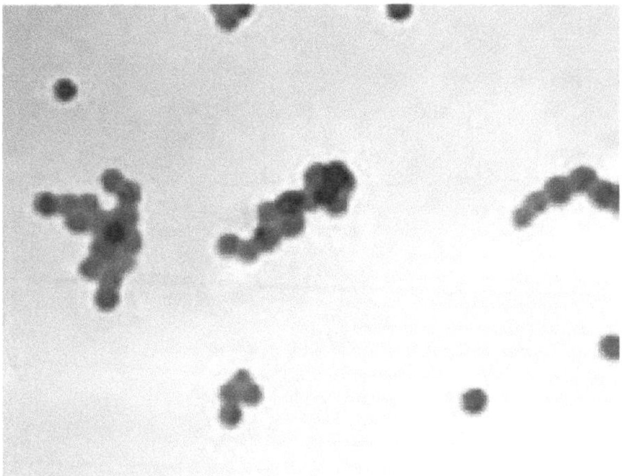

FIGURE 4.11 Transmission electron micrograph of particles of emulsion-polymerized PVDF particles ($D_p = 128$) [6].

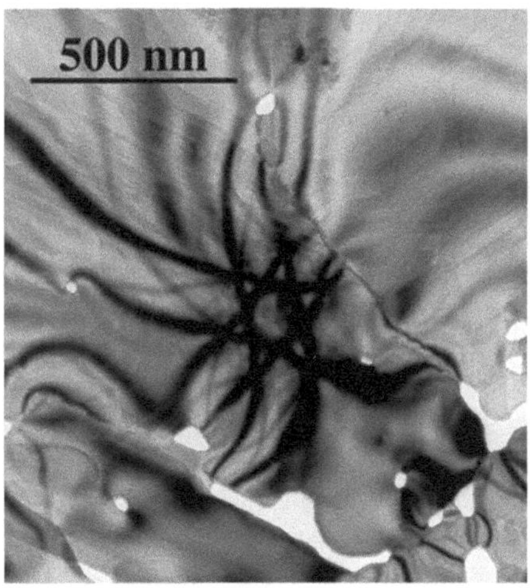

FIGURE 4.12 Bright field TEM micrograph of a thin sapphire film. (http://matwww.technion.ac.il/Kaplan/Gallery/gallery1.html).

An analyzer measures the kinetic energy of the photoelectron, which is equal to the electron's binding energy. Knowledge of the binding energy allows identification of the element. The chemical bond of an atom to other elements shifts the bonding energy of the photoelectron to higher or lower values. This shift in binding energy provides structural information about a molecule.

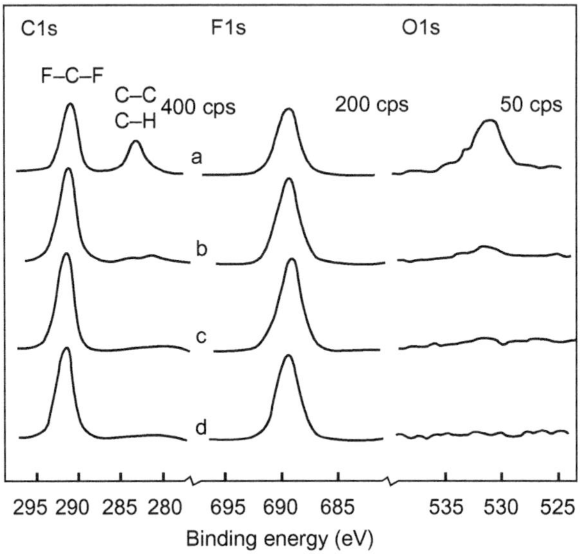

FIGURE 4.13 XPS spectra of polytetrafluoroethylene (a) before cleaning; and after cleaning with an ultrasonic cleaner in (b) methanol; (c) acetone; and (d) n-heptane [13].

Operation and maintenance of ESCA equipment and interpretation of its data are quite complex. Samples intended for ESCA and other surface analysis should be handled carefully because minute contamination can mask the surface structure of the samples. To alleviate this type of complication, the sample surface can be washed with volatile solvents such as methanol, acetone, hydrocarbons, and fluorocarbons using an ultrasound bath. Typically, analysis is conducted before and after the surface wash in studying a sample that has been handled and/or contaminated. Another application of surface wash is removal of loose material that may be weakly bonded to the surface. The interested reader is encouraged to refer to other sources to gain an in-depth understanding of electron spectroscopy for chemical analysis [4,8–12].

A typical spectrum of ESCA shows peaks as a function of binding energy, shown in Fig. 4.13 for polytetrafluoroethylene. C1s and F1s peaks on a "clean" surface indicate that the PTFE surface is comprised of only carbon and fluorine. The energy shift can be curve-fitted by trial and error to determine the functional groups on the surface. The most simplified report that ESCA generates is a survey of the atomic composition of the surface elements with the exception of hydrogen. A helpful tool to investigate surface changes of a polymer is the ratio of other elements to carbon. Table 4.3 shows the effect of sodium etching on the surface composition of a few different fluoropolymers.

TABLE 4.3 Effect of Sodium Etching on the Surface Composition and Adhesion Bond Strength of Fluoropolymers [14]

Polymer	Treatment	Surface Chemical Analysis (%) by ESCA						
		F/C Ratio	Cl/C Ratio	O/C Ratio	Cl	C	F	O
PTFE	–	1.60	–	–	–	38.4	61.6	–
PTFE	Tetra–Etch*[1] (1 minute)	0.011	–	0.20	–	82.2	0.9	16.9
PTFE	N/1 min[2]	0.005	–	0.14	–	87.2	0.4	12.4
PVF	–	0.42	–	–	–	70.4	29.6	–
PVF	Tetra–Etch®[1] (30 minute)	0.21	–	0.026	–	80.7	17.2	2.1
ECTFE	–	0.64	0.27	–	14.1	52.5	33.4	–
ECTFE	Tetra–Etch®[1] (1 minute)	0.16	0.05	0.12	3.8	74.9	12.2	9.1

[1] *Supplied by WL Gore Corporation.*
[2] *Treatment with 1 mole solution naphthalenide in tetrahydrofuran at room temperature.*

4.11 AUGER ELECTRON SPECTROSCOPY (AES)

Auger refers to the emission of a secondary electron after the surface of a solid has been bombarded with electrons. This is a different phenomenon from photoelectron emission (the basis of ESCA) that occurs after bombardment of a solid surface with low-energy X-rays [15]. The energy of the Auger electron depends on the chemical bonding state of the element from which it was emitted. The depth of the escape of an Auger electron is less than 1 nm, with metals having the shortest escape depth and organic material the deepest. Lateral resolution of AES is around 1 μm. Auger spectroscopy thus characterizes the surface of materials.

This method uses a low-energy electron gun, with a power less than 5 keV, to lessen the heating and decomposition of the surface. The number of electrons and their resulting energies are detected by a counter and an energy analyzer. The energy of the electron identifies the element, while the number of emitted electrons indicates the surface concentration of the element.

4.12 ION SCATTERING SPECTROSCOPY (ISS)

When a beam of ions hits a solid surface, projectiles will be scattered back into the vacuum after one or more collisions with target atoms of the top few layers [16]. Measurement of the energy of the backscattered particles can be used to identify the mass of these atoms. The technique is called ion scattering spectroscopy (ISS). The term actually encompasses several techniques depending on the energy of the primary ion beam. Low-energy ion scattering spectroscopy (LEIS) refers to primary energies in the range of 100 eV to 10 keV, medium-energy ion scattering (MEIS) to a range from 100 to 200 keV, and high-energy ion scattering (HEIS) to energies between one and several MeV. The LEIS technique is often called ion scattering spectroscopy (ISS), the term we use below (meaning LEIS), while the HEIS technique is best known as Rutherford backscattering spectroscopy (RBS).

Low-energy ion scattering is attractive as a surface-specific technique. Spectra are usually obtained using noble gas ion beams from 0.5 to 3 keV [17]. Due to strong electron affinity of inert gas ions, the probability of electron transfer is very high in the initial collision with a surface atom. After two or more collisions, most ions will be neutralized, so a detector set to analyze only ions of the same type as those in the incident beam will almost entirely detect ions that have had only one collision with a target atom. Projectiles entering the solid will be discarded since they would need several scattering events to return back to the surface and exit.

The practical use of ISS is determined by its extreme sensitivity to only the top surface layer (for standard experimental arrangements) or two monolayers (for grazing incidence). Typical applications include composition of catalytic surfaces, thin film coatings, adhesion, and arrangement of surface atoms including the localization of adsorbed atoms. Quantification of surface analysis using low energy ions is hampered by the uncertainty of the inelastic losses and the neutralization rate depending on ion trajectories. In addition, overlapping peaks and multiple scattering must be taken into account; thus computer simulation becomes an indispensable tool.

4.13 SECONDARY ION MASS SPECTROSCOPY (SIMS)

Secondary ion mass spectroscopy (SIMS) is a valuable technique for identifying the structure and composition of polymer surfaces and complements ESCA. ESCA spectra for similar materials are difficult to resolve, while SIMS can differentiate among several polymers. This is partly due to the smaller sampling depth required by SIMS. In a typical analysis, the surface of the polymer sample is bombarded by primary ions at low current density, principally intended to minimize alteration of the sample surface by irradiation. The polymer surface generates positive and negative ions that are analyzed using a mass analyzer. The results of detailed analysis provide chemical structure and composition data about the surface. A traditional shortcoming of SIMS, however, is its inability to perform quantitative analysis.

4.14 TIME-OF-FLIGHT SECONDARY ION MASS SPECTROSCOPY (TOF-SIMS)

TOF-SIMS can be utilized to determine the structure of a sample as a function of sampling depth. Time-of-flight is used to measure the mass of an ion, so TOF-SIMS is a means of mass analysis. TOF-SIMS can be useful in determining stratification of resins in coatings or diffusion of atoms into an area underneath a surface [18,19].

The highly surface sensitive technique TOF-SIMS provides detailed chemical information for the outermost 1–2 nm of samples, i.e., it is somewhat more surface sensitive than XPS [20]. The output is essentially a high-resolution mass spectrum from the surface, but TOF-SIMS can also be used as a "chemical microscope" to get ion distribution images with lateral resolution <100 nm for metals, and, for organic compounds, less than about 0.5–1 ~m. The main advantage with TOF-SIMS is that, in addition to elemental information, detailed chemical information is obtained on a molecular level regarding functional groups and molecular fragments such as oligomers, monomers, and radicals. Another advantage is high sensitivity with a detection limit of the order of ppm–ppb. However, the technique suffers from some drawbacks. For example, due to the high surface sensitivity, contamination effects can lead to artifacts. Therefore, care should be taken during preparation, storage, and transport of samples, and it is best to avoid materials such as soft plastic bags due to the risk of contamination from plasticizers and polydimethylsiloxane. Another drawback with TOF-SIMS is that the results are not straightforward to quantify, as XPS results are.

4.15 MASS SPECTROSCOPY OR SPECTROMETRY (MS)

Mass spectroscopy (MS) is applicable to both organic and inorganic substances. Mass spectroscopy is a quantitative technique that allows the study of the structure of organic matter. A sample is degraded into fragments that are identified through fractional mass differences in atoms of the same principal mass number [21]. There are two types of mass analysis. One is high resolution which differentiates between fractional mass differences. The second is mass analysis that only differentiates principle mass numbers. The latter is more common. This method requires extremely low pressure (high vacuum) and the sample fragments must be volatile. Charged fragments are ejected into the vacuum. Strictly speaking, this does not mean these fragments are volatile (which implies evaporation). The determination relies on the pattern of fragmentation of a molecule upon ionization. These patterns are distinct, can be reproduced, and are additive for mixtures. A mass spectrum is a graph that shows the measured values of ion intensities per unit charge (m/e, mass divided by charge). The mass spectrum is unique for each compound (Fig. 4.14). In a high-resolution spectrum, the deviation of the

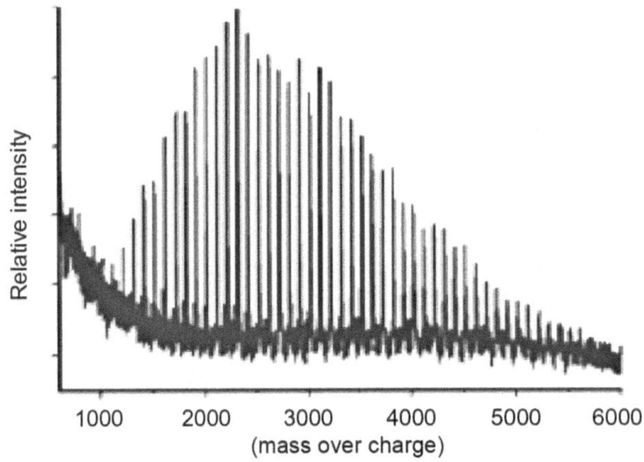

FIGURE 4.14 Matrix-assisted laser description ionized time-of-fight (MALDI-TOF) mass spectrum of poly(methylmethacrylate) (PMMA) [22].

molecular weight of each fragment from an integral value is used to determine the elemental composition.

There are several methods for ionization of organic compounds including particle bombardment, chemical and field ionization, electron impact, field desorption, and laser pulse. For example, in laser micro-mass analysis methods, a laser pulse is used to supply the necessary energy to volatilize a sample from the surface for mass spectroscopy [23]. Mass spectrometry is often used in conjunction with gas chromatography to identify the separated components. The working technology of mass spectrometry is quite complex [24].

4.16 ATOMIC FORCE MICROSCOPY (AFM)

This technique [2] is part a group of methods general known as Scanning Probe Microscopy (SPM). They include Atomic Force Microscopy (AFM), Scanning Tunneling Microscopy (STM), Magnetic Force Microscopy (MFM), Chemical Force Microscopy (CFM), Lateral Force Microscopy (LFM), and Near Field Scanning Optical Microscopy (NSOM) where a surface is imaged at high (and in some cases atomic) resolution by rastering an atomically sharp tip in close, but not direct, contact with the surface back and forth across it. The measurement of the strength of the interaction between the tip and the surface is combined with the measurement of the relative position of the tip to produce an image of interaction strength as a function of position which (depending on the particular technique) represents surface topography or chemistry.

Atomic force microscopy [25] produces high resolution three-dimensional images by scanning a sharp physical probe over a sample

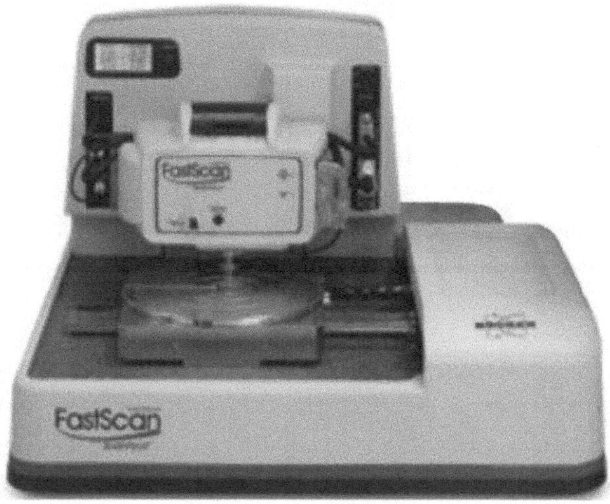

FIGURE 4.15 Photograph of a Bruker Dimension FastScan AFM Instrument. (*Courtesy Bruker Corp*, www.bruker.com, June 2013.)

surface while keeping the tip in close proximity to the sample. The tip is typically integrated into a cantilever beam which allows for both the physical support of the force sensing tip and displacement sensing via an optical beam bounce system. The cantilever/tip assembly is in turn mounted on a three-dimensional scanner. The scanner is a piezo actuator that is physically configured as a tube, which is both rigid and allows for motion in all three directions. A laser is reflected off the backside of the scanner and the reflected light is placed on a position sensitive photodetector. Scanning action is produced by applying a voltage to the appropriate part of the tube to produce motion in X, Y, and Z. The tip-sample proximity is controlled by a feedback loop. The feedback mechanism, in part, determines the operating mode of the instrument.

For example the Bruker Dimension 3000 (Fig. 4.15) and Multimode are conventional optical beam bounce scanning probe microscopes (SPM). Both instruments utilize commercially available cantilever/tip assemblies. Both systems utilize a four segment photodetector for sensing cantilever displacement. A four segment photodetector allows for sensing of both vertical displacement and torsion. Both instruments are equipped with an extender electronics box, which allows for phase (time lag) imaging. The D3000 is a scanning tip instrument. That is, the tip is attached to the scanner and the sample is fixed to a motorized stage. The sample stage can fit up to 150 mm diameter samples that are up to 12 mm thick. The Multimode is a scanning sample instrument, i.e., the tip is fixed and the sample is scanned. The Multimode requires small samples (10 × 10 × 5 mm) that are glued or taped to a sample pack that is attached to the scanner.

The most commonly used operating mode is the *intermittent contact mode,* aka *tapping*. When operated in standard tapping or contact mode, the SPM is more commonly known as an atomic force microscope (AFM). Phase imaging can show changes in elasticity or hardness, but may only show a derivative of the topography. Since the AFM senses heights directly it is trivial to measure step heights and surface roughness (Figs. 4.16 and 4.17). It is also possible to measure long range forces like magnetic and electrical forces. Magnetic and electrical forces require special tips that are sensitive to the forces that are to be measured.

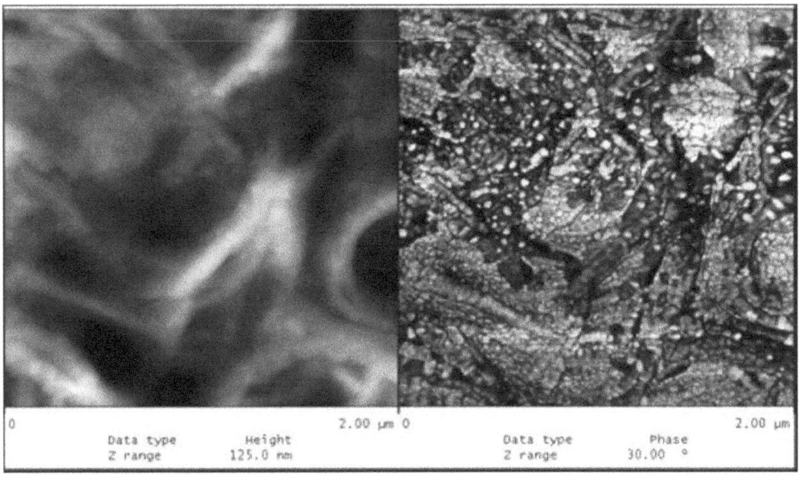

FIGURE 4.16 Height (left) and phase (right) images of an HDPE sample taken from a gallon milk jug. In this case the field of view is 2 micrometers [25].

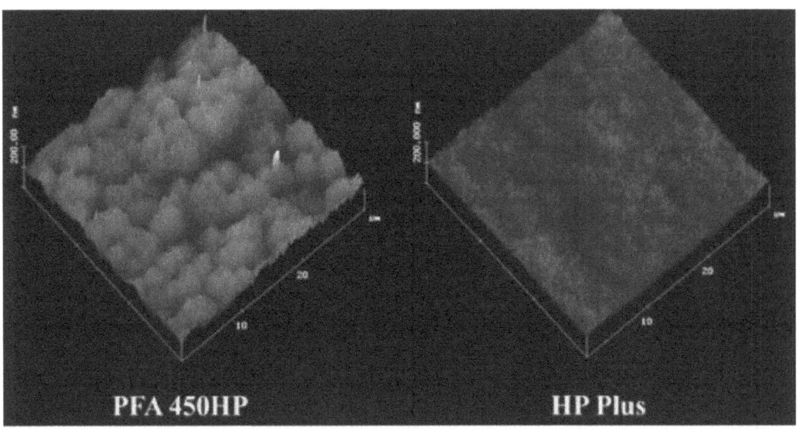

FIGURE 4.17 Comparison of surface roughness of two grades of Teflon® PFA. (*Courtesy DuPont Company*, www.DuPont.com, 2005.)

4.17 ELECTRON ENERGY LOSS SPECTROSCOPY (EELS)

This technique [3] utilizes the inelastic scattering of low energy electrons in order to measure vibrational spectra of surface species: superficially, it can be considered as the electron-analogue of Raman spectroscopy. To avoid confusion with other electron energy loss techniques it is sometimes referred to as: HREELS – high resolution EELS *or* VELS – vibrational ELS.

Since the technique employs low energy electrons, it is necessarily restricted for use in high vacuum (HV) and UHV environments – however, the use of such low energy electrons ensures that it is a surface specific technique and, arguably, it is the vibrational technique of choice for the study of most adsorbates on single crystal substrates.

The basic experimental geometry is fairly simple. It involves using an electron monochromator to give a well-defined beam of electrons of fixed incident energy, and then analyzing the scattered electrons using an appropriate electron energy analyzer.

The resolution of the technique (despite the HREELS acronym!) is generally rather poor; 40–80 cm^{-1} is not untypical. A measure of the instrumental resolution is given by looking at the FWHM (full-width at half maximum) of the elastic peak.

This poor resolution can cause problems in distinguishing between closely similar surface species – however, recent improvements in instrumentation have opened up the possibility of much better spectral resolution ($<10\ cm^{-1}$) and will undoubtedly enhance the utility of the technique.

One of the classic examples of an area in which vibrational spectroscopy has contributed significantly to the understanding of the surface chemistry of an adsorbate, is that of molecular adsorption of CO on metallic surfaces. Adsorbed carbon monoxide usually gives rise to strong absorptions in both the IR and EELS spectra at the (C-O) stretching frequency. The metal-carbon stretching mode (ca. 400 cm^{-1}) is usually also accessible to EELS. More detailed discussions on the bonding of CO to metals can be found in reference sources [26–28].

4.18 GAS CHROMATOGRAPHY (GC)

Chromatography separates a mixture into its components [29], which are then analyzed by one of many detectors. In gas chromatography, the sample is passed in the vapor phase through an appropriated solid bed placed in a column. An oven supplies heat to the column through which the vapor is carried using a constant stream of a gas such as nitrogen or helium. The time required to elute each component (*retention* time) is monitored. Analysis times can run for more than an hour. GC refers to the use of solid absorbent or molecular sieve columns. In the majority of GC columns, the packing is coated with a stationary liquid phase. Separation takes place by partitioning between the liquid coating and the carrier gas.

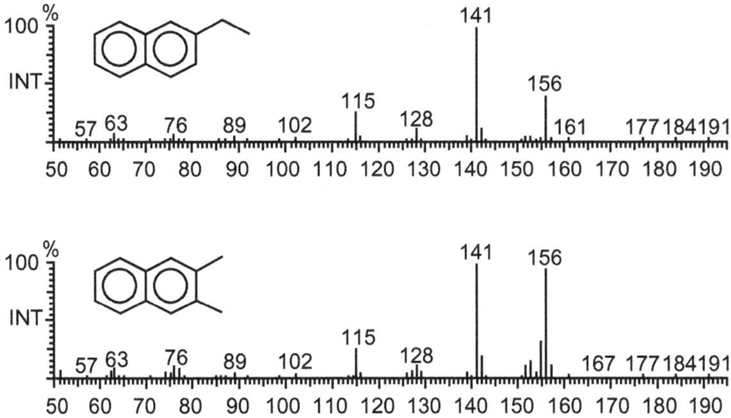

FIGURE 4.18 GC/MS spectra of alkylated polycyclic aromatic hydrocarbons (PAH) isomers [30].

There are a variety of column packing materials and liquid phases. In some cases, a very long capillary column is used without a solid packing and only coated with a liquid on the wall. The oven can be operated isothermally at a given temperature or according to a program where temperature varies as a function of time.

There are several different types of detectors including thermal conductivity and electron capture models. The most common detector is the *flame ionization* variety (FID). A hydrogen flame is utilized to combust the column effluents. Thermal conductivity detectors do not degrade the effluents and are not as sensitive as flame ionization detectors. Electron capture detection is especially sensitive to halogenated compounds.

Retention time of a particular compound is a characteristic of that compound under a set of conditions. To resolve an overlap of retention time, a combination of GC with MS or IR is employed. Figure 4.18 shows examples of GC/MS spectra for two isomers. Note the similarity that is typical of the GC/MS spectra of various isomers. Figure 4.19 shows the chromatogram of contamination (silicone oil) that has resulted in poor adhesion in a thermally bonded package.

4.19 NUCLEAR MAGNETIC RESONANCE (NMR)

Nuclear magnetic resonance (NMR) is [31] a unique technique that depends on the magnetic nature of a few isotopic nuclei, primarily ^{1}H, ^{13}C, ^{19}F, and ^{31}P. In a strong magnetic field, a nucleus precesses about the direction of the magnetic field with a frequency proportional to the field strength. A nucleus is bombarded with radio frequency electromagnetic waves at right angles to the magnetic field. When the spinning nucleus and the radio frequency

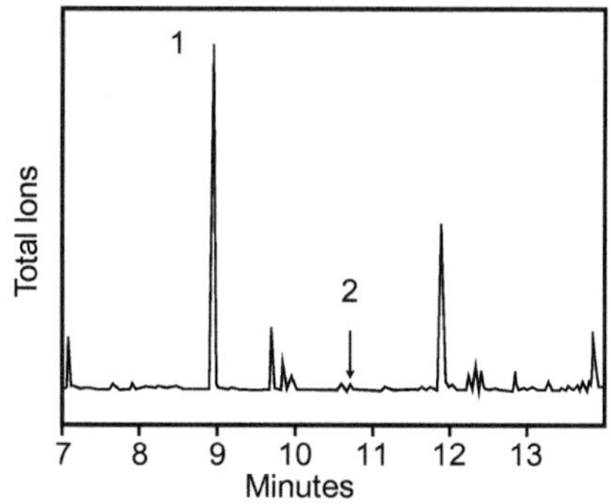

FIGURE 4.19 GC/MS thermal desorption chromatogram of silicon oil causing poor adhesion of thermally bonded package; (peak 1) octamethylcyclotetrasiloxane, (peak 2) decamethylcyclopentasiloxane.

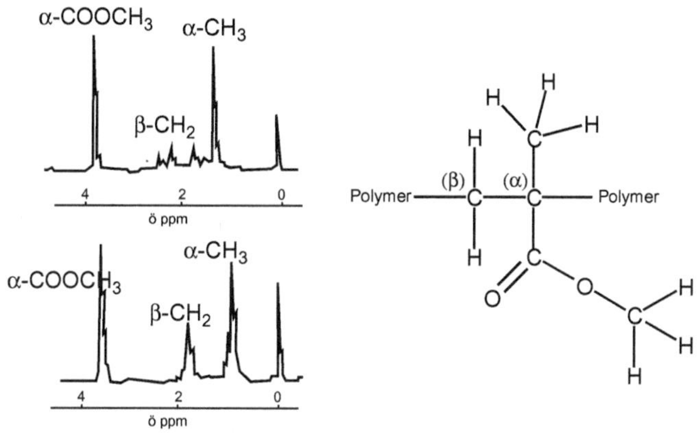

FIGURE 4.20 Example of NMR frequency shift in elucidating chemical structure of poly-methyl methacrylate (PMMA) [32].

become equal, resonance takes place. The energy transfer is the basis of the NMR spectrum of the nucleus and is measured and depicted as a band. The resonance frequency at which the energy absorption occurs depends on the chemical nature of the sample and the environment (e.g., solvent). The nature of the bonds in a molecule can be identified by NMR (Fig. 4.20), and by the use of reference spectra and characteristic frequencies for different types of chemical bonds.

If a strong magnetic field is applied to the nuclei, a difference can be found between alignment in the direction of the applied magnetic field and in the opposite direction. In NMR, the constraint which leads to quantized transitions is applied by the spectrometer. Due to this, the frequency of absorption varies with the applied magnetic field; there is no absolute frequency or wavelength for a given absorption. NMR spectra are not plotted as absorption versus wave number as are IR spectra, but as absorption versus chemical shift, *d*. The chemical shift for proton NMR is the difference between the frequency of absorption of the sample and a standard, tetramethylsilane (TMS), normalized by the frequency of absorption of TMS, which is expressed in parts per million (ppm). Therefore, the above equation is multiplied by 10^6 [33].

In IR we consider two states for a bond: vibrating and non-vibrating. The transition associated with the change from non-vibrating to vibrating leads to the absorption at fixed wave numbers. In NMR, several states are potentially possible, depending on the spin quantum number. The permitted states are given by the allowable values for the magnetic quantum number, $mI = I$, $I-1, \ldots -I$. For $I = 1/2$ there are two states possible, $mI = 1/2$ and $mI = -1/2$. For $I = 1$ three states are possible, $mI = 1$, 0, and -1. For protons, the two allowable states are generally spoken of as parallel and anti-parallel to the applied field.

Polymethyl methacrylate (PMMA) was one of the first polymers studied in depth for tacticity using proton NMR. For syndiotactic PMMA (Fig. 4.20, bottom), three main absorptions are observed: α-CH_3 at 0.91, β-CH_2 at 1.9, and α-$COOCH_3$ at 3.6. For the isotactic polymer (Fig. 4.20, top curve), the α-CH_3 is more deshielded at 1.20, the β-CH_2 becomes a quartet centered at 1.9, and the α-$COOCH_3$ remains a singlet at 3.6. The splittings of the β-CH_2 in what should be a sequence of 1:1:1:1 is due to two types of methylene groups termed erythro, e (more deshielded), and threo, t (less deshielded), corresponding to the bottom and top protons in the molecular sketch in Fig. 4.20. Each of these peaks are split into two peaks by the other, leading to an expected splitting of four equal peaks, with a reported J coupling constant of about 0.2 ppm. The e and t protons are separated by 0.7 ppm, which can be verified by molecular modeling. A higher resolution NMR resolves higher order stereosequences as shown below for isotactic and atactic PMMA. You should compare the information content of the 60 MHz spectrum above to the 500 MHz spectra below. Sixty (60) MHz refers to the natural resonance frequency of a proton for a given magnetic field of the instrument B_0.

4.20 DIFFERENTIAL SCANNING CALORIMETRY (DSC)

Differential scanning calorimetry (DSC) is one of best known techniques in the group known as the thermal analysis methods. Other techniques include differential thermal analysis methods, dynamic mechanical analysis methods, and thermogravimetric analysis methods, all of which are discussed in the following sections.

DSC is a thermal analysis technique used to measure the heat flows related to transitions in materials as a function of time and temperature [34]. These measurements provide qualitative and quantitative information about physical and chemical changes that involve endothermic or exothermic processes or changes in heat capacity. Any event, such as loss of solvent, phase transitions, crystallization temperature, melting point, and degradation temperature of the plastic sample, results in a change in the temperature of the sample. The available DSC systems have a wide range of temperature capability, from −60°C to >1500°C.

Two types of system are commonly used: *power compensation* and *heat flux* DSCs. In the former apparatus, temperatures of the sample and the reference are controlled independently by using separate but identical furnaces. The power input to the two furnaces is adjusted to equalize the temperatures. The energy required for the temperature equalization is a measure of the enthalpy, or heat capacity, of the sample relative to the reference. In heat flux DSC, the sample and the reference are interconnected by a metal disk that acts as a low-resistance heat-flow path. The entire assembly is placed inside a single furnace. The changes in the enthalpy, or heat capacity, of the sample cause a difference in its temperature compared to the reference. The resulting heat flow is small due to the thermal contact between the sample and the reference. Calibration experiments are conducted to correlate enthalpy changes with the temperature differences. In both cases, enthalpy changes are expressed in units of energy per unit mass.

A typical DSC is run isothermally, with the temperature change at a constant rate under an atmosphere of air or another gas. In the isothermal case, the heat flow, or enthalpy change, is plotted against time. The heat flow, or enthalpy, is plotted against temperature or time. Figure 4.21 shows an example of an enthalpy peak generated by an exothermic or endothermic event such as melting of the crystalline phase of a semicrystalline or crystalline polymer. Heat of fusion (H_{f0}) represents the enthalpy change at the crystal melting point of T_0. The weight fraction of the crystalline phase (W) can be determined by comparing the measured heat with H_{f0}. A sample of the polymer is heated in a DSC from a temperature of T_1 to T_2 where the polymer

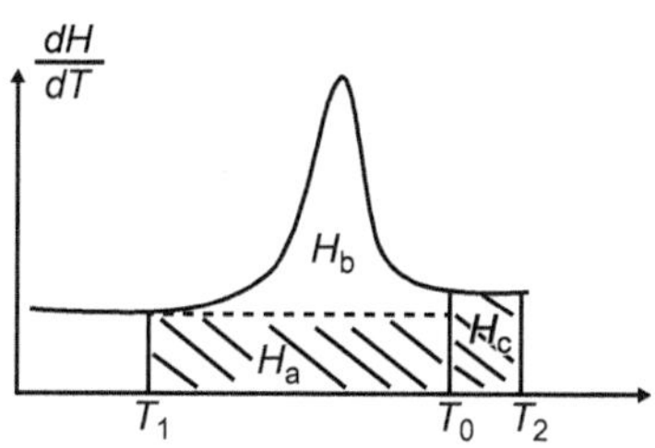

FIGURE 4.21 Example of a DSC peak.

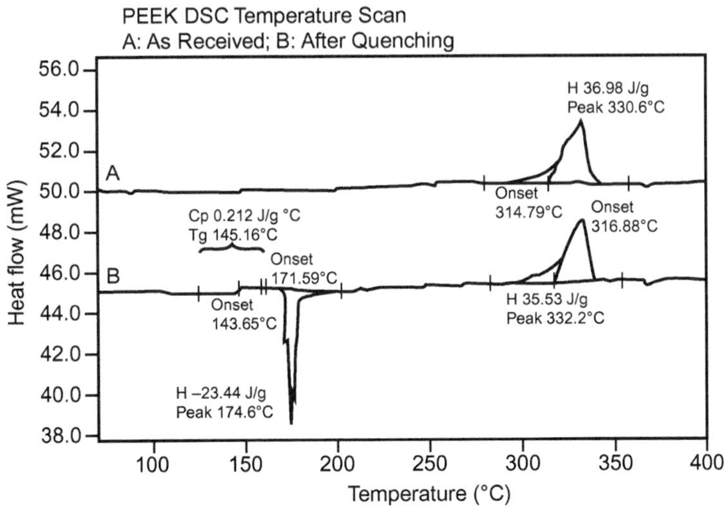

FIGURE 4.22 DSC thermogram for two types of polyetheretherketone [30].

becomes amorphous at a temperature of T_0 prior to reaching T_2, shown by the baseline shift in Figure 4.21. The enthalpy changes are determined according to the following procedure.

H_a represents the simple enthalpy change during the heating of the polymer sample from T_1 to T_0. At T_0, the crystalline phase melts, becoming amorphous. The enthalpy of fusion is expressed in Eq. (4.3). The increase in the enthalpy of the amorphous phase as a result of heating from T_0 to T_2 is designated as H_c. If the enthalpy change calculated from the separation of the DSC curve from the baseline (i.e., area under the curve) is designated as H_T, then Eq. (4.4) can be derived to calculate the weight fraction of the crystalline phase.

$$\begin{aligned} H_b &= W \cdot H_{f0} \\ H_T &= H_a + H_b + H_c \\ H_T &= H_a + W \cdot H_{f0} + H_c \end{aligned} \tag{4.3}$$

$$W = \frac{(H_T - H_a - H_c)}{H_{f0}} \tag{4.4}$$

An example of a DSC thermogram for two types of polyetheretherketone is given in Figure 4.22.

4.21 DIFFERENTIAL THERMAL ANALYSIS (DTA)

DTA measures the temperature difference between a sample and a reference as temperature is increased. A plot of the temperature difference (thermogram)

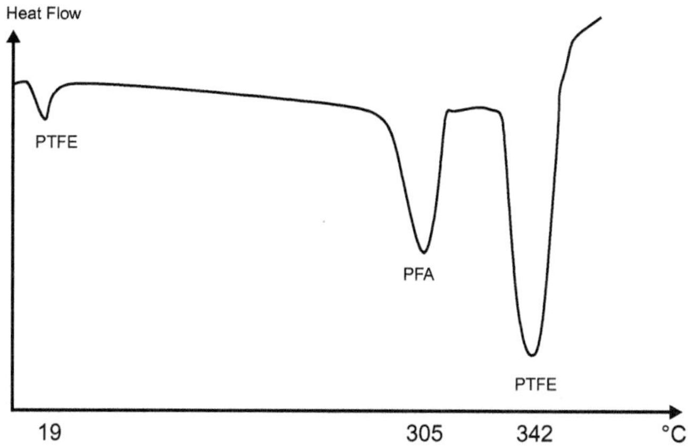

FIGURE 4.23 Example of DTA thermogram for PFA and PTFE.

reveals exothermic and endothermic reactions that may occur in the sample. Temperature for thermal events such as phase transitions, melting points, crystallization temperatures, and others can be determined. Maximum temperature capability of DTA is in excess of 1000°C under air or other gas atmospheres. A typical heating rate for DTA is in the range of 10°C–20°C/min, although slower rates are possible by using a typical optimum sample weight of 50–100 mg. Prior to DTA, the sample should be ground to particles finer than 100 mesh.

Melting of a semicrystalline or crystalline polymer manifests itself as an endothermic peak. The peak temperature is correspondent to the actual melting point of the polymer. As in DSC, the area under the peak is proportional to the crystalline fraction of the sample. Mixtures of polymers can be characterized by DTA because the melting points of individual polymers are, for the most part, unaffected by the mixture. Similar polymers, such as those of high and low density, are distinguishable by DTA, while infrared spectroscopy would not be able to easily resolve such subtle differences. Figure 4.23 shows the thermogram of a mixture of polytetrafluoroethylene (PTFE) and perfluoroalkoxy polymer (PFA). Three peaks are evident. One is at about room temperature (19°C), which is a transition point for PTFE, and two peaks reveal melting points. PTFE and PFA are distinguished by the difference in their melting points.

4.22 DYNAMIC MECHANICAL ANALYSIS (DMA)

Dynamic mechanical analysis (DMA) is generally a more sensitive technique for detecting transitions than the DSC and DTA methods. This is because the properties measured are the dynamic modulus and damping coefficient. Both change significantly when crystalline structures transition to the amorphous phase. The operating principle is that in these transitions, a proportionally larger change takes place in the mechanical properties of a polymer than in its specific heat. Therefore, dynamic mechanical analysis is the preferred method of measurement for glass transition temperature and other minor phase/structure changes of polymers.

The dynamic mechanical analysis method determines [35] elastic modulus (or storage modulus, G′), viscous modulus (or loss modulus, G″), and damping coefficient (tan Δ) as a function of temperature, frequency, or time. Results are usually in the form of a graphical plot of G′, G″, and *tan Δ* as a function of temperature or strain. DMA may also be used for quality control and product development purposes.

The forced non-resonance technique is one of the simpler DMA methods [36]. In the majority of commercially available DMAs, a force is applied to a sample, and the amplitude and phase of the resultant displacement are measured. These instruments utilize a linear actuator in which the applied force is calculated from knowledge of the input signal to the electromagnetic coils in the driver. An alternative is the use of a force transducer to measure the applied load, with the sample held between this transducer and the magnetic driver. These are the two types of arrangements that are found with the forced non-resonance technique. In each case, the sample is driven at a frequency below that of the test arrangement. Typically, the frequency range of instruments is from 0.001 to 1,000 Hz. Any measurements below 0.01 Hz take too long for most analytical experiments, especially if data are required as a function of temperature. Resonance often occurs at frequencies greater than 100 Hz, depending upon the sample stiffness.

Figure 4.24 shows an example of a DMA output for polytetrafluoroethylene. Samples of PTFE were analyzed by DMA in shear mode at a length-to-thickness ratio of 4:1. Figure 4.24(a) shows the DMA output of stress versus time versus temperature, which has been converted to Fig. 4.24(b) by the *time-temperature superimposition* technique.

The time–temperature superimposition technique allows the prediction of material properties that would normally require measurements over many months or years. To collect the necessary data, measurement of a time-dependent variable is made at different temperatures. The curves are shifted mathematically along the time axis until some overlap occurs and a continuous curve is formed covering several decades of time; this curve is called the *master curve*. The master curve can be used to determine the time-dependent property as a function of time. Figure 4.24(c) shows total strain as a function of time and temperature for PTFE.

The choice of geometry will be dependent on the sample under investigation. For example, thin films can only be measured accurately in tension. High quality dynamic mechanical testers perform well in tension and should apply the required pre-tension forces automatically, including those associated with large modulus changes that may occur at the glass transition.

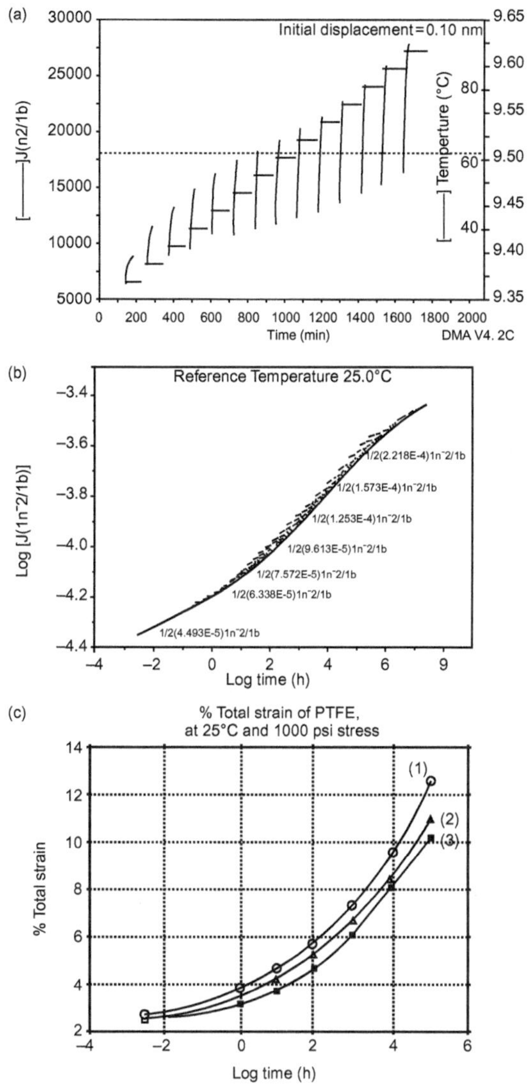

FIGURE 4.24 Examples of DMA graphs for three grades of PTFE. (*Courtesy DuPont Fluoroproducts.*)

Pre-tension is necessary in order to maintain the sample under a net tension to prevent buckling that would otherwise occur. Tension should be the first choice for any sample less than one millimeter thick. Samples thicker than one millimeter will likely be too stiff for the instrument in tension, making the bending mode preferable. Materials that creep excessively, such as polyethylene, may be difficult to test in tension due to creep under the pretension force.

Readers interested in a more in-depth understanding of data obtained from DMA measurements can refer to polymer rheology textbooks. Some available books published in this area include those listed in Refs 37–39.

4.23 THERMOGRAVIMETRIC ANALYSIS (TGA)

Thermogravimetric analysis (TGA) is a powerful technique for the measurement of the thermal stability of materials including polymers. In this method, changes in the weight of a specimen are measured while its temperature is increased. Moisture and volatile contents of a sample can be measured by TGA. The apparatus consists of a highly sensitive scale to measure weight changes and a programmable furnace to control the heat of the sample. The balance is located above the furnace and is thermally isolated from the heat. A high precision hang-down wire is suspended from the balance down into the furnace. At the end of the hang-down wire is the sample pan, the position of which must be reproducible. The balance must be isolated from the thermal effects (e.g., by use of a thermostatic chamber) to maximize the sensitivity, accuracy, and precision of weighing. Addition of an infrared spectrometer to TGA allows analysis and identification of gases generated by the degradation of the sample.

The TGA apparatus is equipped with a micro-furnace that can be rapidly cooled. The heating element is made of platinum (reliable up to 1000°C). An external furnace with a heating element made of an alloy of platinum and 30% rhodium can extend the temperature range to 1500°C.

A modern apparatus is usually equipped with a computer that calculates the weight-loss fraction or percentage. A commercial TGA is capable of >1000°C, 0.1 μg balance sensitivity, and a variable controlled heat-up rate under an atmosphere of air or another gas. The heat-up rate capability of TGA can vary from 0.1°C to 200°C/min.

Figures 4.25 and 4.26 show the TGA spectra for the FEP resins, DSCs of which are shown in Figures 4.27 and 4.28. A comparison of these figures indicates deterioration in the thermal stability of FEP after incorporation of pigment. Figure 4.29 represents the TGA thermogram for a PTFE (31% wt) compounded with carbon black (18% wt) and silica (50.5% wt). The 0.5% difference is due to the evolved volatile gases.

4.24 ADHESIVE BOND TESTING

Adhesive tests are used for a variety of reasons including [40]:

1. Comparison of properties (tensile, shear, peel, flexural, impact, and cleavage strength; durability; fatigue; environmental resistance; conductivity; etc.)
2. Quality checks for a "batch" of adhesives to determine whether the adhesives are still up to standard.

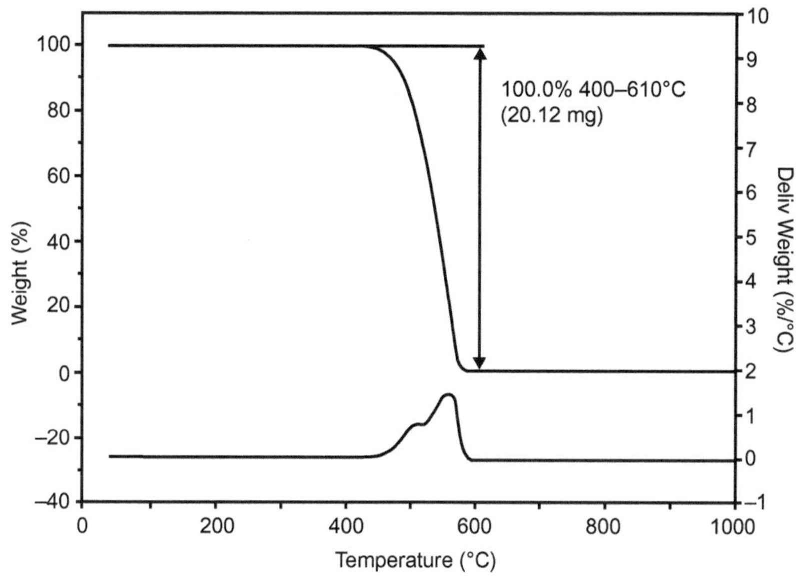

FIGURE 4.25 TGA thermogram of a neat FEP resin. (*Courtesy DuPont Fluoroproducts.*)

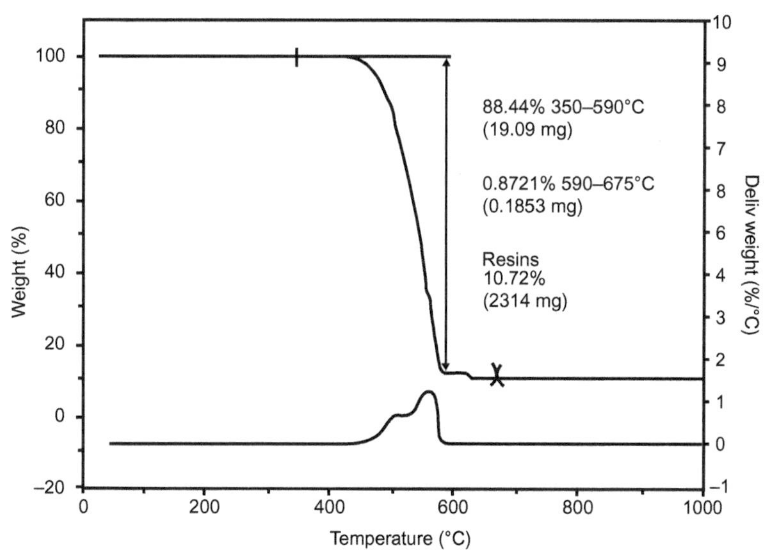

FIGURE 4.26 TGA thermogram of a pigmented FEP resin. (*Courtesy DuPont Fluoroproducts.*)

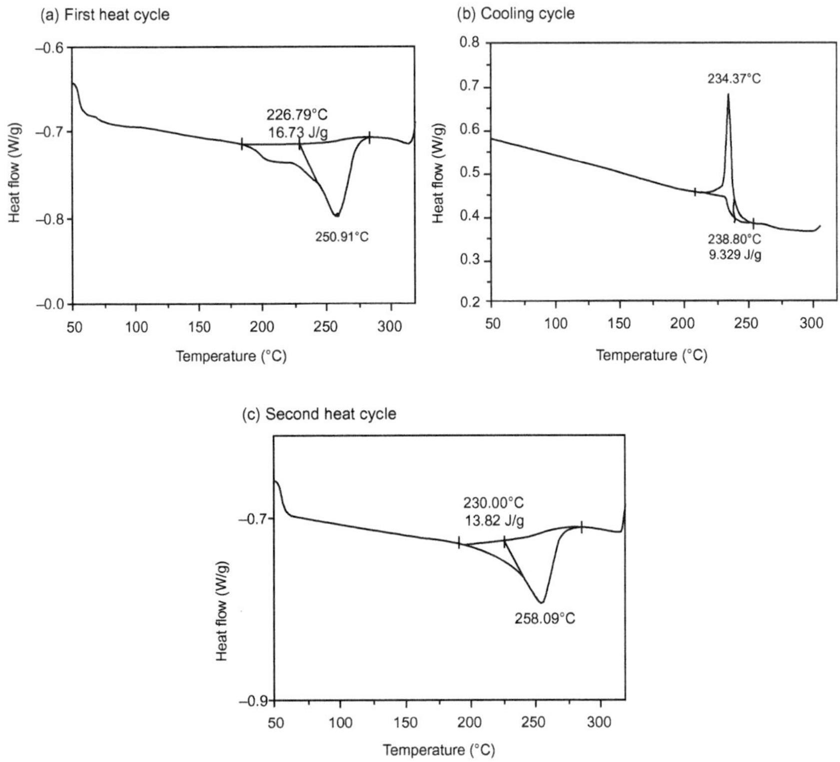

FIGURE 4.27 DSC thermograms of a neat FEP resin (melt flow rate = 30 g/10 min). (*Courtesy of DuPont Fluoroproducts.*)

3. Checking the effectiveness of surface and/or other preparation.
4. Determination of parameters useful in predicting performance (cure conditions, drying conditions, bond-line thickness, etc.).

4.24.1 Tensile

Pure tensile tests are those in which the load is applied normal to the plane of the bond line and in line with the center of the bond areas (Fig. 4.30). ASTM D897-08 (see below for this and all standards discussed) is one of the oldest methods still in the ASTM book on adhesives. The specimens and grips called for require considerable machining and, because of the design, tend to develop edge stresses during the test. Because of these limitations, D897 is being replaced by D2095-96(2008) on rod and bar specimens. These specimens, prepared according to ASTM D2094-00(2006), are simpler to align and, when correctly prepared and tested, more properly measure tensile adhesion [42].

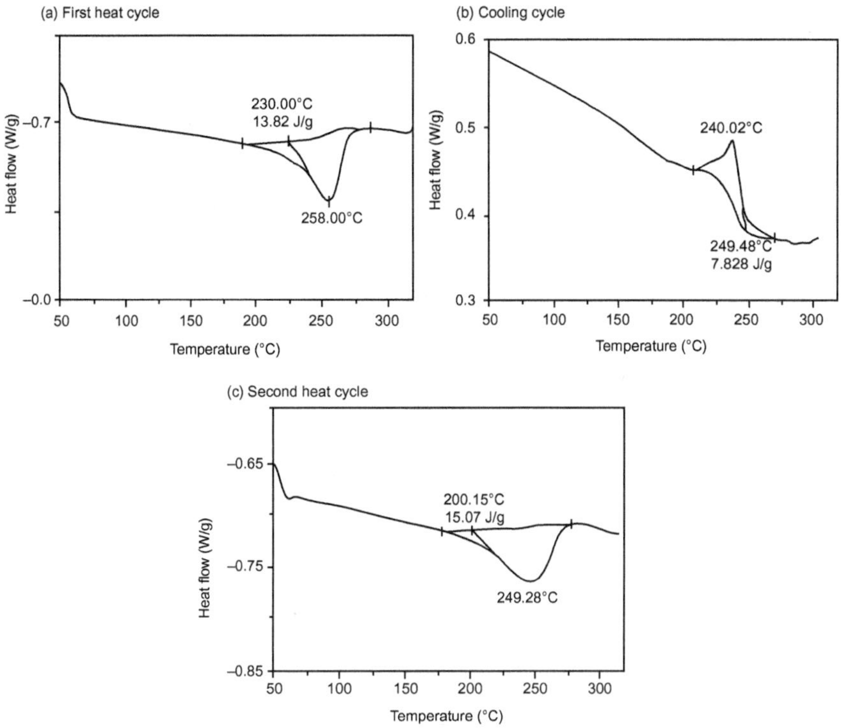

FIGURE 4.28 DSC thermograms of a pigmented FEP resin (melt flow rate = 30 g/10 min). (*Courtesy DuPont Fluoroproducts.*)

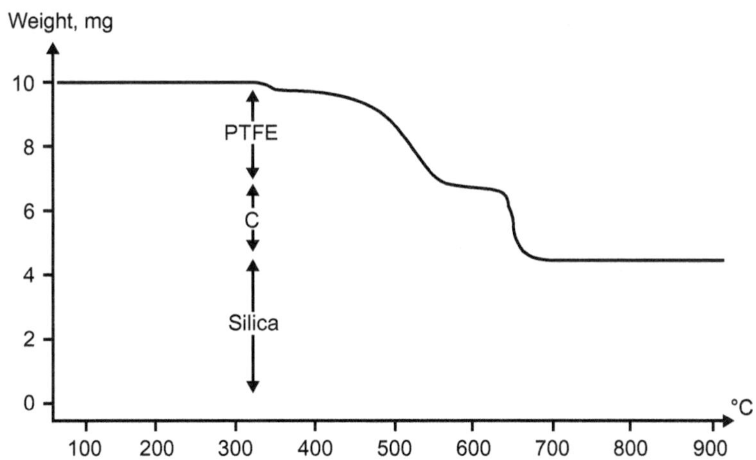

FIGURE 4.29 TGA thermograms of a pigmented PTFE resin [32]. (*Courtesy DuPont Fluoroproducts.*)

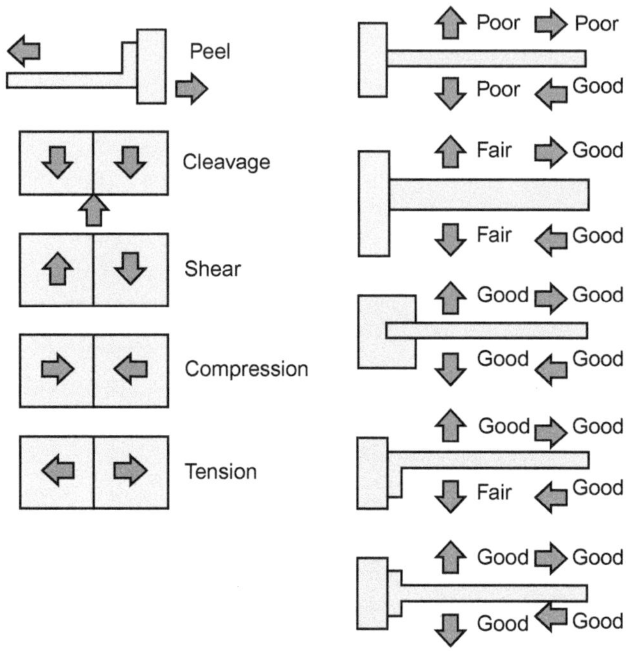

FIGURE 4.30 Different types of stress applied to a joint [41].

Tensile tests are among the most common tests used for evaluating adhesives, despite the fact that, where possible to use joint designs that load the adhesive in other than a tensile mode. Most structural materials have high tensile strengths when compared to the tensile strengths of structural adhesives. One of the advantages of the tensile test over other methods is that it yields fundamental and uncomplicated tensile strain, modulus, and strength data [40].

4.24.2 Shear

Pure shear stresses are those which are imposed parallel to the bond and in its plane (Fig. 4.30). Single-lap shear specimens do not represent pure shear, but are practical and relatively simple to prepare. They also provide reproducible, usable results. The preparation of this specimen and method of testing are described fully in ASTM D1002-10. Two types of panels for preparing multiple specimens are described [42].

Shear tests are very common because samples are simple to construct and closely duplicate the geometry and service conditions for many structural adhesives. As with tensile tests, the stress distribution is not uniform and, while it is often conventional to give the failure shear stress as the load divided by the bonding area, the maximum stress at the bond line may be considerably higher than the average stress. The stress in the adhesive may also differ from pure shear. Depending on such factors as adhesive thickness and adherend stiffness, the failure of the adhesive "shear" joint can be dominated by either shear or tension [40].

Methods other than ASTM D1002-10 are in use. ASTM D3163-01(2008) describes an almost identical test configuration, except for thickness. This method helps alleviate the problem of adhesive extruding out from the edges of the sample. ASTM D3165-07 describes how a specimen can be prepared to determine the strength properties of adhesion in shear by tension loading of laminated assemblies. The double-lap shear test offers the advantage of reducing the cleavage and peel stresses found in the single lap shear test [40].

4.24.3 Peel

Peel tests, intended to be used with flexible adhesives, are designed to measure the resistance to highly localized stresses (Fig. 4.30). Peel forces are therefore considered as being applied to linear fronts. The more flexible the adherend and the higher the adhesive modulus, the more nearly the stressed area is reduced to linearity. The stress then approaches infinity. Since the area over which the stress is applied is dependent on the thickness and modulus of the adherend and the adhesive, and is therefore very difficult to evaluate exactly, the applied stress and failing stress are reported as linear values, i.e., pounds per linear inch. Probably, the most widely used peel test for thin-gauge metal adherends is the T-peel test (ASTM D1876-08). In this test, all the applied load is transmitted to the bond. This type of peel thus tends to provide the lowest values of any peel test [42].

With elastomeric adhesives peel strength is dependent on bond thickness. The elongation characteristics of these adhesives permit a greater area of the bond to absorb the applied load as the bond thickness increases. The T-peel test is probably the most widely used peel test since it uses only one thickness of metal. The Bell peel test is designed to be peeled at a constant radius around a 1-inch (2.52 cm) steel roll and, for this reason, provides more reproducible results. ASTM D1781-98(2012) uses a metal-to-metal climbing drum in an attempt to achieve this same constant peel radius by peeling around a 4-inch diameter rotating drum. While the fixtures used with the bell and drum-peel tests help stabilize the angle of peel, the ideal of a fixed radius of peel is not achieved because the high modulus of the metal tends to resist close conformation to the steel roll or drum. In both methods, considerable energy is used in deforming the metal so that they provide higher peel values for a given adhesive than the T-peel method [42].

ASTM D3167-10 is a test for determining the floating-roller peel resistance of adhesives. The specimens for this test are made by bonding a flexible material to a comparatively rigid one. The method is of particular value for acceptance and process control testing. It may be used as an alternative to ASTM D1781-98(2012) (Climbing Drum Test). This method should be considered more sever, since the angle of peel is greater.

ASTM D903-98(2010) uses a 180° peel to determine the peel or stripping strength. In this method, one of the adherends must be sufficiently flexible to enable it to, essentially, fold back on itself.

REFERENCES

[1] Evans C. and Associates. 810 Kifer Road, Sunnyvale, CA 94086, <www.surface-science.com/tutorial/>; 2004.

[2] UK Surface Analysis Forum. <www.uksaf.org/home.html/>; 2013.

[3] University of Alberta, Alberta Center for Surface Eng. and Sci, <www.ualberta.ca/ACSES/ACSES3/Techniques/iss.htm/ht>; 2013.

[4] Chan CM. Polymer surface modification and characterization. Munich: Hanser Publishers; 1993.

[5] Stewart IM. In: McCall JL, Mueller W, editors. Microstructural analysis: tools and techniques. New York: Plenum Press; 1973. pp. 281–5.

[6] Tsuda N. Fluoropolymer emulsion for high-performance coatings. Osaka, Japan: Daikin Industries Ltd.; 2001.

[7] <www.botany.utexas.edu/facstaff/facpage>

[8] Siegbahn K, et al. ESCA: atomic, molecular, and solid state structure studied by means of electron spectroscopy, Acta Regiae Sco. Sci. Upsaliensis, Ser. IV, vol. 20. Stockholm: Almqvist and Wiksells; 1967.

[9] Siegbahn K, et al. ESCA applied to free molecules. Amsterdam: North-Holland; 1969.

[10] Wheeler DR, Pepper SV. J Vac Sci Technol 1982;20:226.

[11] Briggs D. In: Brewis DM, editor. Surface analysis and pretreatment of plastics and metals. New York: Macmillan; 1982, p. 73.

[12] Briggs D. Ency. polymer science, 16. John Wiley and Sons; 1988.

[13] Chan CM. Polymer surface modification and characterization. Munich: Hanser Publishers; 1994.

[14] Brewis DM. Surface modification of fluoropolymers for adhesion, Paper presented at the Fluoropolymers Conference. Loughborough University, UK; 1992.

[15] Davis LE, MacDonald NC, Palmberg PW, Riach GE, Weber RE. Handbook of Auger electron spectroscopy. 2nd ed. Eden Prairie, MN: Physical Electronics Industries; 1976.

[16] Univ. of Alberta, Center for Surface Engineering and Science, <www.ualberta.ca/ACSES/ACSES3/Techniques/iss.htm>

[17] Taglauer E, Heiland W. Appl Phys 1976;9:261.

[18] Brown A, Vickerman JC. Surf Interface Anal 1984;6:1.

[19] Briggs D, Wotton AB. Surf Interface Anal 1982;4:109.

[20] Ernstsson Ref M, Wậrnheim T. In: Johansson I, Somasundaran P, editors. Handbook for cleaning and decontamination of surfaces. 2007. pp. 747–89.

[21] McGraw-Hill encyclopedia of chemistry. 2nd ed. Parker SP, editor-in-chief. New York: McGraw-Hill, Inc.; 1992.

[22] Available from: <www.public.asu.edu/~ionize/polymer.html/>.

[23] Ezrin M. Plastics failure guide: cause and prevention. New York: Hanser Publishers; 1996.

[24] Marshall AG, Verdun FR. Fourier transforms in NMR, optical, and mass spectrometry. New York: Elsevier; 1989.

[25] Analytical Instrument Laboratory, North Carolina State University, <www.aif.ncsu.edu/scanning-microscopy-laboratory/>; June 2013.

[26] Cotton FA, Wilkinson G. Advanced inorganic chemistry. 5th ed. pp. 58–62.

[27] Hoffman R. Solids & surfaces: a chemist's view of bonding in extended structures. pp. 71–4.

[28] <www.chem.qmul.ac.uk/surfaces/scc/scat5_4.htm/>; June 2013.

[29] GC/MS/MS Analysis of Alkylated Polycyclic Aromatic Hydrocarbons, VARIAN Application Notes, GCMS44:0695, <www.varianinc.com/image/vimage/docs/products/chrom/apps/>

[30] Elmer P. Measurement and calculation of kinetic parameters for the crystallization of poly (ether ether ketone) (PEEK). Ther Anal Newsl <www.thermal-instruments.com/Applications/petan066.pdf/>

[31] Bovey FA. High resolution NMR of macromolecules. New York: Academic Press; 1972.

[32] Krause A, Lange A, Ezrin M, In: K. Rubytranslated by, Plastics analysis guide: chemical and instrumental methods. Munich: Hanser Publishers; 1979.

[33] Campbell D, White JR, Pethrick RA. Polymer characterization, physical techniques. 2nd ed. International Thomson Publishing Services Ltd; 1999.

[34] Available from: <www.mri.psu.edu>.

[35] Plastics Technology Laboratories, Inc., 50 Pearl Street, Pittsfield, MA 01201, <www.PTLI.com/>

[36] Available from: <www.Triton-Technology.Co.UK>.

[37] Bird RB, Stewart WE, Lightfoot EN. Transport phenomena. 2nd ed. New York: Wiley; 2001.

[38] Dealy JM, Wissbrun KF. Melt rheology and its role in plastics processing: theory and applications. Springer; 1990.

[39] Cogswell FN. Polymer melt rheology: a guide for industrial practice. Woodhead Publishing; 1994.

[40] Anderson GP. Analysis and testing of adhesive bonds. New York: Academic Press; 1977.

[41] Gaston T. Building a better adhesive bond. Machine Design <www.machinedesign.com/>; November 6, 2003.

[42] DeLolis NJ. Adhesives, adherends, adhesion. Huntington, NY: Robert E. Krieger Publishing Co.; 1985.

Chapter 5

Theories of Adhesion

5.1 INTRODUCTION

Historically, mechanical interlocking, diffusion, adsorption and surface reaction, and electrostatic theories have been postulated to describe the mechanisms of adhesion. More recently, other theories have been put forward for the adhesive bonding mechanism (Table 5.1). It is often difficult to fully ascribe adhesive bonding to an individual mechanism. A combination of different mechanisms is responsible for bonding within a given adhesive system. The extent of the role of each mechanism would then vary for different systems. An understanding of these theories will be helpful to those who plan to work with most aspects of adhesion.

An important facet of an adhesion bond is the locus of the proposed action or the scale on which the adhesive and adherend interact. Table 5.1 shows a scale of action for each mechanism, which is intended to aid in the understanding of these mechanisms. Of course, adhesive–adherend interactions always take place at the molecular level, which is discussed later in this chapter.

The *microscopic* parameter of interest in *mechanical interlocking* is the contact surface of the adhesive and adherend. The specific surface area (e.g., surface area per unit weight) of the adherend is an example of one such measure. Surface roughness is the means by which interlocking is thought to work, detected by optical or electron microscopy. In an *electrostatic* mechanism, the surface charge is the *macroscopic* factor of interest. The charge in question is similar to that produced in a glass rod after rubbing it with a wool cloth. *Diffusion* and *wettability* involve, respectively, molecular and atomic scale interactions.

Readers who wish to gain an in-depth understanding of the interaction forces, adhesion mechanism, and thermodynamics of adhesion are encouraged to consult *Fundamentals of Adhesion*, edited by Lieng-Huang Lee [1]. This reference provides a qualitative and quantitative treatment of adhesion, complete with derivation of force interaction equations.

5.2 MECHANICAL THEORY

According to mechanical theory, adhesion occurs by the penetration of adhesives into pores, cavities, and other surface irregularities of the surface of the

Surface Treatment of Materials for Adhesive Bonding. DOI: http://dx.doi.org/10.1016/B978-0-323-26435-8.00005-8

TABLE 5.1 Theories of Adhesion

Traditional	Recent	Scale of Action
Mechanical interlocking	Mechanical interlocking	Microscopic
Electrostatic	Electrostatic	Macroscopic
Diffusion	Diffusion	Molecular
Adsorption/surface reaction	Wettability	Molecular
	Chemical bonding	Atomic
	Weak boundary layer	Molecular
	Acid–base	Molecular

substrate or adherend. The adhesive displaces the trapped air at the interface. Therefore, it is concluded that an adhesive penetrating into the surface roughness of two adherends can bond the two. A positive contribution to the adhesive bond strength results from the "mechanical interlocking" of the adhesive and the adherends. Adhesives frequently bond better to porous abraded surfaces than to smooth surfaces. This theory is, however, not universally applicable because good adhesion also takes place between smooth surfaces.

Enhanced adhesion after abrading the surface of an adherend may be due to one or more of the following: mechanical interlocking, formation of a clean surface, formation of a highly reactive surface, and an increase in contact surface area. While the surface does become rougher by mechanical interlocking, it is believed that changes in physical and chemical properties of the adherend's surface increase an adhesive's strength [2]. One can debate whether mechanical interlocking is responsible or, rather, an increase in the adhesive contact surface enhances other mechanisms. More thorough wetting and more extensive chemical bonding are expected consequences of increased contact surface area.

There is supportive data in the literature that relates joint strength and bond durability to increased surface roughness. There are also contrary observations indicating that increased roughness can lower joint strength [3].

It must be remembered that mechanical abrasion on ductile surfaces such as aluminum and titanium can, most certainly, be deleterious to the formation of a strong bond and, particularly, for environmental resistance: just look at the pictures (Figures 3 and 4 in Ref [4]).

5.3 ELECTROSTATIC (ELECTRONIC) THEORY

This theory proposes that adhesion takes place due to electrostatic effects between adhesive and adherend [5–8]. An electron transfer theoretically takes place between the adhesive and the adherend as a result of unlike electronic band structures. Electrostatic forces in the form of an electrical double layer are thus formed at the adhesive–adherend interface. These forces account for the resistance to separating. This theory gains support from the fact that electrical discharges have been observed when an adhesive is peeled from a substrate [2].

The electrostatic mechanism is a plausible explanation of the polymer–metal adhesion bond. In non-metallic systems, the contribution of the electronic mechanism to the work of adhesion has been calculated and been found to be diminutive when compared to that of chemical bonding [9,10].

5.4 DIFFUSION THEORY

This theory suggests that adhesion is developed through the interdiffusion of molecules in and between the adhesive and adherend. The diffusion theory is primarily applicable when both the adhesive and adherend are polymers with relatively long-chain molecules capable of movement. The nature of materials and bonding conditions will influence if, and to what extent, diffusion takes place. The diffuse interfacial (interphase) layer typically has a thickness in the range of 10 Å–1000 Å. Solvent cementing or heat welding of thermoplastics is considered to be a result of diffusion of molecules [2].

No stress concentration is present in the interface because no discontinuity exists in the physical properties. Cohesive energy density [CED, Eq. (5.1)] can be used to interpret diffusion bonding, as defined by Eq. (5.2). Bond strength is maximized when solubility parameters between adhesive and adherend are matched.

$$\mathrm{CED} = \frac{E_{coh}}{V} \tag{5.1}$$

$$\delta = \sqrt{\frac{E_{coh}}{V}} \tag{5.2}$$

E_{coh} is the amount of energy required to separate molecules to an infinite distance, V is the molar volume, and δ is the solubility parameter.

A relevant example is adhesion of polyethylene and polypropylene to a butyl rubber. The adhesive bond is weak when two polymers are bonded at temperatures below the melting point of polyolefin. Bond strength grows sharply when the adhesion process takes place above the melting temperature of polyethylene (135°C) and polypropylene (175°C). Figure 5.1 illustrates the bond strength (peel strength) as a function of bonding temperature. An inference can be made that at elevated temperatures, interdiffusion of polyolefins and butyl rubber increases, leading to higher bond strength.

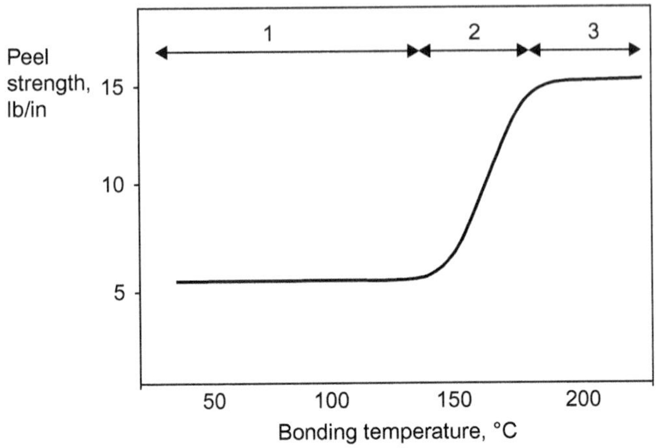

FIGURE 5.1 Peel strength of polypropylene and butyl rubber vs. bonding temperature [11]. 1. Adhesive failure; 2. Adhesive/cohesive failure; and 3. Cohesive failure.

5.5 WETTING THEORY

This theory proposes that adhesion results from molecular contact of two materials and the surface forces that develop between them. The first step in bond formation is to develop interfacial forces between the adhesive and the substrates. The process of establishing continuous contact between the adhesive and the adherend is called *wetting*. For an adhesive to wet a solid surface, the adhesive should have a lower surface tension than the critical surface tension of the solid, which is the reason for the surface treatment of plastics.

Van der Waals forces are extremely sensitive to the distance (r) between molecules, decreasing by the inverse of the seventh power ($1/r^7$) of the distance between two molecules, and the cubic power of the distance between two adherends. In most cases, these forces are usually too insignificant to account for the adhesive bond strength.

Figure 5.2 illustrates [12] complete and incomplete wetting of an adhesive spreading over a surface. Good wetting results when the adhesive flows into the valleys and crevices on the substrate surface. Poor wetting occurs when the adhesive bridges over the valley and results in a reduction of the actual contact area between the adhesive and adherend, and thus a lower overall joint strength [2]. Incomplete wetting generates interfacial defects, thereby reducing the adhesive bond strength. Complete wetting achieves the highest bond strength.

Most organic adhesives readily wet metal adherends. On the other hand, many solid organic substrates have surface tensions lower than those of common adhesives. The criteria for good wetting requires the adhesives to have a lower surface tension than the substrate. This explains, in part, why

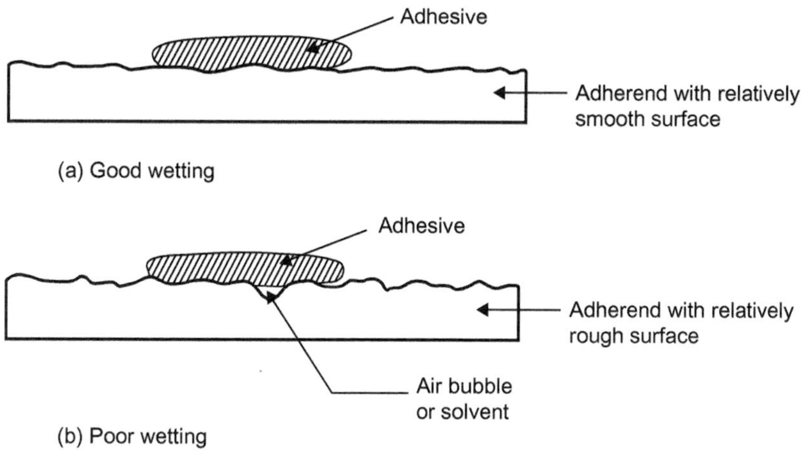

FIGURE 5.2 Examples of good and poor wetting by an adhesive spreading across a surface. Modified after Schneberger [2].

organic adhesives such as epoxies have excellent adhesion to metals, but offer weak adhesion on untreated polymeric substrates such as polyethylene, polypropylene, and fluoroplastics [2]. Surface energy of plastic substrates is increased by various treatment techniques (see Ch. 6) to allow wetting.

5.6 CHEMICAL BONDING

This mechanism attributes the formation of an adhesion bond to surface chemical forces. Hydrogen, covalent, and ionic bonds formed between the adhesive and adherends are stronger than the dispersion's attractive forces. Table 5.2 shows a list of these forces and their magnitudes. In general, there are four types of interactions that take place during chemical bonding: covalent bonds, hydrogen bonds, Lifshitz–van der Waals forces, and acid–base interactions. The exact nature of the interactions for a given adhesive bond depends on the chemical composition of the interface.

Covalent and *ionic bonds* (Table 5.2) are examples of *chemical bonding* that provide much higher adhesion values than the secondary forces provide. Secondary valence bonding is based on the weaker physical forces exemplified by hydrogen bonds. These forces are more prevalent in materials that contain polar groups such as carboxylic acid groups than in nonpolar materials such as polyolefins. The interactions that bind the adhesive and adherends together may also receive contributions from mechanical interlocking, diffusion, or electrostatic mechanisms.

TABLE 5.2 Examples of Energies of Lifshitz–Van Der Waals Interactions and Chemical Bonds

Type	Example	E (kJ/mol)
Covalent	C–C	350
Ion–ion	$Na^+ \cdots Cl^-$	450
Ion–dipole	$Na^+ \cdots CF_3H$	33
Dipole–dipole	$CF_3H \cdots CF_3H$	2
London dispersion	$CF_4 \cdots CF_4$	2
Hydrogen bonding	$H_2O \cdots H_2O$	24

The definitions of intermolecular interactions are listed here:

- *Dipole* (polar molecule): a molecule whose charge distribution is represented by a center of positive charge and a center of negative charge that do not coincide.
- *Dipole–dipole forces*: intermolecular forces resulting from the tendency of polar molecules to align themselves such that the positive end of one molecule is near the negative end of another.
- *Hydrogen bonding*: a special type of dipole–dipole interaction that occurs when a hydrogen atom, bonded to a small, highly electronegative atom (most commonly F, O, N, or S), is attracted to the lone electron pairs (most commonly on an F, O, N, or S atom) of another molecule.
- *London dispersion forces* (dispersion forces): intermolecular forces resulting from the small, instantaneous dipoles (induced dipoles) that result from the varying positions of the electrons during their motion around nuclei. The importance of London dispersion forces increases with atom size and number of electrons.
- *Polarizability*: defined as the ease with which the electron cloud of an atom or molecule is distorted. In general, polarizability increases with the size of an atom and the number of electrons in an atom.

Covalent chemical bonds may form across the interface and are likely to occur in crosslinked adhesives and thermoset coatings. These are usually the strongest and most durable. They do, however, require that mutually reactive chemical groups exist. Some surfaces such as previously coated surfaces, wood, composites, and some plastics contain various functional groups, which under appropriate conditions can produce chemical bonds with the adhesive material. There are ways to intentionally generate these conditions by surface treatment of plastics with techniques like corona or flame treatment.

Organosilanes are widely used as primers on glass fibers to promote the adhesion between the resin and the glass in fiberglass-reinforced plastics. They are also used as primers or integral blends to promote adhesion of resins to minerals, metals, and plastics. During application, silanol groups are produced, which react with the silanol groups on the glass surface or with other metal oxide groups to form strong ether linkages. Coatings containing reactive functional groups such as hydroxyl or carboxyl moieties tend to adhere more tenaciously to substrates containing similar groups. Chemical bonding may also occur when a substrate contains reactive hydroxyl groups, which react with the isocyanate groups from the incoming coating in thermoset polyurethane coatings.

Most likely, chemical bonding accounts for the strong adhesion between an epoxy coating and a substrate with a cellulose interface. The epoxy groups of an epoxy resin react with the hydroxyl groups of cellulose at the interface.

5.6.1 Acid–Base Theory of Adhesion

A special type of interaction, the acid–base theory, is a fairly recent discovery. It is based on the chemical concept of Lewis acid and base. The acid/base definition was proposed separately by J. N. Bronsted and G. N. Lewis. Restatement of these definitions by Lewis in 1938 led to their popularity and acceptance. The Lewis definitions are: "An acid is a substance which can accept an electron pair from a base; a base is a substance which can donate an electron pair" [13]. By this definition, every cation is an acid, in addition to chemical compounds such as BF_3 and SiO_2. Conversely, anions and compounds like NH_3, PH_3, and $C_6H_5CH_2NH_2$ are bases. According to acid/base theory, adhesion results from the polar attraction of Lewis acids and bases (i.e., electron-poor and electron-rich elements) at the interface. This theory is attributed to the work by Fowkes et al. [14–17], Gutmann [18], and Bolger and Michaels [19].

In BF_3, the higher electronegativity of fluorine atoms preferentially displaces the shared electrons away from the boron atom. A bipolar molecule is thus created that has a positive charge on the boron side and is negative on the fluorine sides, making it a Lewis acid. On the other hand, NH_3 has a negative nitrogen end that renders it a Lewis base. The positive boron end of BH_3 and negative nitrogen end of NH_3 interact.

A special case of acid–base interaction is hydrogen bonding, such as water molecules that exhibit both acidic and basic tendencies. Table 5.2 shows that the hydrogen bond strength, while substantially less than ionic and covalent bond energies, is one of the more significant secondary interactions. The reader can refer to inorganic chemistry texts [20] to learn about Lewis acids and bases and their chemical reactions. It is sufficient to say that

TABLE 5.3 Drago *E* and *C* Parameters for a Variety of Molecular Bases [12]

Bases	C_b	E_b
Pyridine	13.09	2.39
Ammonia	7.08	2.78
Methylamine	11.41	2.66
Dimethylamine	17.85	2.33
Trimethylamine	23.6	1.652
Ethylamine	12.31	2.80
Diethylamine	18.06	1.771
Triethylamine	22.7	2.03
Acetonitrile	2.74	1.812
p-Dioxane	4.87	2.23
Tetrahydrofuran	8.73	2.00
Dimethyl sulfoxide	5.83	2.74
Ethyl acetate	3.56	1.994
Methyl acetate	3.29	1.847
Acetone	4.76	2.018
Diethyl ether	6.65	1.969
Isopropyl ether	6.52	2.27
Benzene	1.452	1.002
p-Xylene	3.64	0.851

the interactions between compounds capable of electron donation and acceptance form the foundation of the acid–base theory of adhesion.

Techniques have been developed to measure the acid–base properties of polymer surface using techniques such as electron spectroscopy for surface analysis (ESCA), also called X-ray photoelectron spectroscopy (XPS). For example, the polymer swelling method, in conjunction with X-ray photoelectron spectroscopy, was shown to provide an attractive means of determining the acid–base properties of polymers in the solid state. Study of polymer swelling, for instance, by two solvents of known acid–base properties, allowed determination of Drago's *E* and *C* parameters for the candidate polymer (see Tables 5.3 and 5.4). Examples of the use of this method include a surface treated homopolymer of polypropylene [22].

TABLE 5.4 Drago *E* and *C* Parameters for a Variety of Molecular Acids [21]

Acids	C_a	E_a
Iodine	2.05	2.05
Iodine monochloride	1.697	10.43
Thiophenol	0.405	2.02
p-tert-Butylphenol	0.791	8.30
p-Methylphenol	0.826	8.55
Phenol	0.904	8.85
p-Chlorophenol	0.978	8.88
tert-Butyl alcohol	0.614	4.17
Trifluoroethanol	0.922	7.93
Pyrrole	0.603	5.19
Isocyanic acid	0.528	6.58
Sulfur dioxide	1.652	1.88
Antimony pentachloride	10.49	15.09
Chloroform	0.325	6.18
Water	0.675	5.01
Methylene chloride	0.02	3.40
Carbon tetrachloride	0.00	0.00

The degree of acid–base interaction is measured by the enthalpy of adduct formation. Drago proposed [21] the following relation:

$$-\Delta H_{ab} = E_a \cdot E_b + C_a \cdot C_b \tag{5.3}$$

where $-\Delta H_{ab}$ is the enthalpy of adduct formation per mole. E_a and C_a are empirically determined parameters for the acid. E_b and C_b are empirically determined parameters for the base. The magnitudes of the *E* parameters can be interpreted as measures of the susceptibility of the molecule for electrostatic interaction; the magnitudes of the *C* parameters can be interpreted as measures of the susceptibility for covalent interactions (similar to the concept of "hard" and "soft" acids and bases.)

5.7 WEAK BOUNDARY LAYER THEORY

This theory, first described by Bikerman, states that bond failure at the interface is caused by either a cohesive break or a weak boundary layer [23].

Weak boundary layers can originate from the adhesive, the adherend, the environment, or a combination of any of the three. Weak boundary layers can occur in the adhesive or adherend if an impurity concentrates near the bonding surface and forms a weak attachment to the substrate. When failure takes place, it is the weak boundary layer that fails, although failure appears to take place at the adhesive–adherend interface [2].

Polyethylene and metal oxides are examples of materials that may inherently contain weak boundary layers. Polyethylene has a weak, low molecular weight constituent that is evenly distributed throughout the polymer. This weak boundary layer is present at the interface and contributes to low-stress failing when polyethylene is used as an adhesive or an adherend. Some metal oxides are weakly attached to their base metals. Failure of adhesive joints made with these materials occurs cohesively within the oxide. Certain oxides, such as aluminum oxide, are strong and do not significantly impair joint strength. Weak boundary layers, such as those found in polyethylene and metal oxides, can be removed or strengthened by various surface treatments. Weak boundary layers formed from the bonding environment are very common. When the adhesive does not wet the substrate, as shown in Fig. 5.2, a weak boundary layer (air) is trapped at the interface, causing a reduction in joint strength [2].

5.8 DEFINITION OF FAILURE MODES

A hypothetical adhesion bond is illustrated in Figure 5.3. Assume that the bond is tested in the tensile mode in which the two adherends are pulled apart in the direction perpendicular to the bond. There are different possibilities for the occurrence of failure. The surfaces involved in bond failure are called the *locus* of failure.

If the bond failure occurs between the adhesive layer and one of the adherends, it is called *adhesive failure* [Fig. 5.3 (a)]. A failure in which the separation occurs in a manner whereby both adherend surfaces remain

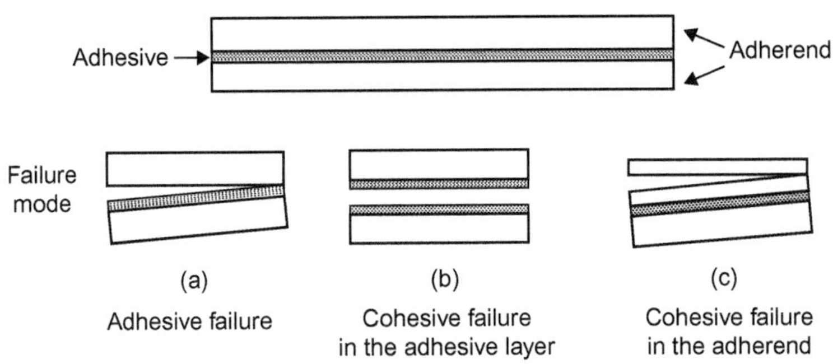

FIGURE 5.3 Schematics of adhesive bond failure modes.

covered with the adhesive is called *cohesive failure* in the adhesive layer [Fig. 5.3 (b)]. Sometimes the adhesive bond is so strong that the failure occurs in one of the adherends away from the bond. This is called a cohesive failure in the adherend. Bond failures often involve more than one failure mode and are ascribed as a percentage to cohesive or adhesive failure. This percentage is calculated based on the fraction of the area of the contact surface that has failed cohesively or adhesively.

It is important to determine the exact mode(s) of bond failure when a problem occurs. Determination of the failure mode allows action to be taken to correct the true cause and save time and money. Some of the methods reviewed in Chapter 4 can be applied to analyze the locus of failure.

Tables 5.5–5.7 show the analyses of several bonds between a substrate and a polyvinyl fluoride film using an acrylic adhesive. All surfaces were

TABLE 5.5 Surface Chemical Analysis (ESCA) in a Cohesive Failure of Adhesive Bond [22]

	Atomic Concentration (%)				
	F	O	N	C	Si
As-is adhesive	nd	26.0	2.1	71.6	nd
As-is film	29.3	6.6	nd	64.4	nd
Polyvinyl fluoride film facing the substrate	nd	24.9	2.5	72.6	nd
Substrate facing the polyvinyl fluoride film	nd	25.0	2.1	72.9	nd

nd = not detectable.

TABLE 5.6 Surface Chemical Analysis (ESCA) in a Cohesive Failure of Polyvinyl Fluoride [22]

	Atomic Concentration (%)				
	F	O	N	C	Si
As-is adhesive	nd	26.0	2.1	71.6	nd
As-is film	29.3	6.6	nd	64.4	nd
Polyvinyl fluoride film facing the substrate	31.0	4.0	nd	63.2	1.7
Substrate facing the polyvinyl fluoride film	30.0	5.4	nd	62.6	2.0

nd = not detectable.

TABLE 5.7 Surface Chemical Analysis (ESCA) in an Adhesive Failure of Polyvinyl Fluoride [24]

	Atomic Concentration (%)				
	F	O	N	C	Si
As-is adhesive	nd	26.0	2.1	71.6	nd
As-is film	29.3	6.6	nd	64.4	nd
Polyvinyl fluoride film facing the substrate	31.6	2.1	nd	66.4	nd
Substrate facing the polyvinyl fluoride film	nd	26.4	3.2	70.5	nd

nd = not detectable.

analyzed by electron spectroscopy for surface analysis (ESCA). ESCA yields chemical analysis of organic surfaces in atomic percentage, with the exclusion of hydrogen, which is undetectable by this technique. To determine the type of bond failure, ESCA results for the failed surfaces are compared with those of the adhesive and polyvinyl fluoride film.

In a pure cohesive failure, the two surfaces involved should have virtually identical chemical compositions, which is nearly the case in Tables 5.5 and 5.6. In a 100% adhesive failure, one surface should have the same chemical composition as the adherend and one should have the same as the adhesive. The examples presented in Tables 5.5 and 5.6 illustrate cohesive failure cases for polyvinyl fluoride (an adherend) and the adhesive. Table 5.5 gives an example of an adhesive failure. One can see, from the chemical composition, that the two surfaces have been separated in a "clean" manner.

5.9 MECHANISMS OF BOND FAILURE

Adhesive joints may fail adhesively or cohesively. Adhesive failure is interfacial bond failure between the adhesive and the adherend. Cohesive failure occurs when a fracture allows a layer of adhesive to remain on both surfaces. When the adherend fails before the adhesive, it is known as a *cohesive* failure of the substrate. The various modes of failure are illustrated in Figure 5.3. Cohesive failure within the adhesive or one of the adherends is the preferred type of failure because with this type, the maximum strength of the materials in the joint has been reached. In analyzing an adhesive joint that has been tested to destruction, the mode of failure is often expressed as a percentage cohesive or adhesive failure. The ideal mode of failure is a 100% cohesive failure in the adhesive layer [2,25].

TABLE 5.8 Environmental and Other Factors Affecting Adhesive Joints [26]

Capability of bonding specific adherends
(see "internal chemical factors" below)
Service requirements
Stress (tension, shear, impact, peal, cleavage, fatigue)
Chemical factors
External (effect of chemical agents, including solvents, acids, alkalis, etc.)
Internal (effect of adherend on adhesive, e.g., extruded placticizers in certain plastics and rubbers; effect of adhesive on the adherend, e.g., crazing, staining)
Environmental factors
Weathering
Light (important only with translucent adherends)
Oxidation
Moisture
Salt spray
Temperature extremes (including thermal cycling)
Biological factors (bacteria, fungi, vermin, rodents)
Specialized functional requirements (thermal or electrical conductance, etc.)
Production requirements
Application method
Bonding range
Blocking
Curing condition (time versus temperature)
Storage stability
Working life
Coverage
Cost
Health and safety hazards

Failure mode should not be used as the only criterion for a useful joint. Some adhesive–adherend combinations may fail by *adhesive* failure, but exhibit greater strength than a similar joint bonded with a weaker adhesive that fails cohesively. The ultimate strength of a joint is a more important criterion than the mode of joint failure. Nevertheless, an analysis of failure mode can be an extremely useful tool in determining whether the failure was due to either a weak boundary layer or improper surface preparation [2].

The exact cause of premature adhesive failure is very difficult to determine. If the adhesive does not wet the surface of the substrate completely, the bond is certain to be less than optimal. Internal stresses occur in adhesive joints because of a natural tendency of the adhesive to shrink during setting, and due to differences in physical properties of the adhesive and substrate. The coefficient of thermal expansion of adhesive and adherend should be as close as possible to minimize the stresses that may develop during thermal cycling or after cooling from an elevated-temperature cure. Fillers are often used to modify the thermal expansion characteristics of adhesives and limit internal stresses. Another way to accommodate these stresses is to use relatively elastic adhesives [2].

The types of stress acting on completed bonds, their orientation to the adhesive, and the rate at which they are applied are important factors in determining the durability of the bond. Sustained loads can cause premature failure in service, even though similar unloaded joints may exhibit adequate strength when tested after aging. Some adhesives break down rapidly under dead load, especially after exposure to heat or moisture. Most adhesives have poor resistance to cleavage loads. A number of adhesives are sensitive to the rate at which the joint is stressed. Rigid, brittle adhesives sometimes have excellent tensile or shear strength but have very poor impact strength. Operating environmental factors, such as those shown in Table 5.8, are capable of degrading an adhesive joint in various ways. If more than one environmental factor (e.g., heat and moisture) is acting on the sample, their combined effect can be expected to produce a synergistic result of reducing adhesive strength. Whenever possible, the candidate adhesive joints should be evaluated under simulated operating loads in the actual environment the joint will encounter [2].

REFERENCES

[1] Lee LH. Fundamentals of adhesion. New York: Plenum Press; 1991.

[2] Petrie EM. Plastics and adhesives as adhesives. In: Harper CA, editor. Handbook of plastics and elastomers. New York: McGraw-Hill; 1975.

[3] Allen KW. Int J Adhes 1993;13:67.

[4] Bishopp JA. Surface pretreatment for structural bonding. In: Cognard P, editor. Handbook of adhesives and sealants, Vol. 1: adhesives and sealants: basic concepts and high tech bonding. Elsevier; 2005.

[5] Deraguin BV, Smilga VP. J Appl Phys 1967;38:4609.

[6] Deraguin BV, Toporov YP. Physiochem Asp Polym Proc Int Symp 1983;2:605.

[7] Cross JA. Surf Contam Genesis Detec, Control Proc Symp, 1. 1979.

[8] Possart W. Int J Adhes 1988;8:77.

[9] Roberts AD. J Phys D 1977;10:1801.

[10] Roberts AD. Adhesion. vol. 1. London; 1977. p. 207.
[11] Wu S. Polymer interface and adhesion. 1st ed. New York: Marcel Dekker, Inc.; 1982.
[12] Polymer adsorption, Lecture 5, Colloidal dispersions. American Chemical Society, <http://colloidaldispersions.com>; 2005.
[13] Parker SP, editor-in-chief. Encyclopedia of chemistry. 2nd ed. New York: McGraw-Hill, Inc.; 1992.
[14] Fowkes FM. Acid-base interactions in polymer adhesion. In: Mittal KL, editor. Physicochemical aspects of polymers surfaces, vol. 2. New York: Plenum Press; 1983.
[15] Fowkes FM. Attractive forces at solid-liquid interface: wetting, SCI Monograph No. 25. London; 1967.
[16] Fowkes FM, Maruchi S. Org Coat Plast Chem Prep 1977;37:605.
[17] Fowkes FM, Mostafa MA. Ind Eng Chem, Prod Res Dev 1978;17:3.
[18] Gutmann V. Donor-acceptor approach to molecular interaction. New York: Plenum Press; 1978.
[19] Bolger JC, Michaels AS. In: Weiss P, Dale Cheever G, editors. Molecular structure and electrostatic interaction of polymer-solid interface, Interface conversion for polymer coatings. New York: Elsevier; 1968.
[20] Douglas B, McDaniel DH, Alexander JJ. Concepts and models of inorganic chemistry. 2nd ed. New York: John Wiley & Sons, Inc.; 1983.
[21] Morrison ID, Ross S. Colloidal dispersions: suspensions, emulsions, foams. Wiley Interscience; 2002.
[22] Watts JF, Chehimi MM. Int J Adhes Adhes 1995;15:91–4.
[23] Bikerman JJ. Causes of poor adhesion. Ind Eng Chem 1967;59(9):40–4.
[24] Schmidt JJ. DuPont fluoroproducts, Private Communication; 2003.
[25] Cagle CV. Adhesive bonding, techniques and applications. New York: McGraw-Hill; 1968.
[26] Landrock AH. Adhesives technology handbook. Westwood, NJ: Noyes Publications; 1985.

Part II

Surface Treatment Methods and Techniques

Chapter 6

Material Surface Preparation Techniques

6.1 INTRODUCTION

Surface treatment and preparation for adhesive bonding is complicated due to the material-specific nature of the required methods. This chapter and the next two have been arranged in a comprehensive fashion to facilitate learning and referral. This chapter describes the surface preparation and treatment techniques in a general way. The emphasis is on method rather than the specific material (adherend). Chapters 7 and 8 focus on material-specific surface treatment processes and conditions for metallic and plastic/elastomer substrates, respectively.

This chapter has been organized to separate the sections for metallic and non-metallic (plastics and elastomers) materials. Sections 6.3–6.6 describe the surface treatment techniques of metals; Section 6.7 describes those of plastics and elastomers. Fluoroplastics are extremely difficult to treat and adhere, and are identified as a special case. Section 6.9 addresses methods to assess the effectiveness of surface treatment. Finally, Sec. 6.10 focuses on the important topic of the treatment age (i.e., the elapsed time between surface treatment and the actual bonding), which may affect the bond strength and durability.

6.2 GENERAL CONSIDERATIONS

Since adhesive bonding is a surface phenomenon, preparation prior to adhesive bonding is the keystone of success. Surface preparation (or *surface pretreatment*) is carried out to render adherend surfaces receptive to the development of strong, durable adhesive joints. It is desirable, although not always practical, to have the basic adherend material exposed directly to the adhesive, with no intervening layer of oxide film, paint, chromate coating, chromate-free coating, phosphate coating, or silicone release agent. Such layers are called *weak boundary layers*, and in their presence the adhesive never directly contacts the adherend surface [1]. The key to good surface

Surface Treatment of Materials for Adhesive Bonding. DOI: http://dx.doi.org/10.1016/B978-0-323-26435-8.00006-X

pretreatment of many metallic adherends is to generate stable, controlled oxide growth on their surfaces.

Selection of the exact surface preparation method for a particular adherend requires careful evaluation. A number of factors, some obvious and some not, influence the choice. The size of component parts and the availability of equipment and facilities are obvious considerations. Less obvious factors include the rapid depletion of active chemicals in an immersion bath, or the accumulation of foreign materials in the bath, which give rise to weak boundary layers [1].

To place surface preparation in proper perspective, the adherend-to-organic material (i.e., adhesive) interface must be considered from design through fabrication. The interdependent factors or joint design, adhesive selection, and processing all must be considered. Optimum surface preparation is of little value if an unsuitable adhesive is used, if the bond is not properly processed, or if the joint design involves peel or cleavage stress [2].

Proper surface preparation should ensure that the weakest link in an adhesive joint is within the adhesive or organic material layer, not at the interface with the adherend. In adhesive bonds, this type of fracture is known as *cohesive failure* (see Chapter 5 for a discussion of bond failure) when a layer of adhesive remains on both adherends. When fracture of failure occurs at the adhesive-to-adherend interface, it is called *adhesive failure*. Many ASTM test methods prescribe reporting the mode of failure of fracture as a certain percentage cohesive and a certain percentage adhesive' failures. From a surface preparation viewpoint, the ideal mode of failure of a bonded joint or test specimen is one with one hundred percent cohesive failure [2].

Table 6.1 shows the different surface treatment processes for metallic and non-metallic substrates in addition to a short description of the effect of the treatment on the material surface.

6.3 SURFACE TREATMENT OF METALS

Preparing the surface of a metallic sample involves multiple steps, all of which are not always applied. It is impossible to obtain a quality adhesive bond without cleaning (and abrading) the metal surface. Metals have high-energy surfaces and absorb oils and other contaminants from the atmosphere. These steps are listed (with the frequency of use in parentheses):

1. Cleaning – using a solvent or other chemical. (Always)
2. Removal of loose materials – mechanical (e.g., grit blasting); also increases contact surface. (Sometimes)
3. Improvement of corrosion resistance. (Almost always)
4. Priming – applying a material to surface. (Sometimes)
5. Surface hardening – mechanical or chemical – to strengthen the surface. (Occasionally)

TABLE 6.1 Surface Treatment of Materials to Enhance Adhesion

Substrate	Treatment Method	Effect of Treatment
Metals	Degreasing	Cleaning of the surface
Metals	Grit blast	Loose material (weak boundary) removal from the surface and increase in contact surface area
Metals	Acid etch/liquid pickling	Surface oxidation
Metals	Anodizing	Surface oxidation
Plastics	Corona treatment	Weak boundary layer removal and surface oxidation
Plastics	Flame treatment	Weak boundary layer removal and surface oxidation
Plastics	Chemical etching	Weak boundary layer removal and surface oxidation
Fluoroplastics	Chemical etching	Surface defluorination and oxidation

Metal surfaces are best cleaned by vapor degreasing with effective aqueous systems; such methodologies have been developed. This treatment is followed by gritblasting to increase the adhesive contact surface area by roughening the metal surface. Chemical etching removes weakly bonded oxides from the metal surface and forms an oxide that is strongly bonded to the bulk of the part. A useful step is the *priming* of the part's surface, which can improve the wettability of the surface and protect it from oxidation.

The optimum surface preparation to provide durability and uniform quality is ordinarily a chemical immersion or spray process. In the case of very large parts, the application is often carried out by the use of reagents in paste form. Low energy surfaces, mainly plastics, require entirely different surface treatments, which often alter the chemical nature of their surfaces.

The chosen process is desirable only if accommodated by the production process of the entire system. Cost must also be considered and balanced against the requirement for reliability, maintainability, and critical roles of the joint.

Table 6.2 illustrates the importance of proper surface preparation for five different metal adherends, using two different adhesives and a variety of surface treatment techniques [3]. Figure 6.1 shows how various surface treatments impact the durability of aluminum bonds under adverse conditions. Surface treatment is particularly important for the durability of aluminum. The exact ranking order of the results may change in studies of this type, depending on the adhesive primer and the type of alloy being used [4].

TABLE 6.2 Effect of Metal Substrate Surface Preparation in Adhesive Bonded Joints [3]

Adherend	Treatment	Adhesive	Shear Strength, MPa
Aluminum	As received	Epoxy	3.06
Aluminum	Vapor degreased	Epoxy	5.77
Aluminum	Grit blast	Epoxy	12.1
Aluminum	Acid etch	Epoxy	19.0
Aluminum	As received	Vinyl-phenolic	16.8
Aluminum	Degreased	Vinyl-phenolic	19.9
Aluminum	Acid etch	Vinyl-phenolic	35.7
Stainless steel	As received	Vinyl-phenolic	36.0
Stainless steel	Degreased	Vinyl-phenolic	43.5
Stainless steel	Acid etch	Vinyl-phenolic	49.7
Cold-rolled steel	As received	Epoxy	20.0
Cold-rolled steel	Vapor degreased	Epoxy	19.9
Cold-rolled steel	Grit blast	Epoxy	29.6
Cold-rolled steel	Acid etch	Epoxy	30.8
Copper	Vapor degreased	Epoxy	12.3
Copper	Acid etch	Epoxy	16.1
Titanium	As received	Vinyl-phenolic	9.35
Titanium	Degreased	Vinyl-phenolic	21.4
Titanium	Acid etch	Vinyl-phenolic	46.5
Titanium	Acid etch	Epoxy	21.8
Titanium	Liquid pickle	Epoxy	22.9
Titanium	Liquid hone	Epoxy	26.9
Titanium	Hydrofluorosilicic acid etch	Epoxy	27.6

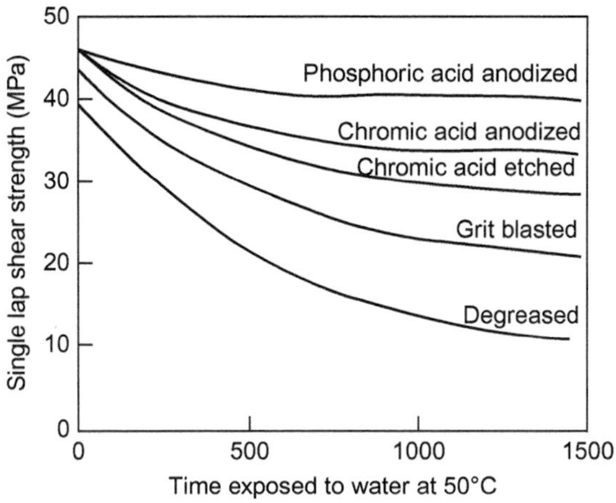

FIGURE 6.1 Effect of surface pretreatment on the performance of aluminum-alloy epoxy joints subjected to accelerated aging in water at 50°C [4].

6.4 CLEANING (DEGREASING) METALS

Before adhesive bonding, it is essential to thoroughly clean the adherends. Unclean adherends will be unreceptive to optimal adhesion regardless of the quality of materials used, or the stringent control of the application process. Proper surface preparation is extremely important in assuring strong and lasting bonds. For many adherends, surface preparation requirements go far beyond simple cleanliness.

After a cleaning process used to remove obvious surface contamination (such as soil, grease, oil, and finger marks), specific chemical or physical treatments are usually required to produce a surface receptive to adhesion. Polytetrafluoroethylene (PTFE) and other fluorinated polymers are good examples of surfaces having this requirement. Adhesion to clean, pure, PTFE is close to zero and it is only after a severe chemical treatment, which alters the surface chemically and physically, that the surface is receptive to adhesion.

Magnesium is similar, requiring considerable treatment beyond simple cleaning. Strong initial adhesion is possible with clean magnesium. Due to the chemically active nature of the metal, however, the permanency of this adhesion is poor. With many organic materials, such as adhesives, reactions occur at the metal-to-organic material interface, producing by-products having low cohesive properties. These by-products subsequently degrade the initial adhesion to an unacceptable level. For this reason, it is necessary to use chemical or anodic treatments after cleaning magnesium. Such treatments result in the formation of thin inorganic chemical films, which adhere tenaciously to magnesium, and are at the same time receptive to organic materials such as adhesives. The treatment film laid down in this manner also functions as a barrier between the metal and organic material, preventing any chemical reaction between them. The term "cleaning" has come to mean removal of dirt, contaminants, and oils, and also specific chemical surface treatments for adhesive bonding [2].

6.4.1 General Sequence of Cleaning

Any surface preparation requires completion of one, two, or all three of the following operations [2]:

- Solvent cleaning
- Intermediate cleaning
- Chemical treatment.

Priming, discussed in Section 6.5, may also be carried out in some cases to ensure superior durable bonds under particularly adverse environments.

6.4.1.1 Solvent Cleaning

Solvent cleaning is the process of removing soil from a surface with an organic solvent without physically or chemically altering the material being cleaned. This includes methods such as vapor degreasing, spraying, immersion, and mechanical or ultrasonic scrubbing. Solvent cleaning can be an end in itself, as in the case of vapor degreasing of an aluminum honeycomb core before bonding. It may also be a preliminary step in a series of cleaning and chemical treatment operations [2].

The four basic solvent cleaning procedures are [2]:

1. Vapor degreasing
2. Ultrasonic vapor degreasing
3. Ultrasonic cleaning with liquid rinse
4. Solvent wipe, immersion or spray.

Vapor Degreasing. Vapor degreasing is a solvent cleaning procedure for the removal of soluble soils, particularly oils, greases, and waxes, as well as chips and particulate matter adhering to the soils, from a variety of metallic and nonmetallic parts. The principle of vapor degreasing is scrubbing the part with hot solvent vapors. These vapors condense on the part at a sufficient rate to form liquid flow, which dissolves and washes the soil away as the condensed solvent drains by gravity. Sometimes it is not possible to wash the heavy soil away by the vapor degreasing. In such cases acetone "scrub" is the first step to remove soil, followed by vapor degreasing to remove residual oils, greases, etc. Vapor degreasing requires both the proper type of solvent and degreasing equipment. The solvents used must have certain properties, including the following [2]:

1. High solvency of oils, greases, and other soils.
2. Nonflammable, nonexplosive, and nonreactive under conditions of use.

3. High vapor density compared to air and low rate of diffusion into air to reduce loss.
4. Low heat of vaporization and specific heat to maximize condensation and minimize heat consumption.
5. Chemical stability and noncorrosiveness.
6. Safety in operation.
7. Boiling point low enough for easy distillation and high enough for easy condensation (for recycling and reuse of dirty solvent or regeneration of clean solvent from used solvent).
8. Conformance to air pollution control legislation.

The eight common vapor degreasing solvents have been:

1. Methyl chloroform (1,1,1-trichloroethane)
2. Methylene chloride (dichloromethane)
3. Perchloroethylene (tetrachloroethylene)
4. Trichloroethylene
5. Trichlorotrifluoroethane
6. Trichlorotrifluoroethane–acetone azeotrope
7. Trichlorotrifluoroethane–ethyl alcohol azeotrope
8. Trichlorotrifluoroethane–methylene chloride azeotrope.

However, 1,1,1-trichloroethane and methylene chloride have been banned and are no longer used. Alternatives with less deleterious environmental and health impact have been developed, examples of which can be seen in Table 6.3.

These materials have a wide range of boiling points, from a low of 39°C for methylene chloride to a high of 121°C for perchloroethylene. Trichloroethylene and perchloroethylene are the solvents most commonly used for vapor degreasing, particularly the former. Considerable detail on the equipment required and the vapor degreasing process is given in Ref. 2 The discussion of chlorinated solvents is from an historic and technical point of view, because the use of nearly every one of them has been banned. The health, safety, and environmental risks of this class of materials are far too high to justify their use in spite of the excellent results that their applications can yield.

Ultrasonic Vapor Degreasing [2]. Vapor degreasers are available with ultrasonic transducers built into the clean solvent rinse tank. The parts are initially cleaned either by the vapor rinse or by immersion in a boiling solvent. They are then immersed for ultrasonic scrubbing, followed by rinsing with vapor or spray plus vapor. During ultrasonic scrubbing, high-frequency inaudible sound waves (over 18,000 cycles per second) are transmitted through the solvent to the part, producing rapid agitation and cavitation. The cavitation (repeated formation and implosion, or collapsing, of the bubbles in the solvent) transmits considerable energy to the parts and any surface contaminants. Particulate materials, insolubles, and strongly adhered soils are quickly

TABLE 6.3 Examples of Environmentally-Friendly Vapor Degreasing Solvents

Trade Name	Composition	Manufacturer
Vertrel CMS	Methanol, Nitromethane, trans-1,2-Dichloroethylene, Decafluoropentane, 1,1,1,3,3-Pentafluorobutane	DuPont
Vertrel SMT	1,1,1,2,2,3,4,5,5,5-decafluoropentane, trans-Dichloroethylene, Methanol, Nitromethane	DuPont
Rho-Tron 225 TM		RHO-CHEM
EnSolv	n-Propyl Bromide	Enviro Tech
HFE-7100	Methyl Nonafluoroisobutyl ether and Methyl Nonafluorobutyl ether	3M
AK-225T	3,3-Dichloro-1,1,1,2,2-pentafluoropropane, 1,3-Dichloro-1,1,2,2,3-pentafluoropropane, Ethanol	Asahi Glass Co.
Asahiklin AE3000	1,1,2,2 tetrafluoroethyl-2,2,2 trifluoroethyl ether	Asahi Glass Co.
Asahiklin AE3000AT	Trans-1,2 dichloroethylene, 1,1,2,2 tetrafluoroethyl-2,2,2 trifluoroethyl ether	Asahi Glass Co.
Asahiklin AE3000ATE	Trans-1,2 dichloroethylene, 1,1,2,2 tetrafluoroethyl-2,2,2 trifluoroethyl ether, ethanol	Asahi Glass Co.
Asahiklin AE3100E	1,2,2 tetrafluoroethyl-2,2,2 trifluoroethyl ether, ethanol	Asahi Glass Co.
Methyl Nonafluoroisobutyl Ether	Methyl Nonafluoroisobutyl Ether	Methyl Nonafluoroisobutyl Ether
Methyl Nonafluorobutyl Ether	Methyl Nonafluorobutyl Ether	Methyl Nonafluorobutyl Ether
Asahiklin AK-225 AES	3,3-Dichloro-1,1,1,2,2-pentafluoropropane, 1,3-Dichloro-1,1,2,2,3-pentafluoropropane, Ethanol, Nitromethane	Techspray
EnSolv-GCS	n-Propyl Bromide	Enviro Tech

removed from a part, even on remote surfaces and blind holes. The ultrasonic frequency and intensity for optimum cleaning must be selected by test. They depend on the type of part being cleaned, soil removed, and particular solvent in the system. Some ultrasonic degreasers have variable frequency and power controls. The most common frequency range for ultrasonic cleaning is from 20,000 to 50,000 cycles per second. Power density may vary widely, but 2, 5, and 10 watts per square inch are common.

Ultrasonic Cleaning with Liquid Rinse [2]. Ultrasonic cleaning is a common procedure for high-quality cleaning, utilizing ultrasonic energy to scrub the parts and a liquid solvent to rinse away the residue and loosened particulate matter. This procedure, rather than using the vapor degreasing technique for pre-cleaning and final rinsing, utilizes the manual application of liquid solvents. The process is not limited to any particular solvents; organic solvents need not be used. It is widely applied with aqueous solutions: surfactants, detergents, and alkaline and acid cleaners. The only real limitations are that the cleaning fluid must not attack the cleaning equipment, fluids must not foam excessively, and the fluids must cavitate adequately for efficient cleaning.

The process is not as efficient as vapor rinse, solvent wipe, immersion, or spray, but is suitable for many surface preparation applications and pretreatments. One or a combination of these techniques may be used. A large number of solvents are recommended. Solvent wiping is the most portable and versatile of these methods, but also the least controllable. There is always a danger of incomplete removal of soil, and the spreading of soil in a uniform manner, so causing its presence not to be readily visible, and contamination of a surface with unclean wiping materials.

For general cleaning, wiping materials should be clean, freshly laundered cotton rags, new cheese cloth, or cellulose tissues. For superclean applications where cleaning must occur in a controlled "clean" room, specially processed lint-free polyurethane foam wiping materials are available (from Sills and Associates, Glendale, CA). The solvent should be used only once, poured onto the wiping material. The wiping material should never be immersed in the solvent. Solvent containers with small openings should be wiped systematically with the solvent-soaked cloth or tissue. The wiping material should be discarded and the surface cleaned again with new solvent and cloth or tissue. This cycle should be repeated until there is no evidence of soil on either the cloth or the cleaned surface.

Although immersion and soaking in a solvent is often sufficient to remove light soil, scrubbing may be required for heavier soils. The most efficient scrubbing method is ultrasonic, discussed above. Other scrubbing techniques include tumbling, solvent agitation, brushing, and wiping. After the parts are soaked and scrubbed, they must be rinsed. The quality of cleaning produced by the immersion process depends primarily on the final rinse. The solvent spray cleaning method is efficient due to the scrubbing effect produced by the impingement of high-speed solvent particles on the surface. The solvent impinges on the surface in sufficient quantity to cause flow and drainage, which washes away the loosened soil. Also, since only clean solvent is added to the surface, and scrubbing and rinsing occur, there is no danger of contamination as there is with the immersion process.

Safety [2]. Four safety factors must be considered in all solvent cleaning operations: toxicity, flammability, hazardous incompatibility, and equipment. The solvents must be handled in a manner preventing toxic exposure of the operator. Where flammable solvents are used, they must be stored, handled, and used in a manner preventing any possibility of ignition. Knowledge of the hazardous incompatibility of the solvents, cleaning equipment, and materials to be cleaned is essential. Safe equipment and proper operation is also critical. Snogren [2] lists maximum acceptable concentrations (MACs), a term synonymous with threshold limit values (TLVs) for a number of solvents used in cleaning. These values are given in parts per million (ppm). Examples are acetone – 1000 ppm; methyl alcohol – 200 ppm; and methyl chloroform – 500 ppm. Obviously methyl alcohol, which requires less concentration to incapacitate than methyl chloroform and acetone, is the solvent with the greatest exposure danger. It should be pointed out that TLVs are merely guides in the control of health hazards and represent conditions under which it is believed that nearly all workers may be repeatedly exposed, day after day, without adverse effect.

All flammable solvents should be stored in metal containers, such as safety cans, and must be applied using metal containers. The container should be grounded during pouring or dispensing. Snogren gives flammability limits in terms of percent by volume in air (lower and upper) for the solvents commonly used. Some of the solvents he lists are nonflammable. The flammable solvents must be used only in ventilated areas to prevent accumulation of vapors and fumes. Other obvious precautions must also be taken. Unstabilized methyl chloroform, trichloroethylene, and perchloroethylene are subject to chemical reaction on contact with oxygen or moisture to form acid by-products. These acids are highly corrosive to metals; therefore only stabilized grades of these solvents should be used. Strong alkalis such as caustic soda may react with trichloroethylene to form explosive mixtures (dichloroacetylene). Fluorocarbon solvents may react violently with highly reactive alkaline earth metals [2].

6.4.1.2 Intermediate Cleaning

Intermediate cleaning [2] is the process of removing soil from a surface by physical, mechanical, or chemical means without altering the material chemically. Small amounts of parent material may be removed in this process. Examples include grit blasting (the most common technique), wire brushing, sanding, abrasive scrubbing, and alkaline or detergent cleaning. Solvent cleaning should always be carried out before this step. Intermediate cleaning operations may be an end in themselves. Examples of such a situation include cleaning stainless steel with uninhibited alkaline cleaner, or detergent scrubbing epoxy laminates.

Grit Blasting. (Contributed by Dr. Laurence W. McKeen.) Grit blasting is the method most commonly used to obtain satisfactory adhesion of fluoropolymer coatings. Grit blasting should precede preheating of ferrous metals to retain the protective oxide formed. With other clean substrates, the order of these two operations is not important (Table 6.4).

Grit blasting is a relatively simple process. Hard grit is propelled by compressed air, or occasionally by high-pressure water, at the substrate needing cleaning or roughening.

Grit blast profiles are commonly measured in microinches or root mean square (RMS) by means of a profilometer. A profilometer drags a diamond stylus across the substrate and measures the depth of the peaks and valleys.

Surface profiles in excess of 100 microinches (2.5 microns) are recommended and 200–250 microinches (5.1–6.5 microns) are frequently employed. On hard substrates, aluminum oxide grit from #40 to #80 is commonly used at air pressures ranging from 80 to 100 psi (5.8–7.3 kg/cm^2) at the gun. Aluminum and brass are commonly used at air pressures ranging

TABLE 6.4 Grit Blast Media Properties

Media	Hardness, Moh	Density, g/cc	Bulk Density, lb/ft^2 (g/cc)	Relative Cost Volume Basis
Walnut shells	1–4		40–80 (0.64–1.28)	19
Silicon carbide	9	3.2	95 (1.52)	50
Aluminum oxide	9	3.8	125 (2.00)	25
Glass bead	6	2.2	100 (1.60)	18
Plastic grit	3–4	1.45–1.52	45–50 (0.72–0.80)	30
Steel shot	6	7.87	280 (4.49)	27
Steel grit	6	7.87	230 (3.68)	
Sand, silica, (silicon dioxide)	7	2.6		11
Sodium bicarbonate	2.5	2.16		

from 80 to 100 psi (5.8–7.3 kg/cm^2). Maximum air pressures on stainless steel may exceed 100 psi (7.3 kg/cm^2).

It should be noted that profiles measured by common profilometers indicate only depth of profile. They do not measure uniformity or coverage of the grit, or the sharpness of the peaks. Full coverage of the grit blast is indicated by lack of gloss on the metal surface when viewed at a flat grazing angle.

There are numerous types of grit. The choice of which grit to use depends upon its intended purpose, the substrate, and the expense. Hardness is an important property of the grit (grit needs to be harder than the substrate, to roughen it). Denser materials have more momentum and impart more energy to the substrate.

Sometimes minimal damage to the substrate is required. This could be due to the relative softness of the substrate, or because the texture of the substrate or pattern machined in the substrate needs to be maintained. Plastic grit, walnut shells, or sodium bicarbonate can clean the substrate or remove the previous coating.

Occasionally, the abrasive is propelled by pressurized water. Sodium bicarbonate slurries have been used to remove fluorocarbon coatings in this fashion.

Aluminum Oxide. Aluminum oxide is usually offered in a size range of 16–240 grit. It is angular in shape. It is the most popular cleaning blast media. Aluminum oxide conforms to major industrial and government standards including MIL A21380B and ANSI B74.12-1982.

Silicon Carbide. Silicon carbide is usually offered in a size range of 16–240 grit. It is angular in shape. Silicon carbide is an extremely hard, sharp, grain that is more friable than aluminum oxide. It is used in blasting of extremely hard materials, and is relatively expensive.

Sand, Silicon Dioxide, or Silica. Sand, silicon dioxide, or silica is considered too smooth and uniform. It breaks down too rapidly to be useful in preparing metal substrates and is thus not recommended, though it is cheap and naturally occurring.

Glass Beads. Glass beads are made from chemically inert sodalime glass. Blasting with glass beads will produce a metallurgically clean surface for parts and equipment. The beads are spheres of uniform size and hardness. Glass beads can meet OSHA standards for cleaning operations. Another advantage of using glass beads is the disposability. Spent glass is environmentally friendly. This can simplify the disposal and reduce cost. Glass beads are often used for stress relief. Mil Spec (MIL-G-9954A) glass beads are available.

Crushed Glass. Crushed glass is available in a range of sizes from coarse to very fine. Crushed glass is an excellent low-cost alternative to various reclaimed blast abrasives. While it breaks down relatively easily, it is silica free with minimal iron content (2%) and produces a luster-white metal finish.

Steel Grit and Shot. Steel grit and shot are generally available with diameters of 0.007–0.078 inches (0.2–2 mm). Steel grit is angular in shape while shot is round. Steel grit and shot have one of the lowest breakdown rates of all blast media and can therefore be recycled and reused. Their high density helps impart more energy to the substrate being cleaned. Steel grit is excellent for use in large blast room applications. It should not be used on stainless steel where iron impregnation is a concern. Cast stainless steel shot is available for nonferrous castings or for other items where ferrous contamination is a problem.

Walnut Shells. Walnut shells are relatively soft. This medium is offered in sizes of 10/14, 14/20, and 20/40. It is soft, friable dried shells or nuts, sometimes called "organic" or agrishell abrasive. It is often used for removing contaminants such as carbon deposits or old paint from delicate parts or soft materials such as aluminum. It is also suitable for blast cleaning with portable equipment.

Plastic Grit. Plastic grit is another soft media typically in sizes ranging from 12 to 60 mesh. It is often made from recycled or waste plastic. Like walnut shells, it is typically used to clean surfaces and remove old paint without harming the substrate.

Baking Soda. Baking soda blasting is unique because of its biodegradable characteristics. Clean up after use is easy because it is water soluble, and can be literally "washed" away. Baking soda is commonly used where one-pass coverage with no recovery is acceptable or desirable, and the substrate is delicate or sensitive. Typical applications for baking soda include graffiti removal, boat hulls, and large printing press rolls. It is often made into waterborne slurry and propelled by high-pressure water.

6.4.1.3 Chemical Treatment

Chemical treatment is the process of treating a clean surface by chemical means. The chemical nature of the surface is changed to improve its adhesion qualities. Solvent cleaning should always precede chemical treatment and, frequently, intermediate cleaning should be used in between. Chemical treatments, such as acid etch procedures, are discussed in Chapter 7 in relation to a number of metals [2].

6.5 PRIMING

An adhesive primer is usually a dilute solution of an adhesive in an organic solvent. The solution is applied to the adherend, producing a dried film with a thickness of 0.0015–0.05 mm. Some of its functions are as follows:

- Improves wetting [5].
- Protects the adherend's surface from oxidation after surface cleaning, extending the time that may elapse between surface preparation and adhesive application. (Such an extension may increase the usable time for

aluminum adherends from 4 hours to up to 6 months, dependent on the chemical treatment method) [2].
- Helps inhibit corrosion (corrosion inhibiting primes – CIPs) [5].
- Modifies the properties of the adhesive to improve certain characteristics such as peel [6].
- Serves as a barrier coat to prevent unfavorable reactions between adhesive and adherend [5].
- Holds adhesive films or adherend in place during assembly. This type of primer retains tack, or develops tack at room or elevated temperatures [6].

The use of primers provide more flexible manufacturing scheduling, high reliability of joints, less rigorous cure conditions, wider latitude in choice of adhesive system, and more durable joints [5]. Primers are usually not fully cured during their initial application. They are dried at room temperature and some are force-dried for 30–60 minutes at 65.5°C. These steps, frequently called "flashing," provide a dry, non-tacky surface that can be protected from contamination and physical damage by wrapping with clean paper, sealing in polyethylene bags, or covering with a non-transferring adhesive-backed paper [2].

When primers are desirable, the manufacturer's literature will ordinarily specify the best primer to use. The primers, like the adhesives, are usually proprietary in nature and made to match the adhesives [6].

6.6 SOL-GEL PROCESS

Sol-gel technology has proven to be an effective and environmentally friendly technology to pretreat metals surfaces for adhesive bonding. Even though originally the sol-gel technique was developed for ceramics processing, it has found many applications outside ceramics. The origins of the sol-gel process date back to the mid 1800s [7,8] when the first observation of the chemical reaction was made. Ebelman in 1846 observed the formation of transparent materials from the slow hydrolysis and polycondensation of silicic acid. Geffcken and Berger in 1939 made the first attempt to synthesize glass from gels, which was focused on making SiO_2 layers [9]. The first multicomponent silicate sol-gel glasses were prepared in 1955 by Della and Rustom Roy [10]. The strongly growing interest of many researchers in the sol-gel method began in the mid 1970s. Sol-gel process for the surface treatments of metals and ceramics for adhesion bonding has been under development since the 1980s.

There are two generic variations of the sol-gel technique [11]. One is called the *colloidal* method the other is called the *polymeric* (or alkoxide) route. The differences between the two stem from the types of starting materials (precursors) that are used. Both routes involve suspending or dissolving the precursor(s) in a suitable liquid, usually water for the colloidal route and alcohol for the polymeric route. The precursor is then activated by the

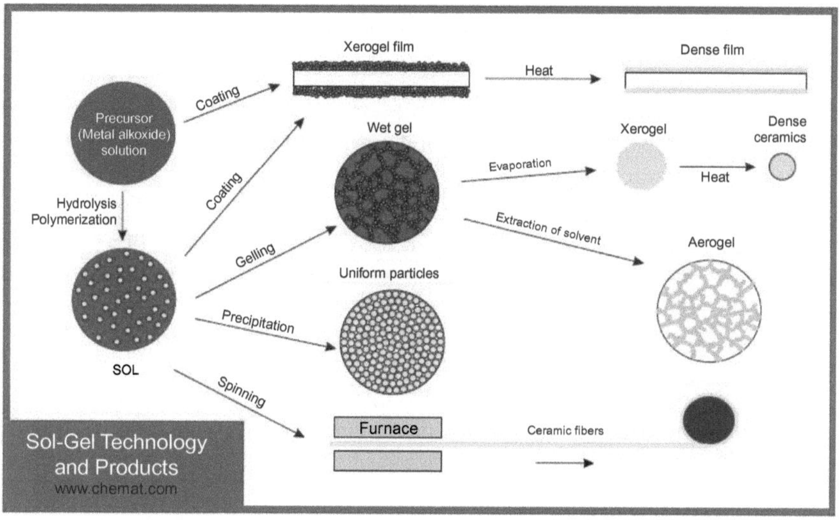

FIGURE 6.2 Basic reaction schemes for sol-gel processes. (*Courtesy Chemat Technology*, www.Chemat.com, June 2013.)

addition of an acid (such as hydrochloric acid) or a base (such as potassium hydroxide). The activated precursors then react together to form a network by condensation. The network grows and ages with time and temperature until it is the size of the container. At this point the viscosity of the liquid increases at an exponential rate until gelation occurs; that is, no more flow is observed. Figure 6.2 shows the basic reaction schemes for the sol-gel process.

In the 1990s Chemat Technology (www.Chemat.com) developed a Sol-gel treatment technology under a grant from United States Environmental Technology [12]. The objective of the project was to demonstrate the feasibility of formation of environmentally benign non-surface pretreatment water-based primers for adhesive bonding. Non-chromate corrosion inhibitors were identified for the water-based primers. A family of water based, non-surface-pretreatment solutions were formulated. The developed primers are useful for treating aluminum alloys, titanium alloys, and steels.

6.7 SURFACE TREATMENT OF PLASTICS

An important and relevant difference between metals and plastics is their surface energy. Polymers have inherently lower surface energy than metals (assuming they are contaminant-free) and tend to form intrinsically poor adhesion bonds without some type of treatment. Adhesion is the mechanical resistance to separation of a bonded system and involves the outer surface of a material. Treatment only impacts the region near the surface and does not alter the bulk properties of the plastic parts.

In the dawn of plastics technology, chemical priming was the only method of surface preparation. Soon after, the first rudimentary machines were developed to treat polymer surfaces, increasing their polarity and surface energy, and making them acceptable for the various laminating, coating, and decorating processes. The flame and corona treatment methods gradually evolved through the 1960s and 1970s. With solvent-based systems predominant, major breakthroughs came slowly. The advent of waterborne, and then energy-cured, systems changed the pace of development. Suddenly, marginal treatment levels (34–38 dynes/cm) caused serious adhesion bond quality problems. Films that had been readily printable exhibited pinholes, fish-eyes, ink lift-off, and other defect types. The remedy was to increase the treatment level to a part surface energy of 40 dynes/cm or higher.

There are chemical, physical, and bulk treatment methods available for adhesion enhancement. Chemical modification techniques include those usually requiring wet or chemical reactions as the primary means of altering the surface such as wet etching, grafting, acid-induced oxidation, and plasma polymerization. Physical surface treatment methods include corona discharge, ion or electron beams, photon beams (laser, ultraviolet light, and X-ray), plasma discharge, and flame oxidation. Bulk methods involve additives, blending, or recrystallization, all of which affect the bulk properties of plastics.

This section describes the significant techniques available for the treatment of plastic surfaces to enhance adhesion (Fig. 6.3). Not all methods have wide commercial application. Some of the techniques are limited in the scope of their use. For example, chemical treatment (acid-induced oxidation) is the most frequently used method to impart adherability to plastic surfaces. Plasma treatment is limited to smaller components and parts. Flame and corona treatments are effective for continuous films (often called webs) and thin sheets of plastic, usually operated at high speeds.

6.7.1 Effect of Treatment on Plastic Surfaces

The changes that occur in the surface of plastics by applying the treatment methods are the result of the four processes of *cleaning, ablation, crosslinking*, and *surface chemical modification* (most commonly oxidation). Common treatment methods include corona, flame and plasma treatment, and chemical etching, all of which serve to increase the surface energy of plastics. The polar component of surface energy should be increased by 15–20 dynes/cm to achieve proper adhesive bonding in polymers such as polyolefin, polyphenylene sulfide, and polyaramide [13]. The other changes in the plastic surfaces include removal of weak materials, strengthening of the surface (by crosslinking), and increased surface roughness.

The increase in surface energy of plastic occurs through the surface oxidation of the polymer chains. For halogenated polymers, such as chlorinated and fluorinated polymers, surface modification involves significant

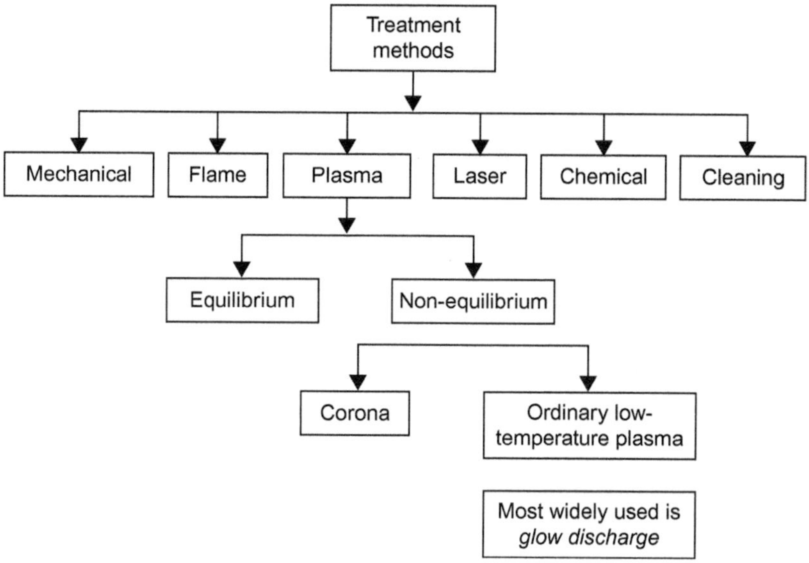

FIGURE 6.3 Classifications of plastic treatment techniques.

dehalogenation, the removal of chlorine and fluorine atoms from the surface molecules. In general, surface treatments can be viewed as a transfer of energy to the surface of plastics. In almost every case, the dissipation of treatment energy on the plastic surface results in more than one change to the surface; these changes are described below.

The *cleaning*, or removal of contamination, including process oils, dirt, waxes, mold release agents, and exuded plasticizers, is an important result of surface treatment. Methods involving chemicals such as solvent cleaning and etching, if not properly used, can leave behind a residue that may interfere with adhesive bond formation. Clean surfaces must be protected due to rapid reacquisition of contamination from the ambient atmosphere.

Plastic surfaces often contain a low molecular polymer such as an oligomer, or loosely bonded matter, creating a *weak boundary layer*. There is some debate about production of low molecular weight polymers as a result of surface treatment [14]. Loosely bonded materials may originate from the plastic part itself and from additives incorporated such as pigments, processing aids, and mold release agents. They may also have been acquired externally as a result of contact with contaminated surfaces or exposure to a contaminated environment. The treatment methods remove loose material from the part surface. One mechanism for removal of material from the part's surface material is *ablation*, which is the degradation and conversion of organic matter into volatile species.

The electrical discharge at the plastic surface during *corona treatment* reacts with the polymer molecules, generating free radicals. The presence of free radicals allows crosslinking and functionalization on the plastic surface with and without chain scission. The radicals rapidly react with ambient oxygen and produce peroxide groups that can decompose and give rise to a variety of polar groups. These groups include hydroxyl (OH), carbonyl (C = O), and carboxylic acid (O = C–OH). The presence of these groups has been confirmed by ESCA (see Ch. 4 for a description of this technique).

Corona treatment is believed to roughen the plastics by the degradation of amorphous regions of the polymer surface [15]. The belief is that corona treatment does not impact the crystalline region of the surface, preferentially attacking the relatively weak amorphous regions. Degradation and subsequent removal of the amorphous material leads to the increased roughening of the surface of plastics such as polyethylene [16–18]. A rough surface provides a much larger adhesive contact area than a smooth surface.

Plasma treatment oxidizes the surface of the polymer in the presence of oxygen. It can thus remove organic contaminants from the surface. Early studies have concluded that the crosslinking of low molecular weight surface species is the mechanism for eliminating a weak boundary layer [19]. More recent research has attributed the effectiveness of plasma treatment to surface cleaning, ablation of surface polymer chains, surface crosslinking of polymer chains, and introduction of polar functional groups that result in increased surface energy [20].

The mechanism of *flame treatment* is the thermal oxidation of the polymer surface. The flame temperature may exceed 2000°C. It can clean the surface and remove the weak boundary layer by vaporizing surface contamination and low molecular weight polymers.

Chemical treatment or etching oxidizes the plastic surface similarly to corona treatment. For instance, chromic acid is used to etch the surface of polyethylene and polypropylene. An increase in etching time and temperature intensifies the surface treatment by increasing the degree and depth of oxidation.

6.7.2 Surface Cleaning

Nonmetallic material, namely plastics, should be degreased, if necessary, with an aqueous detergent solution, followed by a thorough rinse with clean water, and dried. The detergent can be substituted by a solvent. Either solvent or detergent solutions can remove mold-release agents or waxes from the plastic part surface. Effective solvents include methyl ethyl ketone, acetone, and methanol, depending on the plastic type. Resistance of the plastic to the solvent should be considered during solvent selection to prevent dissolving or degrading the plastic during cleaning. The chemical surface cleaning step is unnecessary if a treatment technique, such as plasma treatment, adequately cleans the surface along with surface modification.

6.7.3 Mechanical Treatment (Surface Roughening)

Surface roughening and sanding of plastics accomplishes the same purposes as with metals. Essentially, loose and unstable polymers are removed from the surface, thus increasing the contact surface area. This step is not pertinent if there are decorative and aesthetic considerations that require an even, smooth surface. For instance, if a plastic part is painted or laminated to a thin film, the application of surface roughening and sanding may be impossible or limited. When applicable, plastic surfaces are usually hand-sanded or sand-blasted to impart roughness (increased contact surface area) to the part surface.

6.7.4 Corona Treatment

Corona discharge usually takes place at atmospheric pressure in contrast to low temperature (or cold) plasma, which requires a vacuum. Corona is a stream of charged particles such as electrons and ions that are accelerated by an electric field. A corona is generated when a space gap filled with air or other gases is subjected to a sufficiently high voltage to set up a chain reaction of high velocity particle collisions with neutral molecules, resulting in the generation of more ions. One such species of importance is atomic oxygen (O) generated by the UV light in the corona stream:

$$2O_2^- \rightarrow O + O_3$$

The atomic oxygen (O) reacts with hydrogen and carbon on the film surface, generating $-OH$ and $C{=}O$ groups which form hydrogen bonds with adhesives.

Corona discharge is applied to treat the surface of plastics, rendering them adherable (Fig. 6.4). In this method, the plastic article is exposed to a corona discharge produced by high-frequency, high-voltage alternating current. The corona dosage is calculated from the following equation:

$$D = \frac{P}{CB \times V}$$

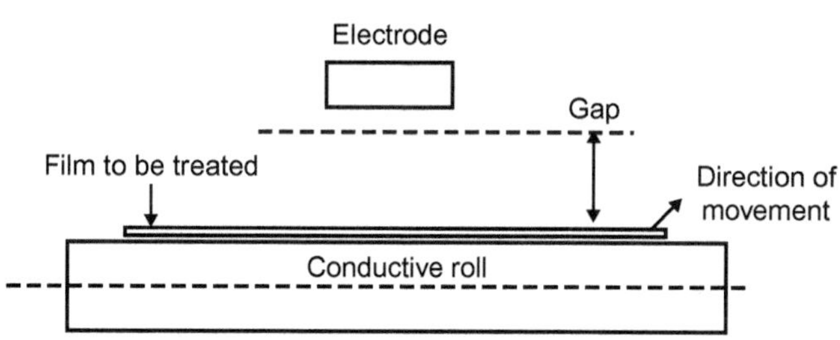

FIGURE 6.4 Conceptual schematic diagram of a film corona treatment system.

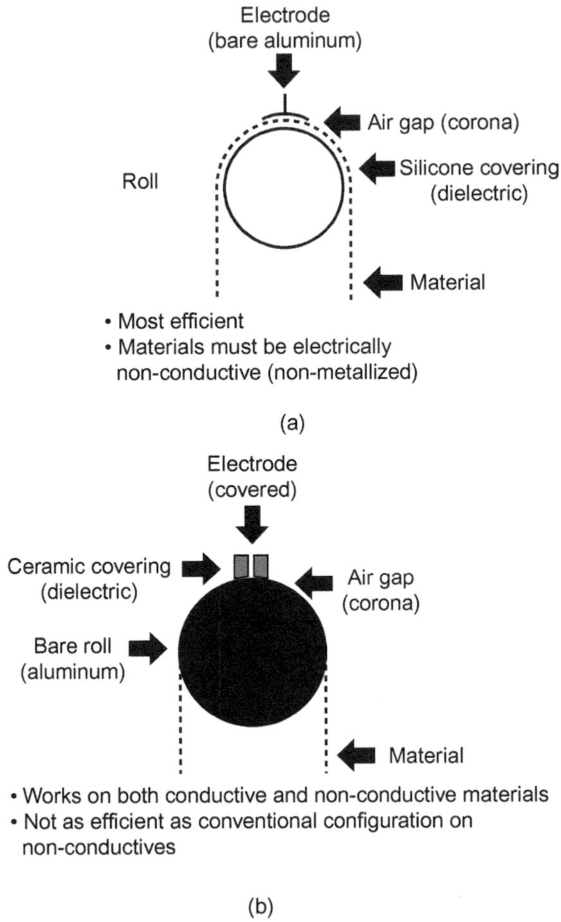

FIGURE 6.5 Configurations of corona treatment equipment. (a) Conventional configuration. (b) Bare-roll configuration [21].

D = Corona dosage, J/mm^2
P = Power, W
CB = Corona treated (electrode) width, mm
V = Web speed, mm/min

There are three types of treating configurations (Fig. 6.5 shows two of them), all consisting of the same parts including an electrode, an electrical insulator (or dielectric), and a return path (or ground). The differences among the three configurations are in the location of the electrode. In a conventional system (Fig. 6.5a), the web passes over a roll that is covered with insulating material such as a silicone rubber. A metal electrode is suspended above the roll so that an air gap of 1.5–2.5 mm exists between the electrode and the insulated roll. High voltage operates across the air gap, ionizing it, and forming a corona discharge curtain between the electrode and the material (e.g., film) that is being treated. The conventional configuration can be used only with nonconductive material. The second configuration is called *bare roll* (Fig. 6.5b) in which the electrode is covered with a dielectric (usually ceramic) and the roll is made of anodized aluminum. In the third configuration, called *double dielectric*, the roll and the electrodes are covered with dielectric material.

The main parameters for the control of the treatment process include the voltage, width of the air gap between electrodes, film/web speed, and the width of the electrodes. Most machines allow treatment of one side of the web (Fig. 6.6) and require two passes for two-sided treatment. There are also machines that are equipped with two sets of electrodes for one pass, two-sided treatment.

The appropriate treatment voltage to turn the air in the electrode gap into a conductor depends mainly on the size of the gap. It is possible to reduce the voltage to about 300 V independent of the gap size (within the operating range) by increasing the current. The color of a stable corona discharge is purple, which resembles a natural gas flame across the gap.

Effect of corona treatment on adhesion of polyethylene to itself (autohesion) has been studied as a function of lamination temperature and density, as shown in Figures 6.7 and 6.8 [22]. Corona treatment has a dramatic effect on the increase of the adhesion bond strength over that achieved with untreated polyethylene. An increase in lamination temperature enhances the bond strength of treated film.

FIGURE 6.6 Photograph of a plastic film treater. (*Courtesy Softal Electric GMBH*, www.softa.de)

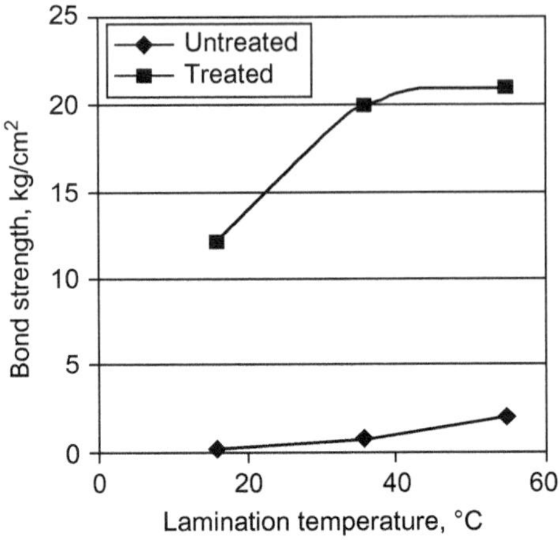

FIGURE 6.7 Adhesive bond strength of polyethylene (PE) to itself as a function of lamination temperature for untreated and corona-treated PE [22].

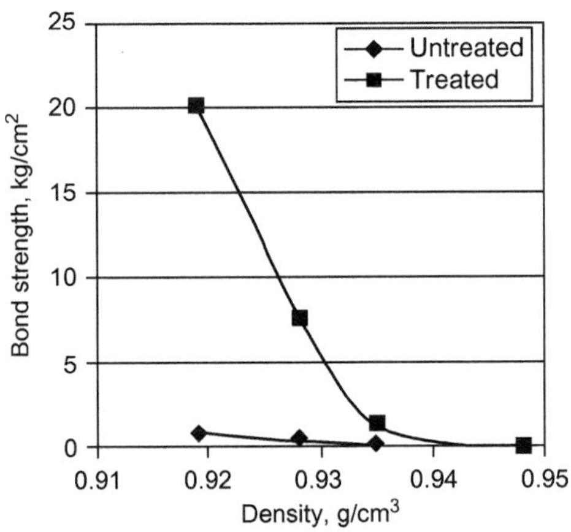

FIGURE 6.8 Adhesive bond strength of polyethylene (PE) to itself vs. density at lamination temperature of 36°C for untreated and corona-treated PE [22].

Effectiveness of corona-treatment depends on the type of polymer. Examples of bond strengths of corona-treated polypropylene and a polyetherimide can be found in Figure 6.9. Polypropylene is affected by the treatment as seen from the increase in adhesive bond strength in contrast to the bond for the polyetherimide, which is unchanged.

The duration of corona discharge treatment in air has a strong influence on the surface energy of the plastic. Figure 6.10 illustrates this relationship for low-density polyethylene (LDPE), polyethylene-propylene copolymer (PEP), and polypropylene (PP). Surface energy increases more slowly as the

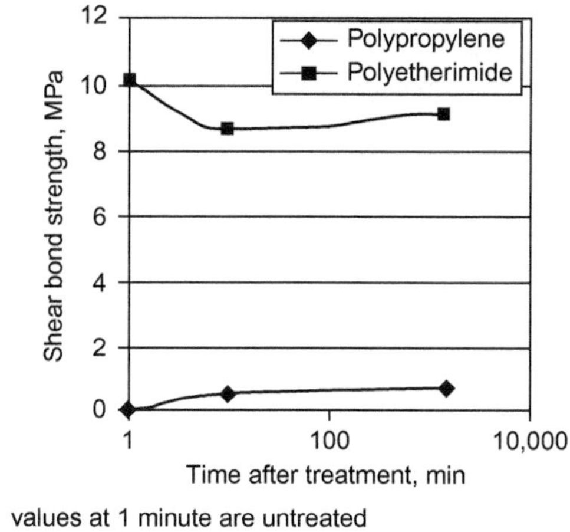

FIGURE 6.9 Bond strength (shear) of corona-treated polypropylene, bonded to aluminum using an epoxy adhesive (Araldite AW 136 H/hardener HY994).

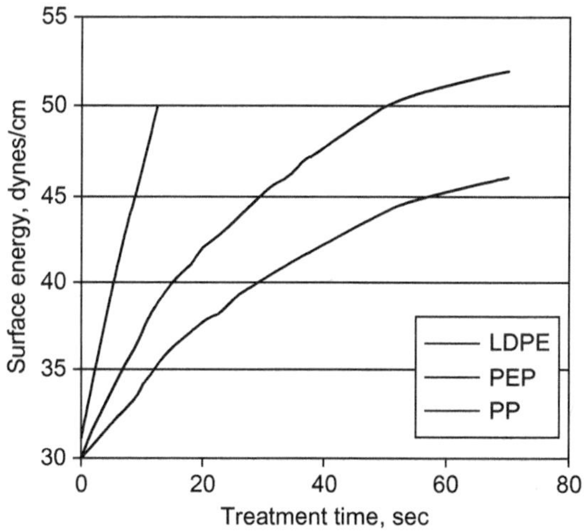

FIGURE 6.10 Surface energy of plastics vs. duration of corona discharge treatment.

crystallinity of the polymer being treated increases (from lower to higher, PE < PEP < PP). Longer exposure time is required to achieve the same level of surface energy.

6.7.4.1 Three-Dimensional Corona Treatment

Three-dimensional (3D) corona treatment, as the name indicates, is aimed at surface modification of objects with a third dimension, compared to a web

(2D). Plastic objects are treated by 3D corona discharge to promote adhesion for printing, painting, coating bonding, and labeling [23–25].

A three-dimensional corona discharge head has a dielectric enclosure with two small electrodes made of medium thickness, bare aluminum or stainless steel wire, connected to a high voltage power supply and fan (Fig. 6.11) [26]. When a high voltage exceeds the air breakdown value (30 kV/cm), an electrical arc forms. This high current arc is blown out from the enclosure by the air stream, and is usually several centimeters long and one millimeter in diameter.

When the arc strikes a solid object (Fig. 6.12), it travels across the surface and returns back into the enclosure [26]. The treatment is achieved by moving the object under the discharge head or by moving the head over the object. The arc is essentially a one-dimensional object (like a wire), and a treatment of 3D surfaces requires at least two heads. It is difficult to treat objects with complicated geometry, especially with internal cavities or channel structures, including tubing. The average power required for one head operation is about 500 W. The effective power dissipated in the arc could be as low as 50 W. Most of the ions and electrons travel between the two electrodes along the arc trajectory where the main transfer of energy occurs.

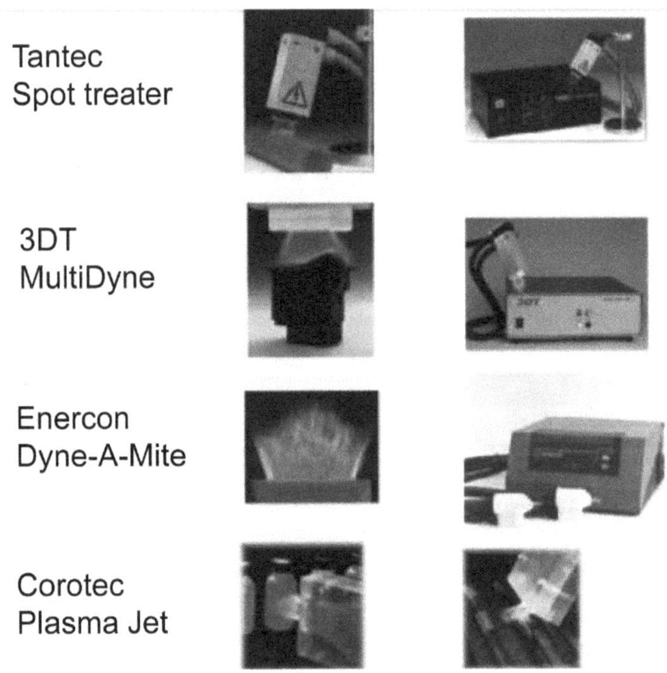

FIGURE 6.11 Examples of 3D corona treater offered by equipment manufacturers [26].

Although the fan cools the arc continuously, it still has a fairly high core temperature. The air stream created by the fan is usually unstable. The arc constantly changes, following the random pulsation of the air, leading to a non-uniform surface treatment. To increase uniformity, the exposure time should be significantly increased, cutting down the production rate. On the other hand, high current density within the arc makes the arc plasma very hot, and exposure time is limited by the thermal damage threshold of the surface.

The plasma effect on the material strongly depends on the exposure time because each material requires a minimum exposure time to activate its surface. The required level of surface modification depends on the application (printing, bonding, coating, etc.) as well as on the applied ink, adhesives, coatings, and curing process. Occasionally, there is not a sufficient window for the necessary exposure time and thermally safe material handling for 3D corona treaters. This is especially true of tough-to-treat materials (plasmaphobic) in which the surface gets burned rather than modified. This is also the case for heat sensitive materials, thin wall plastic objects, wires with thin insulation, fiber optics, thin coating layers, etc. This problem is partly solved by installation of several discharge heads along the process line. If the problem is not solved, other techniques are used to avoid burning.

3D corona treaters also generate ozone. To reduce high ozone concentration, special filters are required to comply with clean room environment regulations for the medical and semiconductor industries. A high-voltage signal applied to the electrodes usually has an audible frequency (60 Hz and up). Electrical breakdown takes place during each half-period of the cycle and produces a small shock wave with distinctive sound. This may generate significant amounts of noise in multiple head discharge systems, requiring noise abatement and hearing protection.

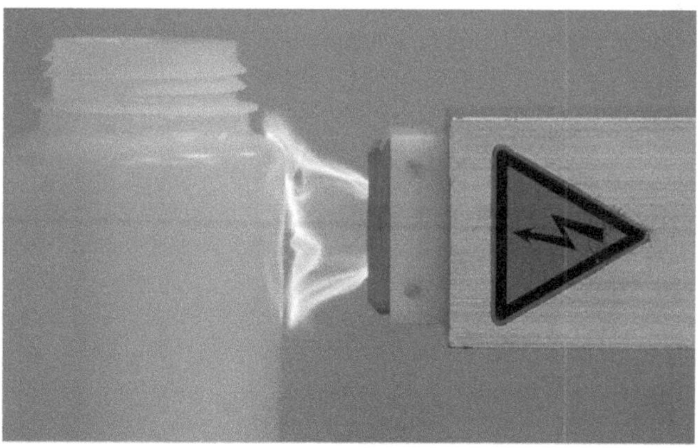

FIGURE 6.12 Three dimensional treatment of a plastic bottle. (*Courtesy Dyne Technology Ltd*, www.dynetechnology.co.uk.)

6.7.4.2 Corona Treatment under Chemical Atmosphere

One of the most important commercial polyolefin films is biaxially oriented polypropylene (BOPP) film, because of its use in a variety of applications such as food packaging, multi-purpose packaging and adhesive-tape packing [1–3]. Biaxially oriented polypropylene (BOPP) films have low hydrophilicity and do not allow printing by ink or wetting by adhesives well. The surface is thus modified to improve wettability and adhesion. The most common method for surface modification of BOPP and other polyolefin films is corona treatment, which offers the benefits of continuous and inline operation and cost effectiveness.

Corona treatment introduces [27] polar functional groups such as hydroxyl, carbonyl, and carboxylic groups onto the film surfaces. The presence of these functional groups on the film surface raises the film surface free energy. Thermodynamic driving forces and the small sizes of these groups, however, result in the overturning or "inversion" of polar groups, which migrate below the film surface [28,29]. This phenomenon is called *hydrophobic recovery*. In just 30 days after treatment, a film like polypropylene can lose the effect of modification significantly, in terms of surface free energy, thus rendering it practically useless for most applications [30].

The mobility of polymer chains at the surface of the film is the reason polar groups invert inwards. To eliminate the time-dependence of corona treatment, the mobility of polymer chains at the surface may be constrained by grafting bulky functional groups [31]. The key issue is, however, that post-corona treatment surface grafting involves additional steps which significantly lengthen the process and increase the cost of the film. Technology has been developed to introduce organic grafting compounds into the atmosphere of the corona process, which is usually air.

Introduction of vapor of acrylic acid (AAc) monomer into the corona region has proved successful. Surface wettability of BOPP films was improved (surface energy increased) by comparison with the normal air-corona treatment (Table 6.5). The hydrophilicity lasted for more than 90 days of aging (Fig. 6.13). The hydrophilicity of the film surface can be tuned by varying the corona energy. The peel strengths of AAc-corona treated films are appreciably higher than those of untreated and air-corona treated films (Table 6.6).

Long before the development of chemistry for surface alteration of plastics films the approach to vary the process atmosphere of corona treatment was attempted for melt processible perfluorinated fluoropolymers such as fluorinated ethylene propylene copolymers (FEP) and perfluoroalkoxy polymer (PFA). Corona discharge treatment of FEP and PTFE films in air has been reported [33] to improve adherability of these films as indicated by higher values of peel strength, although the bond was inconsistent and fairly weak. Later, corona treatment of FEP was reported [34] under an atmosphere of gases other than air (Fig. 6.14). The process conditions included a voltage

TABLE 6.5 Surface Free Energy of Aged BOPP Films at Ambient Temperature [27]

Corona Energy (kJ/m^2)	Surface Free Energy (mJ/m^2)			After Aging (90 Days)	
	Before Treatment	After Treatment			
		Air-Corona	AAc-Corona	Air-Corona	AAc-Corona
153	34.8	48.0	54.1	42.4	51.0
38.2	34.8	49.6	58.4	44.3	53.5
76.4	34.8	55.7	64.0	47.0	56.8

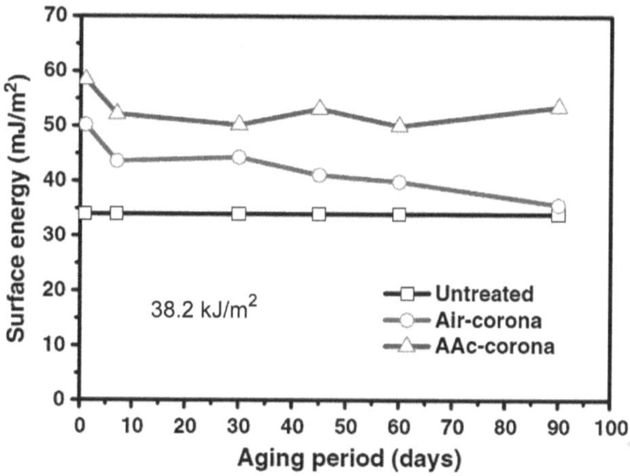

FIGURE 6.13 Surface free energy of aged BOPP films treated with corona energies of 38.2 kJ/m^2 at ambient temperature [32].

in the range of 10,000–30,000 V, pulsing peak voltage of 100,000–500,000 cycles per second, and less than 5% by volume of a suitable gas in nitrogen. A suitable gas must have a vapor pressure of 0.25 mm of mercury at 60°C. Examples included glycidyl methacrylate, tetrahydrofuran, carbon tetrachloride, vinyl butyl ether, and methyl vinyl ketone.

For example, consider a McBride and Wolinski [34] corona-treated FEP film (0.25 mm thick) under the atmosphere (0.5% by volume) of various chemicals. A voltage was applied in the range of 10,000–30,000 with pulsating peak voltages up to 100,000 volts while the film moved at a speed of 1.5

TABLE 6.6 An Average Peel Force of Air-Corona and AAc-Corona Treated BOPP Films Before and After Aging [32]

Sample	Average Peel Force (N)			
Untreated BOPP Film	2.55 ± 0.03			
Corona Treated Film	Air-Corona		AAc-Corona	
	After Treatment	After 90 Days	After Treatment	After 90 Days
15.3 kJ/m^2	3.09 ± 0.02	2.89 ± 0.05	3.58 ± 0.02	3.00 ± 0.06
38.2 kJ/m^2	3.35 ± 0.04	2.98 ± 0.06	3.73 ± 0.02	3.29 ± 0.11
76.4 kJ/m^2	3.63 ± 0.04	3.37 ± 0.14	3.92 ± 0.04	3.69 ± 0.10

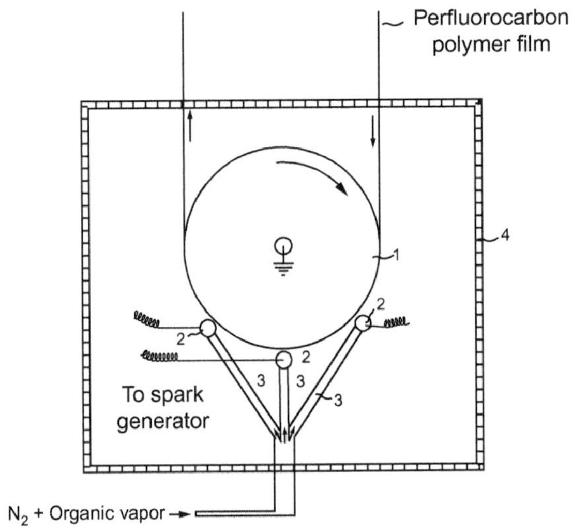

FIGURE 6.14 Schematic diagram of a corona treater under an organic vapor [28]. (1) Metal roll; (2) stationary hollow tube; (3) distributor duct; (4) enclosed chamber.

meters per minute. The treated film was stuck to a strip of cold-rolled steel using an epoxy containing 1% amine hardener. Bond strength was then measured at a peel angle of 90° in a Suter tester, essentially an extensiometer equipped with jaws that can be separated at a controlled speed. Table 6.7 shows the bond strength for various atmospheric chemicals; film treated without a vapor phase chemical developed a weak bond.

TABLE 6.7 Bond Strength of Corona-Treated FEP Under an Atmosphere of Different Chemicals [23]

Atmosphere Chemical	Bond Strength (g/cm)
None	39.7–119
N-vinyl-3-pyrrolidone	3571
Acrylonitrile	2500
p-Chlorostyrene	1190
Toluene-2,4-diisocyanate	1952
Vinyl acetate	1913
Xylene	1389
Hexane	1349
Carbon tetrachloride	1587
Tetraisopropyl titanate	1428

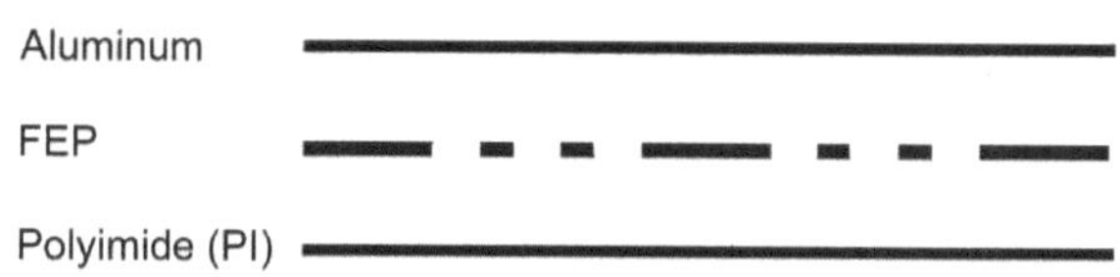

FIGURE 6.15 FEP laminate to aluminum foil and polyimide.

Kreuz and Zytkus [35] reported corona treatment of FEP under an atmosphere of acetone (<5% to 40% by volume in nitrogen) which had the advantage over the previous atmospheric compounds of producing a film that did not block in roll form. Treated FEP could be printed and marked with inks, was heat-sealable, and adhered well to metals. Adhesion of FEP film (25 μm thick) to metal was accomplished by first laminating it to a thin polyimide film (13 μm thick) using a nip roll at 250°C–270°C at a pressure of 270 kPa. The treated FEP surface was then laminated to a 13 μm thick (Fig. 6.15) aluminum foil in a nip roll at a speed of 6–9 m/min.

The importance of the role of acetone in the corona treatment atmosphere is shown in Table 6.8. In this case, a FEP film at a thickness of 13 μm was treated in an atmosphere of acetone (3.1% by volume) and nitrogen. The treated film was heat sealed to a 25 μm thick polyimide (PI) film, both corona-treated and untreated, at different temperatures. Untreated polyimide produced a weaker bond while FEP treated in air (versus acetone/N_2) resulted in almost no bond strength. (Table 6.9)

TABLE 6.8 Bond Strength of Corona-Treated FEP Under an Atmosphere of Acetone [24]

Lamination Temperature (°C)	Post Heat Treatment	Bond Strength, (g/cm)	Failure Interface
230	Yes	137	FEP-Al
230	No	161	FEP-Al
235	No	167	FEP-Al
241	No	127	FEP-Al
240	Yes	182	FEP-Al and PI-FEP
272	Yes	244	PI-FEP
272	No	226	PI-FEP

TABLE 6.9 Effect of Corona Treatment and Heat Sealing Variables on Bond Strength of FEP [25]

FEP Corona Treatment Atmosphere	FEP Film Speed, m/min	Bond Strength, g/cm at Heat Seal Temperature of 315°C		Bond Strength, g/cm at Heat Seal Temperature of 350°C	
		Treated PI	Untreated PI	Treated PI	Untreated PI
Air	15.1	13.5	–	15.5	–
Air	24.2	16.3	–	13.9	–
Acetone	15.1	208	110	416	253
Acetone	24.2	196	50	287	138

6.7.5 Flame Treatment

Flame treatment is a commercial process to render polyolefins and polyethylene terephthalate adherable. The polymer article (e.g., film) is passed over an oxidizing flame formed by an oxygen-rich (relative to stoichiometry) mixture of hydrocarbon gas. Variables affecting the extent of oxidation include the flame characteristics (e.g., excess oxygen) as well as the speed of the article movement. Gas flame contains excited fragments and species such as atomic oxygen (O), NO, OH, and others that can abstract hydrogen from the

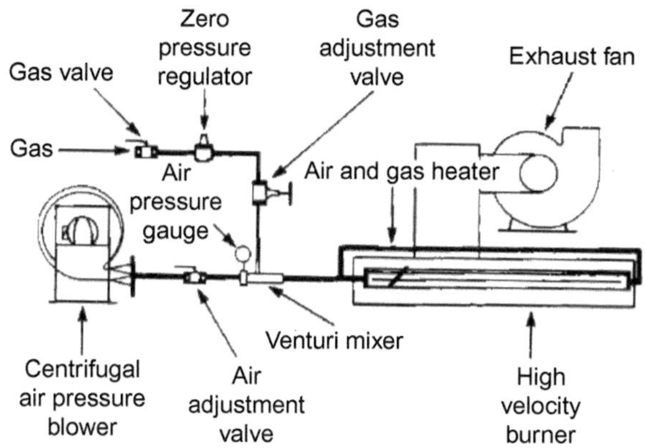

FIGURE 6.16 Flow diagram of flame treatment system [36,37].

surface of the polymer that is replaced by oxygenated functional groups (mostly $-C=O$ and $-OH$). Figure 6.16 depicts the flow diagram of the flame treatment system.

Polyethylene and polypropylene have a surface tension of 30 dynes/cm that must be raised to a minimum of 38 dynes/cm (preferably to 42 dynes/cm) to render their surfaces adherable. Flame treatment is the most widely used method of pretreatment. It is flexible and reliable if carefully controlled. It enables the treatment of uneven and curved surfaces. It uses a mixture of air at 138–345 kPa and a fuel gas at a (low) pressure of 1.7 kPa. The gas can be butane, propane, natural gas (methane), and coal gas.

For the flame to be effective, it must be oxidizing, (represented by a blue flame). Correct flame control is very important. A basic flamer will complete simple work; specially designed flame control systems are recommended for routine operations and long production runs. These systems are fitted with gas and air control valves to compensate for pressure fluctuations, ensuring that the mixture is always at its optimum. Safety devices, such as flame failure shutdown, are fitted as standard. Automatic ignition is also standard. Flame nozzle design is important for these normally single or double row ribbon burners because they give a more stable flame shape and characteristic. "Flame throwers" are inefficient and unreliable. Flame control and position of the item in the flame are critical. Setting up the flamer is also very important (see Fig. 6.17). Over-flaming will damage the surface of the product, while under-flaming can cause failure with the ink adhesion [38].

It is imperative to find the correct flame treatment suitable to the component in question. Key factors include the gas type, gas to air ratio, burner type, gas flow rate, flame distance, and flame passage speed (treatment time). Excessive treatment of the surface results in degradation of the polypropylene and, therefore, poor adhesion. Insufficient treatment, however,

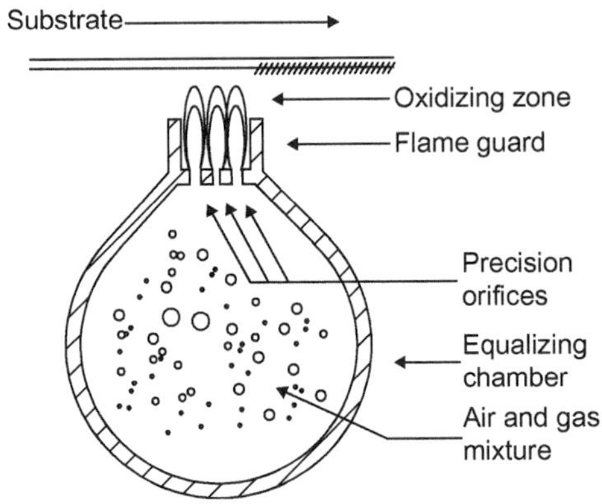

FIGURE 6.17 Effect of distance between part and inner cones of flame [38].

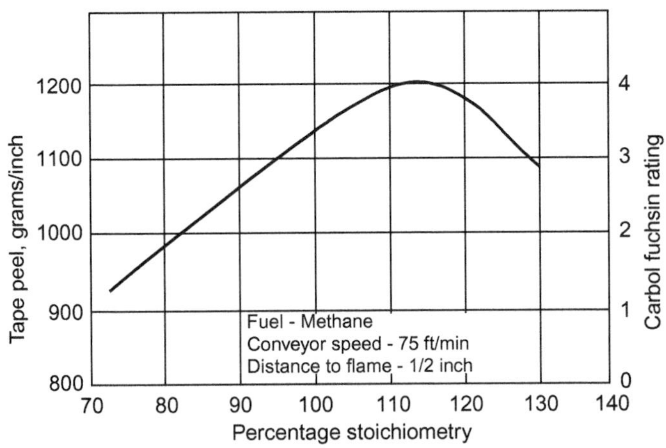

FIGURE 6.18 Example of burner in a flame treater [36,37].

fails to modify the surface adequately, leading to poor adhesion. It is often a rather delicate procedure to find the proper conditions and great care must be taken in the experimentation (Fig. 6.18).

During flame treatment, all hydrocarbon gases combine with oxygen to produce heat, as shown here for methane:

$$CH_4 + 2O_2 \rightarrow 2H_2O + CO_2$$

The above equation indicates complete combustion of the fuel gas methane. The volume of air required for complete theoretical combustion of gas (such as methane) is called the *stoichiometric* volume (100%). A flame containing an excess (>100% stoichiometry) of air is referred to as an *oxidizing*

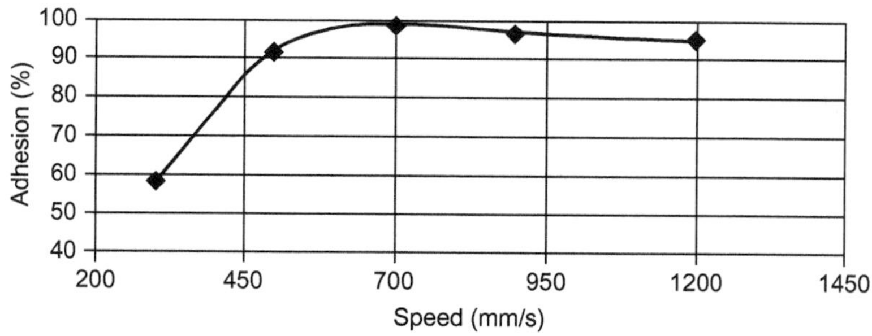

FIGURE 6.19 Effect of air/gas ratio on treatment level [38].

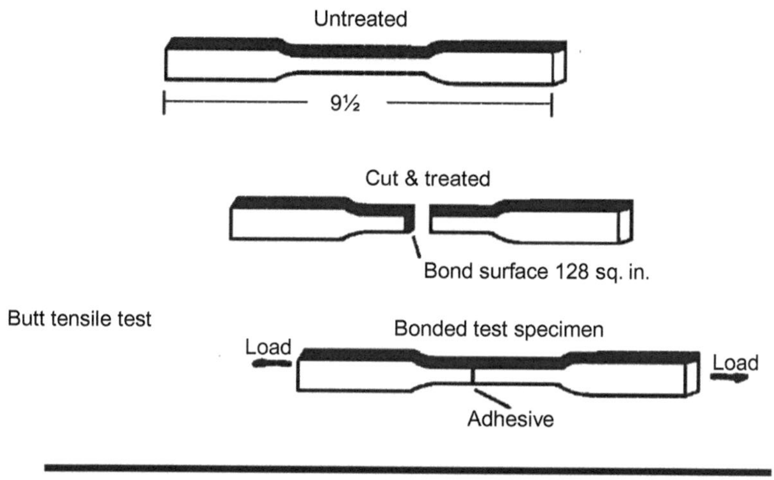

FIGURE 6.20 An example of measured influence of burner speed on the adhesion of molded GMT. Where the optimum is situated depends on the whole system [39].

flame; a flame that does not have sufficient air (<100% stoichiometry) to complete combustion is a *reducing* flame. Figure 6.19 [38] shows the effect of using more or less air than the stoichiometric volume on the tape peel strength of polyethylene or polypropylene. A maximum value develops at about 115% stoichiometry or 15% excess air.

In Figure 6.20 [27], the influence of the flame passage speed over the plastic on the adhesion (shown as percentage of maximum adhesive bond strength) between the adhesive and glass matte-reinforced thermoplastic (GMT) is depicted. The machine supplier often provides assistance with the optimization of the system.

The speed of the conveyer greatly affects the flaming. As a rule, the higher the speed of passage of the item through the flame, the less likelihood there is of damage to the surface. High gloss surfaces are susceptible to blooming,

TABLE 6.10 Effect of Flame Treatment on the Surface Composition and Adhesion Bond Strength of Fluoropolymers

Polymer/ Treatment	Surface Chemical Analysis (%) by ESCA*							Bond Strength, N**
	F/C Ratio	Cl/C Ratio	O/C Ratio	Cl	C	F	O	
PVF								
No	0.41	–	0.011	–	70.4	28.8	0.8	360
Yes	0.41	–	0.065	–	67.6	28.0	4.4	3240
ECTFE								
No	0.61	0.27	–	14.3	53.2	32.5	–	240
Yes	0.25	0.12	0.087	8.0	68.8	17.2	6.0	2980
PTFE								
No	1.60	–	–	–	38.4	61.6	–	420
Yes	1.94	–	–	–	34.0	66.0	–	80

**Electron Spectroscopy for Chemical Analysis.*
***Bond strength in newtons (N) using lap shear test and an epoxide adhesive.*

which reduces the gloss. Higher speeds help reduce blooming. Conveyers are constructed from metal mesh, which must be of sufficient length to allow the mesh to cool. Hot meshes will mark plastic components.

Flame treatment is not effective in the adhesion treatment of perfluoroplastics. The data in Table 6.10 reveals a large increase in the bond strength of PVF and ECTFE after flame treatment. Fluorine-to-carbon ratio (F/C) of PVF remained unchanged but O/C ratio increased significantly. In the case of PTFE, F/C ratio actually increased, which could explain the drop in the bond strength as a result of flame treatment. The flame most likely removed contamination that had previously masked some of the F atoms on the surface.

6.8 SODIUM ETCHING OF FLUOROPLASTICS

Perfluorinated fluoroplastics are chemically unaffected by nearly all commercial chemicals. An exception is highly oxidizing substances such as element forms of sodium, potassium, and other alkaline metals. This is the basis for sodium etching of fluoroplastic parts.

The original method of surface treatment for PTFE adhesive bonding is etching by a sodium solution in anhydrous liquid ammonia [40,41]. The reagent is prepared by simply dissolving metallic sodium in liquid ammonia

TABLE 6.11 Effect of Sodium Etching on the Surface Composition and Adhesion Bond Strength of Fluoropolymers [43]

Polymer	Treatment	Surface Chemical Analysis (%) by ESCA*							Bond Strength, MPa**
		F/C Ratio	Cl/C Ratio	O/C Ratio	Cl	C	F	O	
PTFE	None	1.60	–	–	–	38.4	61.6	–	2.1
PTFE	Tetra-Etch®† (1 min)	0.011	–	0.20	–	82.2	0.9	16.9	21.3
PTFE	N/1 min‡	0.005	–	0.14	–	87.2	0.4	12.4	21.4
PVF	None	0.42	–	–	–	70.4	29.6	–	1.8
PVF	Tetra-Etch®† (30 min)	0.21	–	0.026	–	80.7	17.2	2.1	20.8
ECTFE	None	0.64	0.27	–	14.1	52.5	33.4	–	1.2
ECTFE	Tetra-Etch®† (1 min)	0.16	0.05	0.12	3.8	74.9	12.2	9.1	10.0

**Electron Spectroscopy for Chemical Analysis.*
***Lap shear test.*
†*Supplied by WL Gore Corporation.*
‡*Treatment with a 1 mole solution of sodium naphthalenide in tetrahydrofuran at room temperature.*

to obtain a 0.5%–1% by weight concentration. The solution is dark blue in color and should be thoroughly stirred before use. The surface of the fluoropolymer should be cleaned carefully with an organic solvent such as acetone to remove oils, grease, and other contaminants, which can cause poor treatment and weak bonding. Moisture must be kept from the solution by storing it under positive pressure in protective packaging.

The fluoropolymer must be inserted in the solution for a brief duration ranging from 2 to 10 seconds [42]. Ammonia rapidly volatilizes after the article is removed from the bath. Sodium can be removed by dipping the treated article into ethyl alcohol. An immersion time that is too long actually weakens the adhesive bond. The optimum time depends on the freshness of the etching solution. The treated fluoropolymer has a shiny, dark brown color, which grows into a dull brown after exposure to air. Analysis of the baths shows the presence of fluoride and/or chloride ions, suggesting defluorination and/or dechlorination of the surface during treatment.

Table 6.11 provides the bond values of fluoroplastics before and after etching, obtained by the lap shear test method. Bond strengths of the treated surfaces of PTFE, PVF, and ECTFE are larger than that of the untreated surfaces by an order of magnitude or more.

Etching has a profound effect on the surface chemistry of the fluoroplastics. Results of the effects of sodium etching on the surface chemical composition of fluoroplastics, using electron spectroscopy for chemical analysis (ESCA), have been summarized in Table 6.11. The consistent changes in surface composition of fluoroplastics due to the treatment are a reduction in fluorine and/or chlorine content, as well as an increase in carbon content and oxygen content. Treated PTFE surface is comprised virtually of entirely carbon and oxygen, with a small amount of fluorine.

The change in the surface of the treated fluoroplastics is also reflected in the contact angle. For example, the water contact angle of FEP and PTFE decreases from 109° to 52° (advancing) upon sodium/ammonia treatment [44,45]. Heating the treated fluoropolymer for a lengthy period can reverse the change in the surface. The water contact angle of treated PTFE increases back to 101° after 96 hours exposure to a temperature of 200°C.

Table 6.12 shows the effect of sodium etching on several fluoroplastics' surface composition and lap shear bond strength. In general, the data for fluoroplastics indicate an increase in the adhesive bond strength with increasing fluorine and chlorine content.

Using a commercial dispersion of sodium in naphthalene can reduce preparation time. The shelf life of the bath is sixty days provided that it is stored in a closed container and isolated from air. The polymer is immersed in the solution for 1–5 minutes, followed by rinsing in alcohol or acetone. Bond strength using epoxy adhesives ranges from 7 to 14 MPa in tensile mode. This broad range of adhesive bond strength is obtained in butt tensile, disk tensile, and lap shear test configurations seen in Figure 6.21.

Research has shown that the defluorination depth of the polymer surface extends about 300 nm (Fig. 6.22) as a result of sodium naphthalenide treatment [47]. These workers reported that exposure to X-rays decreased the defluorination depth to 30–150 Å. The structure of the defluorinated layer is highly porous. The suggested adhesion mechanism is mechanical interlocking of adhesive with this porous structure. Bond failure nearly always occurs by stripping the etched layer as opposed to cohesive failure of the adhesive.

Etching solutions for in-house treatment can be purchased from a number of sources. These include Fluoro Etch® by Acton Corp. and Tetraetch® by W.L. Gore & Associates, and Porter. Some companies such as Acton Corp. and Goodrich Corp. (Porter Process), provide a surface treatment service. Operational safety and waste disposal are two issues concerning parties who deal with etching solutions. Suppliers should be contacted for assistance.

6.9 METHODS FOR EVALUATING EFFECTIVENESS OF SURFACE PREPARATION

Before actual bonding, the subjective "water-break" test, or the quantitative and objective contact-angle test, may be carried out. After bonding, the

TABLE 6.12 Effect of Sodium Etching (Tetra-Etch® by W. L. Cole) on the Surface Composition and Adhesion Bond Strength of Fluoropolymers [28]

Polymer/ Treatment	Surface Chemical Analysis (%) by ESCA*							Bond Strength, N_2**
	F/C Ratio	Cl/C Ratio	O/C Ratio	Cl	C	F	O	
PVF								
None	0.41	–	0.011	–	70.4	28.8	0.8	360
10 sec	0.37	–	0.012	–	72.4	26.7	0.9	800
1 min	0.37	–	0.021	–	75.4	23.0	1.6	2080
60 min	0.13	–	0.015	–	87.3	11.4	1.3	3020
PVDF								
None	0.93	–	0.014	–	51.4	47.9	0.7	1580
1 min	0.17	–	0.12	–	77.4	12.9	9.7	2450
60 min	0.11	–	0.14	–	79.5	9.2	11.3	2940
ECTFE								
None	0.61	0.27	–	14.3	53.2	32.5	–	240
1 min	0.24	0.05	0.095	3.7	72.5	17.7	6.9	3300
PTFE								
None	1.6	–	–	–	38.4	61.6	–	420
10 sec	0.01	–	0.13	–	87.6	0.8	11.6	4280
1 min	0.01	–	0.20	–	82.2	0.9	16.9	4260

**Electron Spectroscopy for Chemical Analysis. **Bond strength in Newtons (N) using lap shear test. An epoxide adhesive was used for PVF, PTFE, and ECTFE, a cyanoacrylate adhesive for PVDF.*

effectiveness of surface preparation may be determined by measuring the bond strength and determining the mode of the failure of the adhesive joint.

6.9.1 Dyne Liquids

This method is in common use for estimating the treatment level of plastic surfaces in manufacturing/production environments. It involves using solutions made from a mixture of two chemicals that produce liquids (dyne) with surface tension in the range of 30–70 dynes/cm. The test consists of placing droplets of the various dyne liquids on the treated surface and observing the spreading of the drops in two seconds. Successive liquids with different

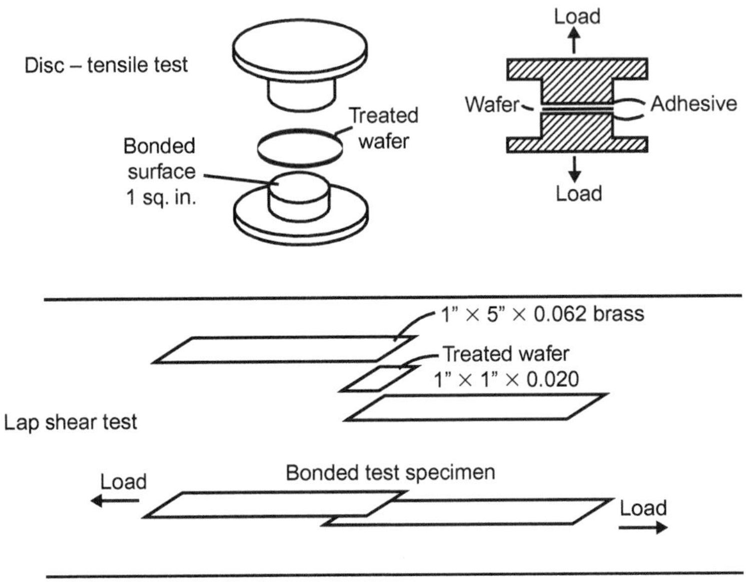

FIGURE 6.21 Test specimens for determination of bondability of treated PTFE [46].

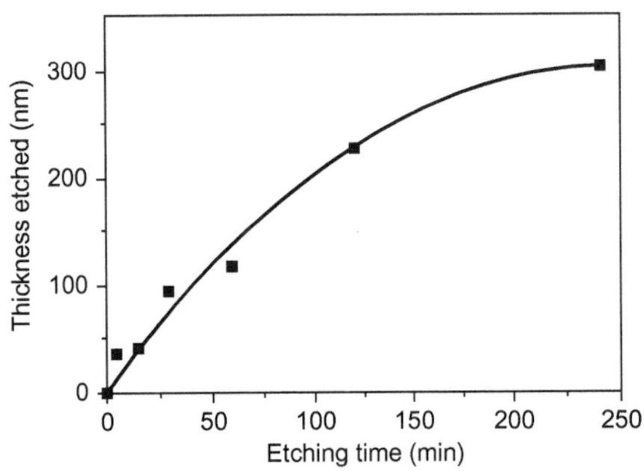

FIGURE 6.22 Apparent defluorination depth as a function of Na/naphthalene treatment time [43].

surface tensions allow narrowing of the surface tension range of the plastic surface. This method is subjective, but it provides a rapid means of assessment of the treatment level, particularly useful in a production environment. There are also pens that operate similarly to dyne liquids. A more quantitative approach is the measurement of contact angle, which decreases with an increase in treatment level. A perfect wetting liquid forms a contact angle of zero on the solid surface.

6.9.2 Water-Break Test

This test depends on the observation that a clean surface (one that is chemically active, or polar) will hold a continuous film of water, rather than a series of isolated droplets. This is known as a water-break-free condition. A break in the water film indicates a soiled or contaminated area. Distilled water should be used in the test, and a drainage time of about 30 seconds should be allowed. Any trace of residual cleaning solution should be removed to avoid a false conclusion. If a water-break-free condition is not observed on the treated surface, it should not be used for bonding. The surface should be recleaned until the test is passed. If failures continue to occur, the treating process itself should be analyzed to determine the cause of the problem [2].

6.9.3 Contact-Angle Test

Wettability may also be determined by measuring the contact angle between the polymer surface and the drop of a reference liquid such as distilled water. A small contact angle indicates that the liquid is wetting the polymer effectively, while large contact angles show that the wetting is poor. Every surface has a critical surface tension, γ_c, of wetting. Liquids with surface-free energies below γ_c will have zero contact angles, and will wet the surface completely. In contrast, liquids with surface-free energies above γ_c will have finite contact angles. The critical surface tension is in units of dynes/cm at 20°C. Contact angles for untreated materials vary from 37° to 48° for relatively polar materials such as nylon to respective highs of 100° and 97° for the nonpolar, unbondable silicone, and polyethylene resins. After exposure to activated argon plasma, contact angles are reduced to 40° for PMMA and to 19° or less for nylon, polystyrene, polyethylene, and RTV silicone [6]. Zissman and others have written comprehensively on surface-tension phenomena relating to adhesion [48,49].

Table 6.13 gives the treatment values recommended for a number of plastics as a function of the adhesive coating system and the manufacturing processes. Notice that, as discussed earlier in this section, waterborne adhesives universally require higher surface energy levels than solvent based systems.

6.10 SURFACE EXPOSURE TIME (SET)

Surface exposure time (SET) is the time elapsed between the surface preparation and the actual bonding. After parts have been subjected to surface preparation, they must be protected from contamination during transportation and storage. The clean surface should never be touched with bare hands or soiled gloves. If more than a few hours are required between cleaning and priming, the parts should be covered, or, for still longer periods, wrapped in clean kraft paper until the priming can be carried out. After priming, the

TABLE 6.13 Suggested Surface Energy (Treatment Level) for Various Plastics Adhesion Substrates [57]

		Printing Processes												Other Processes					
	Process	Flexo and Gravure			Litho			Offset, Letterpress			Screen and Pad			Laminating[6]			Coating		
Substrate	Coating Type	Water	Solvent	UV	Water	Solvent	UV	Water	Solvent	UV	Water	Solvent	UV	Water	Solvent	UV	Water	Solvent	UV
	PE[2]	38	36	38	40	37	40	40	37	42	42	38	44	42	38	42	42	38	44
		44	40	50	46	42	50	46	42	54	48	44	60	50	44	54	48	45	54
	PP[3]	38	36	40	40	38	40	40	37	40	42	38	44	42	38	42	42	38	44
		44	40	50	46	42	50	46	42	54	48	44	60	50	44	54	48	45	54
	PVC[4]	38	36	36	40	37	36	40	38	40	42	38	42	42	38	42	40	38	42
		44	40	50	45	42	52	45	42	52	48	44	60	50	44	54	48	45	54
	PET[5]	44	40	42	46	42	44	46	42	46	48	42	44	46	42	44	42	42	46
		52	46	54	56	46	56	56	46	60	60	48	62	60	48	62	52	48	60
	PS	38	35	42	40	37	42	40	38	42	42	38	42	42	37	42	42	38	44
		44	40	48	45	42	50	46	44	58	48	44	56	52	44	54	50	46	54
	PVDC	40	38	42	42	40	42	42	38	42	42	40	42	42	38	44	42	40	44
		46	42	52	46	42	52	48	44	54	50	45	58	50	44	52	48	46	54
	PU	40	38	38	40	38	38	40	38	42	42	38	42	42	38	42	42	38	44
		46	42	50	46	42	52	45	44	56	50	44	58	50	44	56	48	46	54

ABS	42	40	40	42	40	42	42	38	45	42	40	46	42	40	42	42	38	44
	46	44	52	46	45	52	48	46	52	48	45	56	52	45	56	48	46	54
PTFE	40	34	36	40	35	38	40	38	42	42	38	42	42	38	42	42	40	42
	44	39	52	45	40	52	48	44	60	52	46	60	56	46	56	50	48	54
Silicone	40	35	40	40	38	38	40	38	40	42	38	42	42	38	42	42	40	42
	44	40	50	45	42	52	48	44	56	50	46	60	56	46	56	50	48	54

Notes to Table 6.13:

Polymer name abbreviations: PE = polyethylene, PP = polypropylene, PVC = polyvinyl chloride, PET = polyethylene terephthalate, PS = polystyrene, PVDC = polyvinylidene chloride, PU = polyurethane, ABS = acrylo butyl styrene, and PTFE = polytetrafluoroethylene.

In most cases, if the substrate is somewhere between the low and high dyne levels cited, a satisfactory bond can be obtained. For demanding production conditions (e.g., high web speeds and/or critical quality process work), it is safest to aim for the top of the range. Use Table 6.13 as a general guideline only; each operation has slightly different requirements. These data are for flame or corona treatment in an oxygen-containing atmosphere. (1) Some values theoretical. (2) Any density, any type, including films, coated board, and molded products. (3) All types; cast or molded. (4) More plasticizer generally requires levels at the high end of these ranges. (5) Unquoted. (6) For heat sealing, the surface energies of the mating PE surfaces should match closely; if both faces are treated, heat sealing can usually be accomplished over a broader and lower temperature range.

dried primer surfaces, if not to be bonded immediately, should again be protected by wrapping in kraft paper. Whether or not a primer is used prior to adhesion, these steps should be carried out. The period of time for which the parts can be safely stored in this way will vary depending on the nature of the adherends, the adhesive, the surface preparation method, and the ultimate bond strength required [50,51].

Peel tests, particularly the roller peel test, have been found to be more sensitive to variation in surface preparation than shear tests. Peel tests show that, in general, increasing SET tends to reduce the critical joint strength. If the faying surfaces are protected and the relative humidity is kept at about 50%, up to 30 days may elapse between surface preparation and actual bonding without serious loss in joining strength. Temperatures and relative humidity above normal will cause deterioration in shorter periods of time [51,52].

REFERENCES

[1] Landrock AH. Processing handbook on surface preparation for adhesive bonding, Picatinny Arsenal technical report 4883. Dover, NJ: Picatinny Arsenal; 1975.

[2] Snogren RC. Handbook on surface preparation. New York: Palmerton Publishing Co.; 1974. ISBN 0686482212.

[3] Petrie EM. Plastics and elastomers as adhesives. In: Harper CA, editor. Handbook of plastics, elastomers and composites. 4th ed. New York: McGraw-Hill; 2002. ISBN:0070266867.

[4] Kinloch AJ. Introduction. In: Kinloch AJ, editor. Durability of structural adhesives. London and New York: Kluwer Print on-Demand Publisher; 1983. ISBN: 0853342148.

[5] Akers SC. The function of adhesive primers in adhesive bonding of aircraft structures. In: Bodnar MJ, editor. Applied Polymer Symposia No. 19, Processing for adhesives bonded structures. New York: Interscience Publisher; 1972, pp. 23–8.

[6] DeLollis NJ. (This book is the 2nd edition of a 1970 publication under the name Adhesives for metals: theory and technology) Adhesives, adherends adhesion. Huntington, New York: Robert E. Krieger Publishing Co.; 1980. ISBN:0882759817.

[7] Almeida RM, Marques AC. Crystallization of sol-gel materials. SGT Annual meeting Workshop on Crystallization. Derby, September 6–7; 2007.

[8] Dimitriev Y, Ivanova Y, Iordanova R. History of sol-gel science and technology. J Univ Chem Tech Metall 2008;43(2):181–92.

[9] German Patent 736411 (Granted 6 May 1943) Anti-Reflective Coating (Geffcken W. and Berger E.), Jenaer Glasswerk Schott.

[10] Ro DM, Roy R. Synthesis and sytability of minerals in the system MgO-Al_2O_3-SiO_2-H_2O. Am Min 1955;40:147.

[11] The Welding Institute, Great Abington, Cambridge, United Kingdom, <www.TWI.com>; May 2013.

[12] EPA Rep for Project No., 68D99083 Development of an Environmentally Benign Primer for Metal Adhesive Bonding, September 1, 1999–September 1, 2001,<http://cfpub.epa.gov/ncer_abstracts/index.cfm/fuseaction/display.abstractDetail/abstract/1521>; 2013.

[13] Blitshteyn M. Proceedings of the Society of Automotive Engineers, Paper No. 930052. Detroit, MI; 1993.

[14] Chan CM. Polymer surface modification and characterization. Munich: Hanser Publishers; 1993. Ch. 7.
[15] Kim CY, Goring DAI. J Appl Polym Sci 1971;15:1357.
[16] Briggs D. J Adhes 1982;13:287.
[17] Blythe AR, Briggs D, Kendall CR, Rance DG, Zichy VJI. Polymer (Guildf) 1979;19:1273.
[18] Briggs D, Kendall CR. Int J Adhes Adhes 1982;2:13.
[19] Schonhorn H. Adhesion: fundamentals and practice. New York: Gordon and Breach; 1969.
[20] Garbassi F, Morra M, Ochiello E. Polymer surfaces: from physics to technology. New York: John Wiley & Sons; 1994. Ch. 10.
[21] Stobbe BD. Corona treatment 101: understanding the basics from a narrow web perspective. Label & Narrow Web Industry Magazine; Jun 1996.
[22] Stradal M, Goring DAI. J Adhes 1976;8:57–64.
[23] Web treatment–going solventless. Stuart Greig–Sherman Treaters Limited; 2005.
[24] <www.enerconind.com>.
[25] <www.corotec.com>.
[26] Murokh I. Atmospheric plasma surface treatment technique. El Segundo, CA:Tri-Star Technologies, <http://tri-star-technologies.com/news/articles/AtmosphericPlasmaSurfaceTreatment.pdf>; 2005.
[27] Kalapat N, Amornsakchai T. Surface modification of biaxially oriented polypropylene (BOPP) film using acrylic acid-corona treatment: part I. Properties and characterization of treated films. Surf Coat Technol 2012;207:594–601.
[28] Guimond S, Wertheimer MR. J Appl Polym Sci 2004;94:1291.
[29] Chi-Ming C. Polymer surface modification and characterization. Munich, Vienna, New York: Carl Hanser Verlag; 1994.
[30] Novak I, Florian S. J Mater Sci 2004;39:2033.
[31] Chan C-M, Ko T-M. Surf Sci Rep 1996;24:1.
[32] Kalapat N, Amornsakchai T, Srikhirin T. Surface modification of biaxially oriented polypropylene (BOPP) film using acrylic acid-corona treatment: part II. Long term aging surface properties to be published in Surf Coat Technol 2013;xxx.
[33] Ryan DL. British Patent 890,466, assigned to DuPont Co. (Feb. 28, 1962).
[34] McBride RT, Wolinski LE. US Patent 3,296,011, assigned to DuPont Co. (Jan. 3, 1967).
[35] Kreuz JA, Zytkus EH. US Patent 3,627,624, assigned to DuPont Co. (Sep. 16, 1969).
[36] Markgraf DA. Corona treatment: an overview, published by Enercon Industries Corporation, <www.EnerconInd.com>; 2013.
[37] Markgraf DA. Corona treatment: an overview. Coextrusion Conference Proceedings. Atlanta: TAPPI Press; 1986.
[38] Surface treatment of polyolefins for decorating and adhesive bonding. 101 Secor Lane, Pelham Manor, New York: Ensign Ribbon Burners LLC, <www.erbensign.com> (reprint courtesy Phillips Petroleum Co.)
[39] Quadrant Plastic Composites, <www.quadrantcomposites.com>; 2002.
[40] Purvis RJ, Beck WR. US Patent 2,789,063, assigned to Minnesota Mining and Manufacturing Co. (Apr. 16, 1957).
[41] British Patent 765,284, assigned to Minnesota Mining and Manufacturing Co. (Jan. 1957).
[42] Fluon® Technical Service Notes F9, Finishing Processes, Imperial Chemical Industries, Ltd.

[43] Marchesi JA, Ha K, Garton A, Swei GS, Kristal KW. Adhesion to sodium naphthalenide treated fluoropolymers. Part II. Effects of treatment conditions and fluoropolymer Structure. J Adhes 1991;36:55–69.

[44] Wu S. Polymer interface and adhesion. New York: Marcel Dekker, Inc.; 1982.

[45] Siperko LM, Thomas RR. Chemical and physical modification of fluoropolymer surfaces for adhesion enhancement: a review. J Adhes Sci Tech 1989;3(3):157–73.

[46] Benderly AA. Treatment of Teflon® to promote bondability. J Appl Polym Sci VI 1962;20:221–5.

[47] Rye RR, Arnold GW. Depth dependence of alkali etching of polytetrafluoroethylene: effect of X-ray radiation. Langmuir 1989;5:1331–4.

[48] Zissman WA. Influence of constitution of adhesion. In: Skeist I, editor. Handbook of adhesives. 2nd ed. New York: Van Nostrand Reinhold; 1977.

[49] Gent AN, Hamed GR. Fundamentals of adhesion. In: Skeist I, editor. Handbook of adhesives. 3rd ed. New York: Van Nostrand Reinhold; 1990.

[50] Diversified Enterprises, 101 Mulberry St., Suite 2N, Claremont, NH 03743, (603) 543-0038 or 1-800-833-4644; fax (603) 543-1334; 2004.

[51] Tanner WC. Manufacturing processes with adhesive bonding, pp. 1–21, in Applied Polymer Symposia No. 19. In: Bodnar MJ, editor. Processing for adhesives bonded structures. New York: Interscience Publishers; 1972.

[52] Wegman RF, et al. Effects of Varying processing parameters in the fabrication of adhesion bonded structures. Part 2, Important considerations for the bonding process, Picatinny Arsonal technical report 3999. Dover, NJ: Picatinny Arsonal; 1970.

[53] Katz M, Schmidt FH. EP0467096A1, assigned to DuPont Co. (Jan. 22, 1992).

[54] Mathieson I, Brewis DM, Sutherland I, Cayless RA. J Adhes 1994;46:49.

[55] Brewis DM. Surface Modification of Fluoropolymers for Adhesion, paper presented at the Fluoropolymers Conf. UK: Loughborough University; 1992.

Chapter 7

Surface Preparation of Metals

7.1 INTRODUCTION

The methods listed for surface preparation of metals in Chapter 6 are generally applicable, but the processes required for specific metals are different from the general techniques. The specific preparation and treatment (or pretreatment) techniques described in this chapter have been reported to provide strong reproducible bonds and fit easily into the bonding operation. It must be noted that the methods for preparing metal surfaces are generally much older than those of plastics because of the length of time that metals have been in use. This is not to say that improvements have not been made in recent decades, but the pace of upgrades has been slow in recent years. Improvements have often been driven by environmental rules promulgated by governments to reduce emissions and toxic waste generated by the surface preparation methods of metals. An important example is chromate-free etching of aluminum.

7.2 ALUMINUM

There are a number of methods to treat the aluminum surface prior to adhesion bonding. The choice of the technique depends on the performance requirements of the adhesive bond. Table 7.1 lists the techniques available for the treatment of aluminum. Figure 7.1 shows the durability of various adhesive bonds.

Chemical treatments have been traditionally most effective with aluminum alloys, especially where long-term environmental exposure is required. The sulfuric acid–dichromate etch (*FPL* etch, named after *Forest Products Laboratory*, US Dept. Agriculture) has been used successfully for many decades. The more recently developed techniques are often modifications of the FPL procedure. Other important methods include chromate conversion coating and anodizing. Corrosion-resistant adhesive primer (CRAP), as well as anodic and chromate conversion coatings, help prevent corrosion failure of adhesion [2].

Surface Treatment of Materials for Adhesive Bonding. DOI: http://dx.doi.org/10.1016/B978-0-323-26435-8.00007-1

TABLE 7.1 Methods of Treatment of Aluminum Substrates [1]

"Light" Abrasion	Wire wool or Scotchbrite
"Heavy" Abrasion	Grit-blast with alumina particles
"Heavy" Abrasion	Grit-blast with alumina particles plus silane treatment
Chemical	Chromic-sulfuric acid pickle
Chemical	P2 etch
Electrochemical	Chromic acid anodizing
Electrochemical	Phosphoric acid anodizing
Electrochemical	Sulfuric acid anodizing
Electrochemical	Boric acid-sulfuric acid anodizing
Chemical	Sol-gel procedures
Surface Bombardment	Activated plasma

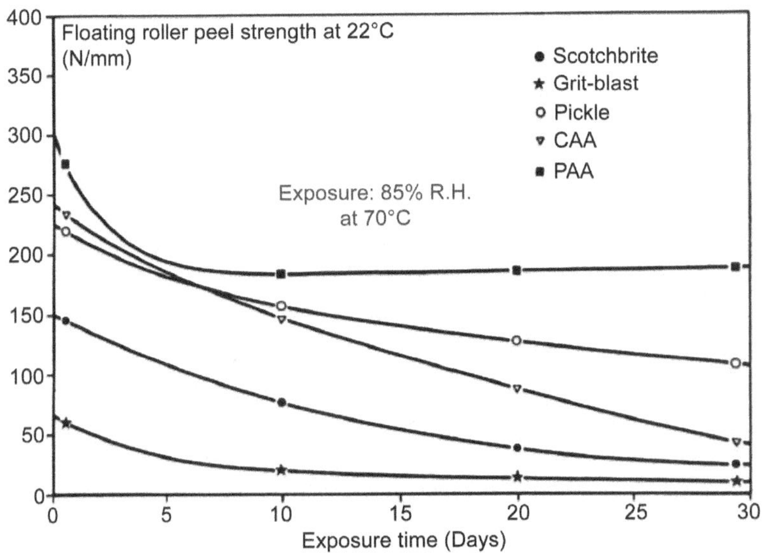

FIGURE 7.1 Durability of adhesive bonds to aluminum treated by various methods. PAA = phosphoric acid anodizing, CAA = chromic acid anodizing, and Pickle = acid treatment [1].

7.2.1 Sol-gel Process

Development of sol-gel process technology has received significant interest since the 1980s. The sol-gel reaction can produce homogeneous inorganic materials with desirable properties of hardness, optical transparency, chemical durability, tailored porosity, and thermal resistance. Originally, sol-gel

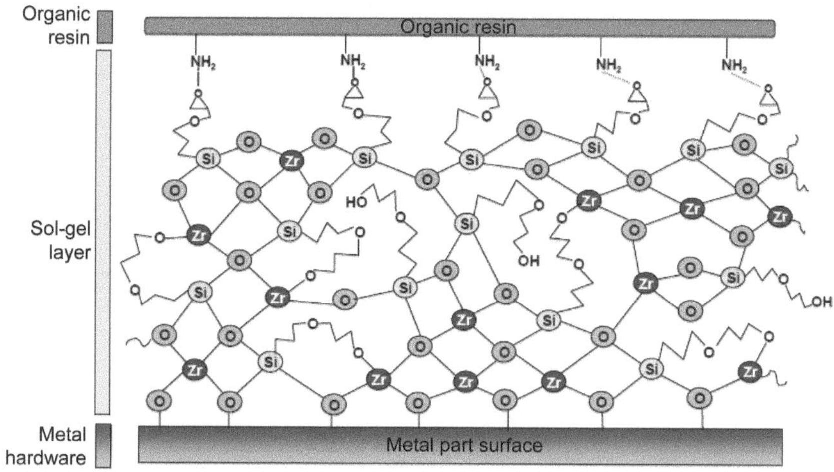

FIGURE 7.2 Nanostructured Boegel (Boeing's original name for its sol-gel technology) Interface group [4].

technology was only aimed at the production of ceramic coatings, foams, fibers, and powders. In 2001 it was discovered sol-gel processing could yield both inorganic and hybrid organic–inorganic materials, further widening the applicability of this technology [3].

Chemistry is at the core of the sol-gel process technologies. The sol-gel process, as the name implies, involves the evolution of inorganic nanoscale networks (Fig. 7.2) through the formation of a colloidal suspension (sol) and the gelation of the sol to form a network in a continuous liquid phase (gel). Different products are obtained by controlled hydrolysis and condensation reactions.

The raw materials in sol-gel processing usually include silicon or metal alkoxide precursors (Fig. 7.3). The most common metal alkoxides are the alkoxysilanes, such as tetramethoxysilane (TMOS) and tetraethoxysilane (TEOS). Water is required for the hydrolysis reactions during which alkoxide (OR) groups are substituted by hydroxyl groups (OH). The reaction is improved by using alcohol as a solvent, because of the immiscibility of water and alkoxides. The alcohol also enhance the hydrolysis reaction by homogenizing the system. The silanol (Si-OH) groups condense to form siloxane bonds (Si-O-Si) in condensation reactions. Water and alcohol are by-products of the reaction. The environmental aspects of the processes can by improved by the selection of raw materials and by-products.

Sol-gel technology was first implemented for several applications by the Boeing Company to treat titanium, followed by treatment of aluminum, titanium, and steel components to be repaired through bonding [5].

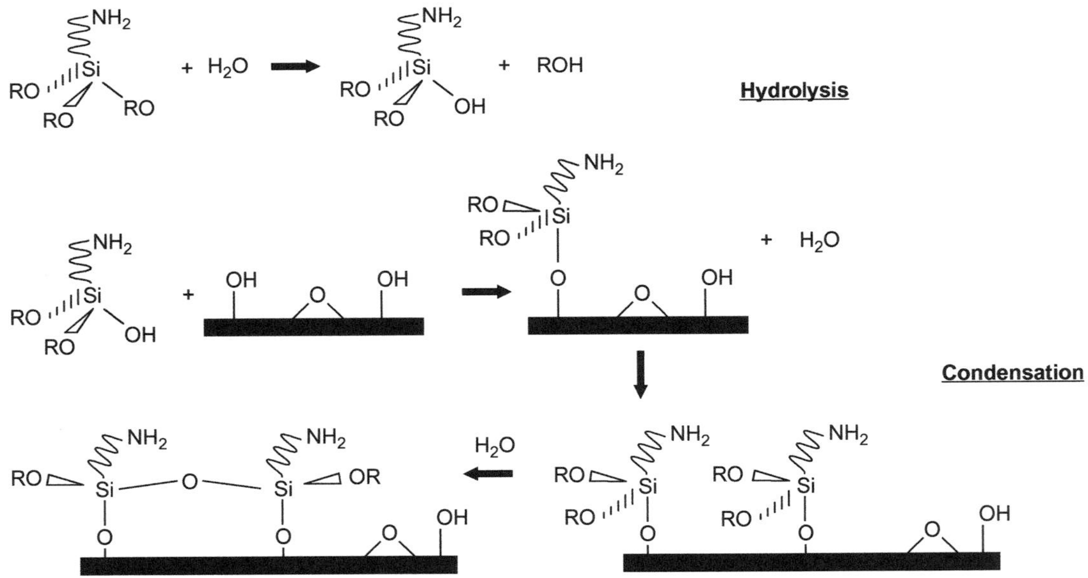

FIGURE 7.3 Basic chemistry of sol-gel process – R = alkyl group [4].

FIGURE 7.4 Demonstration of the sol-gel prebond process on a composite patch repair of fatigue cracks on a B-52 fuselage component [7].

For example, the Boeing sol-gel system [4] (originally designated Boegel) uses a dilute aqueous solution of tetra-*n*-propoxy zirconium (TPOZ) with a silane coupling agent (Fig. 7.2). the actual silane is chosen to give optimum compatibility as well as having the ability to form strong bonds and to enhance the final surface durability. When epoxy based adhesives are to be used, the choice is often γ-glycidoxypropyl trimethoxy silane (Silquest® A-187 offered by Momentive Company, www.momentive.com). Finally, acetic acid is added to the solution to control stability and rate of reaction. This system has been commercialized by Advanced Chemistry & Technology (acquired by 3M Company) as AC®-130 (AC®-130 is a trademark of 3M Company) [6].

With the sol-gel process, it is possible to achieve a reproducible surface that results in durable bonded interfaces using readily available materials. Using appropriate materials and conditions, sanding, used in conjunction with the sol-gel prebond treatment and a bond primer, can yield a robust, durable bond interface system for use at depot sites, repair facilities, or on-aircraft in the field. Figure 7.4 shows a demonstration of this process on a B-52 fuselage component.

7.2.2 Immersion Etch (Optimized FPL Process)

This method is specified in ASTM D2651-01 [8].

1. Remove ink markings and stamped identification by wiping with commercial solvents such as acetone, methyl ethyl ketone, lacquer thinner, and naphtha.

TABLE 7.2 Composition of FPL Etching Solution

Solution Component	Parts by Weight
Sulfuric acid (specific gravity 1.84)	10
Sodium dichromate ($Na_2Cr_2O_7 \cdot H_2O$)	1
Distilled water	30

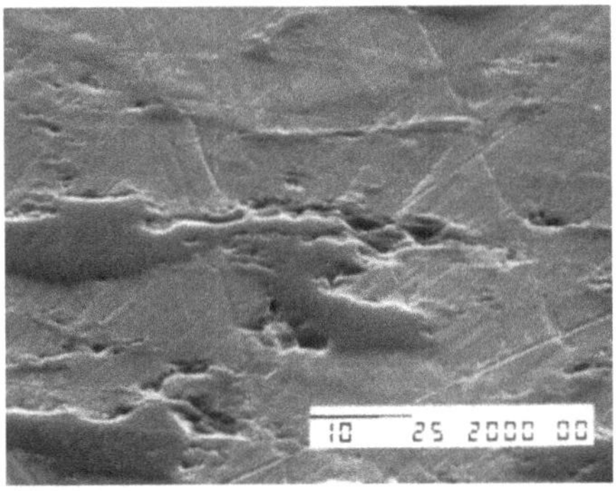

FIGURE 7.5 Scanning electron microscopy (SEM) image of smooth aluminum surface [10].

2. Degrease by vapor degreasing with acetone or toluene, or by immersion in a non-etching alkaline solution for 10 minutes at 70°C–82°C. A typical solution is made by mixing 3.0 parts by weight sodium metasilicate, 1.5 parts sodium hydroxide, and 0.5 parts sodium dodecylbenzene sulfonate such as Nacconol® 90G8 (Nacconol 90G is available from the Stepan Co.), to 133.0 parts water [9].
3. Immerse for 12–15 minutes at 66°C–71°C in the etching solution (Table 7.2).
4. Rinse with water at 60°C–65°C for 30 minutes.
5. Air-dry in an oven, or use infrared lamps, not above 65.5°C.

Figures 7.5 and 7.6 show comparisons of smooth aluminum and treated aluminum surfaces.

7.2.3 FPL Paste Etch

FPL paste etch is used for secondary bonding of parts that contain previously bonded areas, for repair of assemblies, or when the size of parts makes immersion impractical. The parts should be bonded in the temperature range of 21°C–32°C. A paste is prepared by mixing the sulfuric acid–sodium dichromate solution described above with finely divided silica (available from Stepan Chemical Co.) [12] or Fuller's earth (composed mainly of

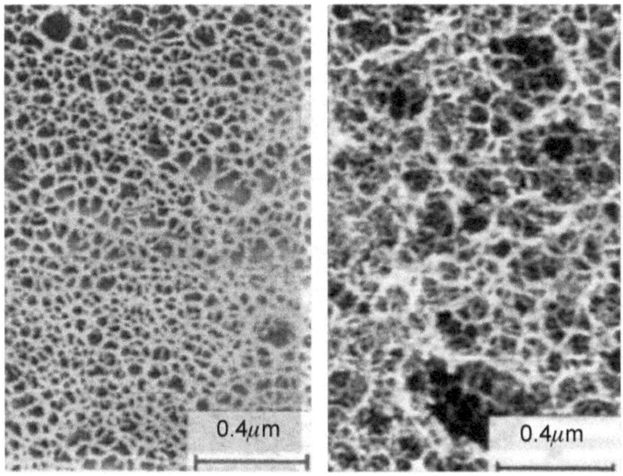

FIGURE 7.6 SEM micrograph of etched aluminum surface by FPL (sulfuric acid–dichromate etch) and P2 (a chromate-free acid solution) methods [11].

alumina, silica, iron oxides, lime, magnesia, and water) and then applied to the surface. The paste is applied by brushing and should not be allowed to dry after application. Polypropylene (or similar) brushes should be used because of their chemical resistance. The paste should be allowed to remain in place for 20–25 minutes. Extra coats may be applied to prevent the paste from drying out or turning green. Clean dry cheesecloth moistened with water should be used to remove all traces of the paste at the end of the exposure period. Water may be sprayed on if desired. Drying should be carried out at a maximum of 66°C. As might be expected, bond strengths obtained by this technique are somewhat lower than those obtained by immersion [8].

7.2.4 Chromate-Free Etch Process

Acid chromate etching solution is not only toxic and hazardous during use, but also highly harmful if released into water supplies. Equally effective chromate-free etching solutions have been developed (Table 7.3). Russell and Garnis [13] found that an etching solution recommended for the precleaning of aluminum prior to resistance-welding gave excellent results. This solution consisted of nitric acid and sodium sulfate (N-S). In later modifications, a "P" etch was developed containing sulfuric acid, sodium sulfate, nitric acid, and ferric sulfate.

The presence of nitric acid resulted in the production of oxides of nitrogen when aluminum was treated. These oxides are toxic and must be vented. In an effort to eliminate the necessity for venting the toxic etching fumes, a new etchant composition called "P2" was developed (see Figs 7.5 and 7.6),

TABLE 7.3 A Comparison of Corrosion Protection Performance of Chromate Free Iridite NCP with Chromate [7]

Corrosion Performance							
Aluminum alloy	5052	6022	3003	1100	6111	6061	A356
Salt spray hours							
Iridite NCP	1000 +	1000 +	1000 +	1000 +	1000 +	1000 +	576
Chromate	1000 +	1000 +	168	432	432	648	1000 +

TABLE 7.4 Composition of P2 Etching Solution

Solution Component	Concentration
Sulfuric acid (6.5–9.5N)	27%–36%
Ferric sulfate	135–165 g/l
Water	30

which does not emit any appreciable fumes and results in good bond strength and improved durability [13–15].

Degreasing or solvent cleaning may be carried out prior to using the P2 etch using the procedure described in Section 7.2.4. The composition of P2 etching solution is given in Table 7.4.

To prepare a one-liter solution, the acid is added to approximately one-half liter of water while constantly stirring. Ferric sulfate is then added and mixing is continued. Next, water is added to bring the volume to 1 liter. The solution is heated to 60°C–65°C and the parts are immersed in this solution for 12–15 minutes. Follow by rinsing in agitated tap water for 2 minutes. A second rinse, also at room temperature, using deionized water is sprayed on the part to rinse off the tap water [13,16]. This sulfuric-acid-ferric-sulfate etch yields bonds at least equal to those made using the sulfuric acid–dichromate (FPL) etch. When used as a deoxidizer prior to phosphoric acid anodizing (PAA) (see below), the results are essentially equal to those using the sulfuric-acid–dichromate etch. In a variation of this process, the final rinse lasts 1–3 minutes in demineralized water at an ambient air temperature of up to 71°C, followed by drying in ambient air up to 71°C [16].

7.2.5 Anodization

The anodization process is sometimes used for bare (nonclad) aluminum machined or chem-milled parts that must be protected against corrosion.

Anodic coatings include chromic acid (CAA), sulfuric acid (SAA), phosphoric acid, boric sulfuric acid (BSAA) anodization processes. The anodizing process involves an electrolytic treatment of metals during which stable films or coatings are formed on the surface of the metals. Anodic coatings can be formed on aluminum alloys in a wide variety of electrolytes, using either alternating or direct current.

Anodizing was first applied on an industrial scale in 1923 to protect aluminum seaplanes parts from corrosion. Early on, chromic acid anodization (CAA) was the process of choice, sometimes called the *Bengough–Stuart* process as documented in British defense specification DEF STAN 03-24/3.

Oxalic acid anodizing was patented in Japan in 1923 and later widely used in Germany, particularly for architectural applications. Anodized aluminum extrusion was a popular architectural material in the 1960s and 1970s, but has since been displaced by cheaper plastics and powder coating. A variety of phosphoric acid processes are among the recent new development in pretreatment of aluminum parts for adhesive bonding or painting. A wide variety of rather complex variations of anodizing processes using phosphoric acid continue to be developed. The trend in military and industrial standards is to classify the anodization processes by coating properties in addition to the identification of process chemistry.

7.2.5.1 Chromic Acid Anodization (CAA)

CAA was the first major commercial pretreatment method for aluminum and has remained in use in spite of requirements for a complicated voltage cycle even though it has been found to be unnecessary. Further it is highly toxic because of chromium and generates hazardous waste. There have been variants of this process over the years. One involves use of sulfuric and chromic acid anodizing processes, patented by Gower and O'Brien [17].

The most widely used anodizing specification is MIL-A-8625. The United States Military Specification (Mil Spec) A-8625 specifies the CAA process, which has been adopted by the American Anodizing Council (AAC) [18]. Table 7.5 shows the types of coatings obtained from anodization processes.

Table 7.6 provides data on different types of chromic acid anodization. Type 1 is chromic acid anodization. Other anodizing specifications include MIL-A-63576, AMS 2469, AMS 2470, AMS 2471, AMS 2472, AMS 2482, ASTM B580, ASTM D3933, ISO 10074, and BS 5599. AMS 2468 is obsolete. None of these specifications defines a detailed process or chemistry, but rather a set of tests and quality assurance measures which the anodized product must meet. BS 1615 provides guidance in the selection of alloys for anodizing. For British defense work, detailed chromic and sulfuric anodizing processes are described by DEF STAN 03-24/3 and DEF STAN 03-25/3 respectively.

TABLE 7.5 Classification of Coatings Obtained from Anodization Processes According to Mil Spec 8625 F

	Thickness
Type I A	
Conventional coatings produced from chromic acid bath	0.5 μ–7.6 μ (microns)
Type I B	0.5 μ–7.6 μ
Low voltage chromic acid anodizing (20 volts)	
Used for 7xxx series alloys	
Type II	1.8 μ–25.4 μ
Conventional coatings produced from sulfuric acid bath	
Type III	12.7 μ–115 μ
Hard coat (uniform anodic coatings)	
• **Class 1**	
◦ Non dyed	
• **Class 2**	
◦ Dyed	

TABLE 7.6 Different Types of Chromic Acid Anodization According to Mil Spec 8625

Type I	Chromic acid anodized coating. This process is used principally for the treatment of aircraft parts. An example is the Bengough-Stewart process where a 30–50 g/l chromic acid bath is maintained at 100°F and the voltage is gradually raised to 50 V. Adjustments are made for high copper, zinc, and silicon alloys. Coating weights must be greater than 200 mg/ft^2. Criteria for corrosion resistance, paint adhesion, and paint adhesion testing must be specified.
Type IB	Low voltage (20 V) chromic acid anodized coating. Typically associated with higher temperature, more concentrated chromic acid electrolytes. Coating weights must be greater than 200 mg/ft^2. Criteria for corrosion resistance, paint adhesion, and paint adhesion testing must be specified.
Type IC	Anodized coating produced in a non-chromic acid electrolyte. As with other Type I coating processes, the treatment is designed to impart corrosion resistance, paint adhesion, and/or fatigue resistance to an aluminum part. Coating weights must fall between 200 and 700 mg/ft^2. Criteria for corrosion resistance, paint adhesion, and paint adhesion testing must be specified.

Color will vary from clear to dark gray depending on alloy. Copper bearing alloys only yield gray colors.

The anodized aluminum layer is grown by passing a direct current through an electrolytic solution, with the aluminum part serving as the anode (the positive electrode). The current releases hydrogen at the cathode (the negative electrode) and oxygen at the surface of the aluminum anode, creating a build-up of aluminum oxide. Alternating current and pulsed current could be used but rarely are used. The voltage required by various solutions ranges from 1 to 300 V DC although the common range is 15–21 V.

Aluminum anodizing is usually performed [19] in an acid solution which slowly dissolves the aluminum oxide. The acid impact is balanced with the oxidation rate to form a coating with nanopores, 10–150 nm in diameter. These pores are what allow the electrolyte solution and current to reach the aluminum substrate and continue growing the coating to greater thickness beyond what is produced by autopassivation. However, these same pores will later permit air or water to reach the substrate and initiate corrosion if not sealed. They are often filled with colored dyes and/or corrosion inhibitors before sealing. Because the dye is only superficial, the underlying oxide may continue to provide corrosion protection even if minor wear and scratches may break through the dyed layer.

Conditions such as electrolyte concentration, acidity, solution temperature, and current must be controlled to allow the formation of a consistent oxide layer (Table 7.7). Harder, thicker films tend to be produced by more dilute solutions at lower temperatures with higher voltages and currents. The film thickness can range from under 0.5 micrometers for bright decorative work up to 150 micrometers for architectural applications.

7.2.5.2 Sulfuric Acid Anodization (SAA)

Mil Spec 8625 specifies Type II, which is sulfuric acid anodization (SAA) and Type III, which is a sulfuric acid hardcoat anodization (Table 7.5). Sulfuric acid is the most widely used solution to produce anodized coating (Table 7.7). Coatings of moderate thickness 1.8 μm–25 μm are known as Type II in North America, as named by MIL-A-8625, while coatings thicker than 25 μm (0.001") are known as Type III, hardcoat, hard anodizing, or engineered anodizing.

Very thin coatings similar to those produced by chromic anodizing are known as Type IIB. Thick coatings require more process control, and are produced in a refrigerated tank near the freezing point of water with higher voltages than the thinner coatings. Hard anodized coatings can be obtained in the thickness range of 25–150 microns. Anodizing thickness increases wear resistance, corrosion resistance, ability to retain lubricants and PTFE coatings, and electrical and thermal insulation. Standards for thin (Soft/Standard) sulfuric anodizing are given by MIL-A-8625 (Types II and IIB), AMS 2471 (undyed), and AMS 2472 (dyed), BS EN ISO 12373/1 (decorative), BS EN 3987 (Architectural). Standards for thick sulfuric anodizing are given by MIL-A-8625 (Type III), AMS 2469, BS 5599, BS EN 2536 and the obsolete AMS 2468 and DEF STAN 03-26/1.

TABLE 7.7 Processing Steps for Different Aluminum Anodization [20]

	CAA[a]	PPA	Bell Helicopter	Fokker-DIN	BSAA
Cleaning Deoxidation	Alkaline	Alkaline	Alkaline	Alkaline	Alkaline
Anodization	FPL[b]	FPL	FPL	CAE	BAC5765
	40–50 g/l	100–120 g/l	60–700 g/l	50 g/l	30.5–52.0 g/l
	CrO_3	H_3PO_4	CrO_3	CrO_3	H_2SO_4, 5.2–10.7 g/l H_3BO_3
	32–38°C	22–33°C	33–37°C	38–42°C	24.5–29°C
	8 V/min to 40 V	3 V/min to 15 V	3–5 V/min to 40 V	5 V/min to 40 V	5 V/min to 15 V
	Hold for 55 min	Hold for 20–25 min	Hold for 30–35 min	Hold 20 min	Hold for 18–22 min
Rinse	1–5 min, 20°C oven dry, $T < 65°C$	5–15 min, <43°C air dry, $T < 71°C$	20–25°C Seal 75–125 ppm CrO_3 82–85°C 7–9 min, air dry	Raise to 50 V Hold for 10 min 5 min, 20–25°C air dry, $T < 60°C$	3–15 min, <35°C air dry, $T = 88°C$

[a] *per MIL-A-8625C*
[b] FPL was the original deoxidizer. Alternative, nonchromate processes include P2 and phosphoric acid deoxidizer (BAC6004-RH).

7.2.5.3 *Phosphoric Acid*

Phosphoric acid anodizing (PAA) is a widely used method, as described in ASTM D3933-98 (2004) [21]. In short, the process involves immersing the part in a 9%–12% solution of phosphoric acid at 19°C–25°C at anywhere from 9 to 16 volts under direct current (DC) for 20–25 minutes. Rinsing, drying, inspecting, and priming (Table 7.8) all follow this step. Durability data obtained by this method is slightly better than that obtained by using an etch [16].

Phosphoric acid anodizing (PAA) was studied by Boeing Aircraft in the early 1960s and commercially introduced in 1974. This process is less critically dependent than etching on processing variables such as the time between treatment and rinsing. It is also possible to use polarized light as a quality control test for the anodizing pretreatment. The oxide layer formed by this process is much thicker and the "whiskers" are longer than with chromic acid etching, although the thickness of the anodic oxide is dependent on the nature of the aluminum alloy being treated. Phosphoric acid anodizing produces surfaces that are more resistant to hydration than those produced with other anodizing methods, including chromic acid [22]. Phosphoric acid anodizing is also known to give more consistent results in durability studies than chromic acid etching.

7.2.5.4 *Boric Sulfuric Acids Anodization (BSAA)*

Chromium (hexavalent) has serious health and safety problems and is closely regulated, hence a number of chromate-free processes have been developed. An important process is boric–sulfuric acid anodization (BSAA) [5] which, while not as popular as PAA or CAA, is used in some aerospace applications by the Boeing Corporation. The BSAA oxide exhibits a porous morphology intermediate to those of PAA and CAA (Fig. 7.7) although its total oxide thickness is closer to that of CAA. With regard to its chemistry, a small amount of sulfur is incorporated in the oxide surface, but no detectable boron. Durability results are similar to those for PAA and CAA. Table 7.7 shows a typical set of processing conditions for BSAA.

TABLE 7.8 Phosphoric Acid Anodization (PAA) Conditions

Solution Component	Value
Phosphoric acid, 85%	9–12 weight %
Temperature, °C	19–25
Voltage, DC	9–11 (single rack) 14–16 (multi-rack)
Anodizing time, min	20–25

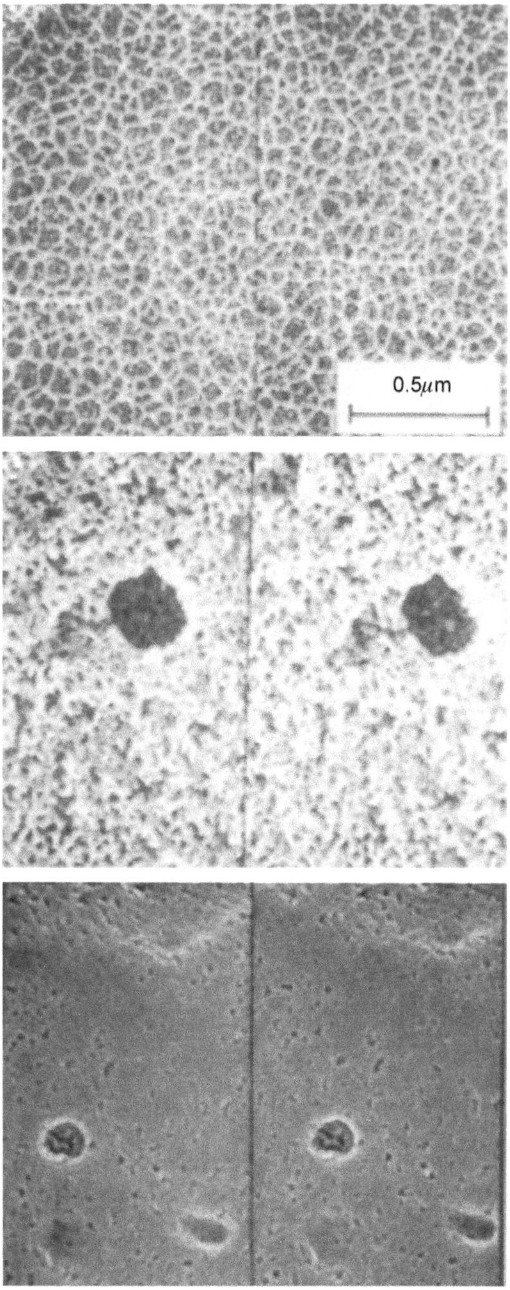

FIGURE 7.7 Scanning electron micrographs of BSAA oxide at different magnifications [20].

7.2.6 Brush Plated Etch (Stylus Method)

This method of electrochemical surface preparation is not broadly known, but has been effectively applied for small batch runs where tanks and other capital-intensive methods might have been used. Almost all of the solutions, "brushes" (or styli), and electrical power supplies are proprietary. When this method is used, the entire "family" of proprietary items should be used (mixed usage of items between vendors should not occur). Additional details can be found in ASTM D2651-01.

7.2.7 Ciba Laser Pretreatment (CLP)

The CLP process for the treatment of aluminum is comprised of a mere two steps (Fig. 7.8):

1. Primer application (incl. evaporation of solvent)
2. Laser treatment.

Laser instruments, from small mobile lasers to high-performance stationary laser units, accommodate a broad range of CLP-uses, including repair to high volume production. Speed of the treatment operation depends on the power of the laser apparatus. For example, a CLP equipped with a high-performance laser is easily able to pretreat a width of 4 cm at a speed of 8 m/min; this is not the upper limit. The choice of the lasers allows CLP with short cycle times. The primer applicator and the laser can be adapted to industrial robots and in this way CLP can be integrated as an in-line pretreatment into the overall production process [24].

CLP was developed for typical substrates used in the automotive industry but as it is equally efficacious in pretreating aluminum, titanium, and stainless steel, it could work for the aerospace parts. The process is a two-step one where the initial stage is to prime and then dry/stove the substrate to be

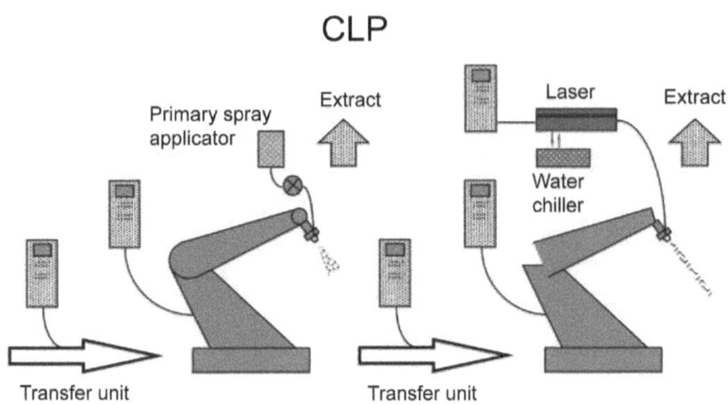

FIGURE 7.8 Schematic diagram of the Ciba laser pretreatment [25].

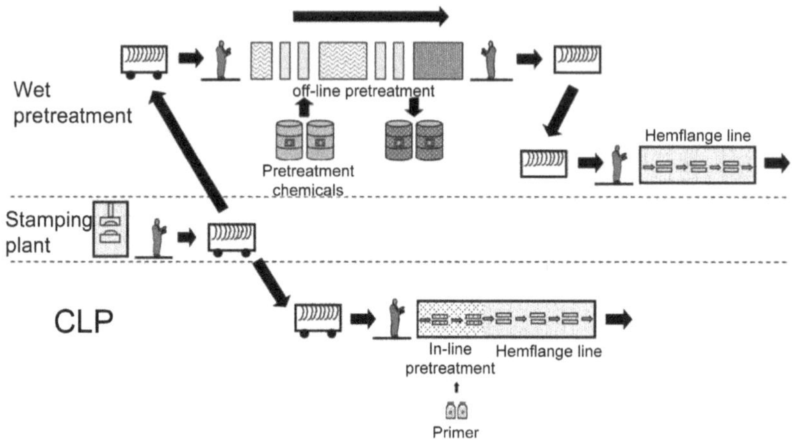

FIGURE 7.9 Comparison of wet treatment and CLP lines [25].

bonded. Although the literature does not give examples of the primer used, it is fairly clear that a suitable surface protection or corrosion protection primer could be utilized; by inference, this could be extended to a silane-based primer. The area to be bonded, and only the area to be bonded, is then exposed to the beam of a suitable laser; laser type, power, and speed of treatment have to be optimized for each type of substrate [26].

A comparison of a wet pretreatment and a CLP line is presented in Figure 7.9. The most important advantage of CLP is that it is environmentally friendly. Wet treatment methods generate toxic by-products which must be disposed in the face of increasingly stricter regulations around the world.

7.3 BERYLLIUM

Beryllium and its alloys must be heated with care. Handling and processing produce dust, chips, scale, slivers, mists, or fumes. Air-borne particles of beryllium and beryllium oxide are extremely toxic with serious latent effects. Abrasives and chemicals used with beryllium must be properly disposed of [2]. One procedure is to degrease with trichloroethylene, followed by immersion in the etching solution (Table 7.9) for 5−10 minutes at 20°C [27].

Rinse in distilled water after washing in tap water and oven-dry for 10 minutes at 121°C−177°C. Caution should be exercised because beryllium reacts quickly with methyl alcohol, fluorocarbons, perchloroethylene, and methyl ethyl ketone/Freon® and can be pitted by long-term exposure to tap water containing chlorides or sulfates [27].

A proprietary coating used to provide a corrosion-resistant barrier is Berylcoat "D", currently available from Materion Corporation, Inc., Mayfield Heights, OH 44124 (http://materion.com).

TABLE 7.9 Composition of Etching Solution for Beryllium

Solution Component	Parts by Weight
Sodium hydroxide	20–30
Distilled water	170–180

TABLE 7.10 Composition of Brass Etching Solution

Solution Component	Parts by Weight
Zinc oxide	20
Sulfuric acid, concentrated	460
Nitric acid, 67% (specific gravity 1.41)	360

7.4 BRASS

Brass is an alloy of copper and zinc. Sandblasting or other mechanical means of surface preparation may be used. The following procedure combines mechanical and chemical treatment [28]:

1. Abrasive blast, using either dry or wet methods. Particle size is not especially critical.
2. Rinse with deionized water.
3. Treat with a 5% solution of sodium dichromate in deionized water.
4. Rinse in deionized water.
5. Dry.

Another method is the following [27,29]:

1. Degrease in trichloroethylene.
2. Immerse for 5 min at 20°C in etching solution (Table 7.10).
3. Rinse in water below 65°C.
4. Re-etch in the acid solution for 5 minutes at 49°C.
5. Rinse in distilled water after washing.
6. Dry in air (temperatures of washing and drying must not exceed 65°C).

7.5 BRONZE

Bronze is an alloy of copper and tin. The surface treatment involving zinc oxide, sulfuric acid, and nitric acid given above for brass is satisfactory for bronze.

7.6 CADMIUM

Cadmium is ordinarily used as a coating on steel. It can best be made bondable by electroplating with silver or nickel. Another procedure is the following [30]:

1. Degrease or solvent clean with trichloroethylene.
2. Scour with a commercial, nonchlorinated abrasive cleaner [31] (such as Ajax® from Phoenix Brands).
3. Rinse with distilled water.
4. Dry with clean, filtered air at room temperature.

It may be desirable to use a primer or sealant. Adhesive choice is particularly important with cadmium coatings [30].

7.7 COPPER AND COPPER ALLOYS

Copper is used in three basic forms: pure, alloyed with zinc (brass), and alloyed with tin (bronze). Copper has a tendency to form brittle amine compounds with curing agents from some adhesive systems. The most successful surface treatments are black oxide (see below), and chromate conversion coatings, which are especially recommended when the adhesive is slightly corrosive to copper. Other treatment methods have also been listed.

7.7.1 Nitric Acid–Sodium Chlorite (Black Oxide)

This method is intended for relatively pure copper alloys containing over 95% copper [8]. It tends to leave a stable surface. It is not recommended for use with adhesives that contain chlorides or for hot bonding to polyethylene. The procedure follows:

1. Degrease.
2. Immerse in the etching solution for 30 seconds at room temperature. The solution is made by mixing 30 parts of nitric acid (70% technical) and 90 parts of water, all by volume.
3. Rinse in running water and transfer immediately to the next solution without allowing the parts to dry. Immerse for 2–3 minutes at 93°C–102°C in a bath (see data in Table 7.11). This solution should not be boiled.
4. Rinse thoroughly in running water until a neutral test is produced when indicated by pH indicator paper.
5. Air-dry.
6. Bond as soon as possible (no later than 12 hours after surface treatment).

7.7.2 Nitric Acid–Ferric Chloride

The procedure is as follows:

1. Degrease (see Section 7.2.2).

TABLE 7.11 Composition of Copper Stabilization Bath

Solution Component	Concentration (g/l)
Sodium chlorite	31.6
Trisodium phosphate	10.5
Sodium hydroxide	5.3

2. Immerse for 1–2 minutes at room temperature in the following solution by weight:
   197 parts water
   30 parts nitric acid (sp gr 1.42)
   15 parts ferric chloride solution (42%)
3. Rinse thoroughly.
4. Dry as quickly as possible.
5. Apply adhesive immediately.

7.7.3 Nitric Acid

The procedure is as follows:

1. Bright dip in concentrated nitric acid at 16°C–21°C for 15 seconds or until all corrosion has disappeared.
2. Rinse thoroughly.
3. Dry as quickly as possible.
4. Apply adhesive immediately.

7.7.4 Acid Etch (Sulfuric Acid–Dichromate–Ferric Sulfate)

The procedure is as follows:

1. Remove surface contamination by sanding, wire brushing, or sand blasting, if necessary.
2. Degrease.
3. Immerse for 10 minutes at 66°C in the solution given in Table 7.12.
4. Rinse (in water at or below room temperature).
5. Dry.
6. Immerse the parts until a bright, clean surface has been obtained in the solution described in Table 7.13.
7. Rinse using cold tap water.
8. Dip in concentrated ammonium hydroxide.
9. Rinse in cold tap water.

TABLE 7.12 Composition of Copper Etching Solution

Solution Component	Parts by Weight
Sulfuric acid (specific gravity 1.84)	75
Ferric sulfate (commercial grade)	1
Distilled water	8

TABLE 7.13 Composition of FPL Etching Solution

Solution Component	Parts by Weight
Sulfuric acid (specific gravity 1.84)	2
Sodium dichromate ($Na_2Cr_2O_7 \cdot 2H_2O$)	1
Distilled water	17

10. Dry quickly.
11. Apply adhesive immediately.

7.8 GOLD

Use methods given for platinum.

7.9 MAGNESIUM AND MAGNESIUM ALLOYS

The surface preparation methods for magnesium alloys are closely associated with corrosion prevention. Magnesium is highly reactive, so corrosion-preventive coatings must be applied for most service applications. The major problem is to apply a sufficient thickness of coating to prevent corrosion, but not so thick that the bond fails cohesively in the coating [8]. (See Chapter 5.)

7.9.1 Alkaline-Detergent Solution

The procedure is as follows:

1. Degrease using procedure described in Section 7.2.2.
2. Immerse for 10 minutes at 60°C–71°C in an alkaline-detergent solution, described in Table 7.14.
3. Rinse thoroughly.
4. Dry at a temperature below 60°C.

TABLE 7.14 Composition of Magnesium Etching Solution

Solution Component	Parts by Weight
Sodium metasilicate	2.5
Trisodium pyrophosphate	1.1
Sodium hydroxide	1.1
Sodium dodecylbenzene sulfonate (Nacconol® 90G)	0.3
Distilled water	95

TABLE 7.15 Composition of Magnesium Etching Solution

Solution Component	Parts by Weight
Chromic oxide (CrO_3)	24
Chromic nitrate	1.1
Distilled water	123

7.9.2 Hot Chromic Acid

1. Degrease as per Section 7.2.2.
2. Immerse for 10 minutes at 71°C–88°C in a solution comprised of 1 part chromic oxide (CrO_3) and 4 parts distilled water.
3. Rinse thoroughly.
4. Dry at a temperature below 60°C.

The methods in Sections 7.9.1 and 7.9.2 can be run consecutively with an aqueous wash in between to give improved bond strengths.

7.9.3 Sodium Hydroxide–Chromic Acid

1. Degrease (as per Section 7.2.2).
2. Immerse for 5–10 minutes at 63°C–79°C in a solution of:
   12 parts by weight water
   1 part sodium hydroxide (commercial grade)
3. Rinse in water at or below room temperature.
4. Immerse for 5–15 minutes at room temperature in the solution whose details are shown in Table 7.15.
5. Rinse thoroughly.
6. Dry at a temperature below 60°C.

7.9.4 Anodic Treatments

Light anodic treatment and various corrosion-preventive treatments produce good surfaces for adhesive bonding. These treatments have been developed by magnesium alloy producers such as Dow Chemical Company, Midland, MI, and others. Details are available from the *ASM Metals Handbook*, Vol. V [32] and MIL-M-45202, Type I, Classes 1, 2, and 3 [33].

7.9.5 Conversion Coatings and Wash Primers [8]

Some dichromate conversion coatings and wash primers designed for corrosion protection can be used for adhesive bonding. Preliminary tests should be carried out to determine the suitability of the process before its acceptance. Details can be found in the *ASM Metals Handbook*, Vol. V [32] and MIL-M-3171 [34].

7.10 NICKEL AND NICKEL ALLOYS

The steps for surface treatment of nickel and its alloys are described below.

7.10.1 Abrasive Cleaning [2]

1. Solvent clean, preferably using vapor degreasing.
2. Abrade with 180–240-grit paper, or grit-blast with aluminum oxide 40-mesh abrasive.
3. Solvent clean again, according to step 1.

7.10.2 Nitric Acid Etch [27]

1. Vapor degrease in trichloroethylene.
2. Etch for 4–6 seconds at room temperature (~20°C) in concentrated nitric acid (sp. gr. 1.41).
3. Wash in cold and hot water, followed by a distilled-water rinse.
4. Air-dry at 40°C.

7.10.3 Sulfuric–Nitric Acid Pickle [2,35]

1. Immerse parts for 5–20 seconds at room temperature in the solution shown in Table 7.16.
2. Rinse in cold water.
3. Immerse in a 1%–2% ammonia solution for a few seconds.
4. Rinse thoroughly in distilled water.
5. Dry at temperatures up to 65.5°C.

TABLE 7.16 Composition of Nickel Etching Solution

Solution Component	Amount
Sodium chloride	30 g
Sulfuric acid (60°Be′)	1.5 l
Nitric acid (40°Be′)	2.25 l
Distilled water	1.0 l

7.10.4 Cleaning and Treatment of Plated Parts [27]

Thinly nickel-plated parts should not be etched or sanded [36]. Two unique problems are often encountered when preparing and bonding plated metals:

- Vigorous mechanical treatment may penetrate and destroy the plating.
- Plating-to-base adhesion may be inadequate.

Since the plating process produces a clean, bondable surface, freshly plated surfaces often do not require additional preparation. Keep in mind, however, that plating changes surface properties such as adhesion, porosity, and surface stress of the metal deposit. Current density, composition of the plating bath (including brightener content), and bath temperature affect the bondability of the plated surface.

When mechanical pretreatment is used, abrade the surface lightly. Fine grades of sand or abrasive paper will minimize penetration. The correct chemical treatment depends upon the type of metal that is deposited during the plating process.

If the plating does not adhere tightly to the base metal, the plating process itself should be investigated. Application of Chemlok® 205 or 207 primer (0.1–0.2 mil, 2.5–5 μm) often improves adhesion to plated metals.

A recommended practice is light scouring with a nonchlorinated commercial cleaner, rinsing with distilled water, drying at below 49°C, and then priming as soon as possible.

7.10.5 Nickel-Base Alloy Treatments

Monel® (nickel–copper), Inconel® (nickel–iron–chromium), and Duranickel® (primarily nickel) are the major types of nickel-base alloys. According to Keith, et al. [37], procedures recommended for stainless steel will give at least satisfactory results with these alloys.

7.11 PLATINUM

7.11.1 Abrasive Cleaning [2]

1. Solvent clean, preferably by vapor degreasing (avoid immersion and wiping unless these steps can be followed by vapor rinsing or spraying).
2. Abrade lightly with 180–240-grit paper.
3. Solvent clean again, as in step 1.

7.11.2 Abrasive Scouring [2]

1. Scrub surfaces with distilled water and nonchlorinated scouring powder.
2. Rinse thoroughly with distilled water (surfaces should be water-break-free).

7.12 SILVER

Use the methods given above for platinum, or use the following methods.

7.12.1 Chromate Conversion Coating [2]

1. Solvent clean.
2. Immerse for 5–10 seconds at room temperature in the solution shown in Table 7.17.
3. Rinse in distilled water.
4. Dry at temperatures up to 66°C.

7.12.2 Degrease/Abrade/Prime [38]

1. Vapor degrease.
2. Abrade lightly with emery cloth.
3. Prime-coat with an adhesive such as Lord Corporation's Chemlok (silicone-type polymer), diluted with 5–10 volumes of methanol or ethanol.

TABLE 7.17 Composition of Chromate Conversion Solution for Silver [2]

Solution Component	Amount
Kenvert No. 14 powder (supplied by Gallade Chemical, www.galladechem.com)	10.6 g
Distilled water	3.6 l

7.13 STEEL

Compared to aluminum, relatively little research has been carried out on optimum surface treatments for steel preparatory to adhesive bonding. Steel alloys are usually used where high temperatures and other rigid requirements are expected. As a result, problem areas such as corrosion, thermal degradation of adhesives and primers, and moisture effects become important. Surface preparation is even more critical in bonding steel than in many other metals. Initial adhesion may be good, but bonds degrade rapidly under rigid environmental exposure. For this reason, primers are desirable. Many steel alloys will form surface oxides rapidly, therefore drying cycles after cleaning are critical. Alcohol rinses after water rinses tend to accelerate drying and reduce undesirable surface layer formation. Mild steel (carbon steel) may require no more extensive treatment than degreasing and abrasion to give excellent adhesive bonds. Tests should be carried out with the actual adhesive to be used to determine whether a chemical etch or other treatment is essential [30,39,40].

Brockmann [41] suggests that in contrast to aluminum and titanium alloys, where the surfaces are usually treated by chemical methods, etching procedures for different types of steel, except for stainless steel, are not recommended. According to Brockmann, the best results are usually obtained by using mechanical roughening techniques such as grinding with corundum (Al_2O_3) as the grit material. Brockmann states the only chemical treatment required is etching for stainless steel in an aqueous solution containing oxalic acid and sulfuric acid.

DeLollis [42] suggests that the most common method for preparing steel surfaces for bonding is probably sandblasting, preceded by vapor degreasing to remove any oily film present. Since a clean, steel surface is easily oxidized, it is particularly important that the compressed air source used in sandblasting is dry and free of oil. While some methods call for rinsing in a clean volatile solvent such as methyl ethyl ketone or isopropyl alcohol, it should be recognized that these solvents frequently contain small amounts of water that may cause subsequent rusting (xylene or toluene are preferable solvents). Furthermore, when volatile solvents evaporate from a metal surface, the surface is chilled. In a humid environment, such an effect may result in condensation of moisture, which nullifies the cleaning effort. After sandblasting, the best treatment is to blow the particulate matter off the surface with clean, dry air. Solvent wiping should be used only as a last resort.

DeLollis [42] suggests that vapor honing or vapor blast can be used for small steel parts without danger of warping or any significant change in tolerance. A wetting agent and rust inhibitor are usually present in the abrasive-water suspension used in this method. These agents must be rinsed off thoroughly in clean water. After this step, a clean, dry, water-compatible solvent must be used to remove the water. Finally, clean, dry compressed air or nitrogen should be used to blow off the solvent. Abrasive treatments may result in warping of thin, sheet-metal stock. In this type of application, any of the acid-etch solutions used with stainless steel can be used with carbon steels. They must be used carefully, however, since they would react more rapidly with carbon steel than with stainless steel. Chemical treatments for carbon steel are to be used only when abrasive equipment is not available, and are described below:

7.13.1 Acid Treatment [39]

1. Prepare one of the solutions shown in Tables 7.18 and 7.19.
2. Degrease in a vapor bath of trichloroethylene.
3. Immerse for 10 minutes in the orthophosphoric acid solution (Table 7.18) at 60°C or for 5–10 minutes at 20°C in the hydrochloric acid solution (Table 7.19).
4. Remove the black residue formed on the metal with a clean, stiff brush while holding the metal under cold, running distilled or deionized water.
5. Dry the metal by heating at 120°C for 1 hour. If the steel part cannot be stored immediately in an area of low humidity (less than 30% RH), apply the adhesive immediately after cleaning.

7.13.2 Potassium Iodide–Phosphoric Acid Method [43]

This is a surface treatment process based on the oxidation of the steel to produce a continuous epsilon-oxide (Fe_2O_3). It is reported to give strong durable bonds and to improve the salt-spray and humidity resistance of polymer-coated steel.

TABLE 7.18 Composition of Etching Solution for Steel

Solution Component	Parts by Weight
Orthophosphoric acid (specific gravity 1.73)	1
Ethyl alcohol, denatured	1–2

TABLE 7.19 Composition of Etching Solution for Steel

Solution Component	Parts by Weight
Hydrochloric acid, concentrated	1
Distilled water	1

1. Vapor degrease.
2. Alkaline-clean in a solution containing 3% trisodium phosphate and 3% sodium carbonate for 5 minutes at 82°C.
3. Rinse in deionized water.
4. Immerse in a solution of 50 g potassium iodide (KI) per liter of 1:1 concentration phosphoric acid and water (volume/volume) at 99 ± 11°C for 2–10 minutes.
5. Rinse in deionized water.
6. Dry at 11°C–71°C.

7.13.3 ASTM Suggested Methods

ASTM D2651-01 [8] lists the following mechanical and chemical methods for surface preparation of carbon steel.

7.13.3.1 Mechanical Method

Carbon steel may be prepared using the same general methods as for stainless steel (Section 7.14). Sand blast, grit blast, or vapor hone may be used. Thin sheet steel may be warped due to the peening action of the blasting abrasives.

7.13.3.2 Method for Stainless Steel

Thin parts and sheet materials may be prepared using methods for stainless steel. The major problem is in preventing oxidation of the surfaces. Oxidation can occur in a very short time, even before the parts are dry. Parts must be bonded immediately. The adhesive or primer must be applied immediately after drying.

7.13.3.3 Nitric–Phosphoric Acid Etchant

The nitric–phosphoric etchant is environmentally acceptable. It produces a smut-free surface that can be treated by other chemical processes, by immersion for 5 minutes at 23°C in the solution given in Table 7.20.

7.14 STAINLESS STEEL

Stainless steel, or corrosion-resistant steel (CRES), has a high chromium content (11% or higher) as the primary alloying element. There have been a large number of surface preparation methods reported in the literature. An excellent report prepared in 1968 by Battelle Memorial Institute workers [44] gives details of twelve surface preparation methods and discusses the relative advantages and disadvantages of each. In addition to mechanical methods, strong acids and strong alkalis are used. A wet-abrasive blast with a 200-grit abrasive, followed by thorough rinsing to remove the residue, is

TABLE 7.20 Composition of Etching Solution for Steel

Solution Component	Volume (%)
Nitric acid, 40° Baume	5
Phosphoric acid–85% weight	30
Surfactant (Triton® X200)	0.01
Deionized water	64.99

an acceptable procedure for some uses, but does not produce high bond strengths. Strong acid treatments are usually used for general bonding to produce strong bonds with most adhesives. Passivation in either nitric acid solution or concentrated sulfuric acid-saturated sodium dichromate solution produces high-strength bonds, but with low or marginal peel strengths. Such joints may fail under vibration stress, particularly when a thin stainless steel sheet is bonded with low-peel-strength adhesives.

The acid-etch process outlined below can be used to treat Types 301 and 302 stainless steel. This process results in a heavy black smut formation on the surface. This material must be removed if maximum adhesion is to be obtained. The acid-etch process produces bonds with high peel and shear strength. The 400 series of straight chromium stainless steels should be handled in the same manner as the plain carbon steels. The precipitation hardening (PH) stainless steels each present an individual problem. Processes must be adopted or developed for each one [2,28,39].

7.14.1 Acid Etch (for Types 301 and 302 Stainless Steel) [45]

This method has been used successfully with modified epoxy adhesive (such as 3M's Scotch-Weld®), achieving excellent stressed durability results at 60°C and 95% RH. The procedure is as follows:

1. *Degrease:* Wash with acetone to remove grease, oil, markings, and solvent-removable soils. Follow with vapor degreasing using a suitable, safe solvent.
2. *Sulfuric acid etch:* Immerse the parts for 4 minutes at 60°C in a solution of 25–35% by volume of sulfuric acid (sp. gr. 1.84) in deionized water. Do not start timing until gassing is evident. A piece of 1020 steel may be rubbed across the parts to start the etching process.
3. *Rinse:* Use running tap water at service-water temperature for 2 minutes.
4. *Smut removal:* Immerse parts for 5 minutes at 66°C in a solution of 22–28 parts-by-weight (pbw) of sulfuric acid (specific gravity 1.84) and 2–3 pbw of sodium dichromate in deionized water.

5. *Rinse:* Use running tap water at service-water temperature for 2 minutes.
6. *Dry:* Dry the parts at 60°C for 30 minutes in a preheated air-circulating oven.
7. *Packaging:* Wrap the parts in clean Craft paper until ready to bond.

7.14.2 Oxalic–Sulfuric Acid Process for Maximum Heat Resistance [46]

1. Immerse for 10 minutes at 90°C in the solution detailed in Table 7.21.
2. Remove the solution and brush the black smut deposit with a good stiff brush (nylon is suggested).
3. Rinse with clean, running water.
4. Dry in a stream of hot air.
5. Return to the desiccator.

Lord Corp. [47] recommend soaking the parts in a commercial steel cleaner of high alkalinity, before immersing in the oxalic–sulfuric acid solution.

7.14.3 Bromophosphate Treatment

This treatment is a modification of the iodo-phosphate treatment described above for carbon steel. No chloride ion is used in the method. KBr is used instead of KI, and 1–5% by volume of concentrated H_2SO_4 should be added to the treating solution in the case of difficult-to-etch CRES steels to ensure the removal of the existing surface layers [43].

1. Vapor degrease.
2. Alkaline-clean in a solution of 3% trisodium phosphate and 3% sodium carbonate for 5 minutes at 82°C.
3. Rinse in deionized water.
4. Immerse in a solution of 50 g potassium bromide (KBr) per liter of 1:1 concentrated phosphoric acid and water for 2–10 minutes at 93°C ± 11°C. In the case of difficult-to-etch CRES steels, to ensure removal of the existing surface layer, use 50 g KBr per liter of 1:9:10 concentrated sulfuric acid, concentrated phosphoric acid, and water.

TABLE 7.21 Composition of Etching Solution for Stainless Steel

Solution Component	Amount
Sulfuric acid, concentrated	10 ml (18.4 g)
Oxalic acid	18.5 g
Distilled water	150 ml

5. Rinse in deionized water and dry at 15.6°C–71°C.

This method can be adapted to spraying techniques.

7.14.4 ASTM Suggested Methods

ASTM D2651-01 [8] lists the following mechanical and chemical methods for surface preparation of stainless steel.

7.14.4.1 Mechanical Methods

Mechanical methods such as wire brushing, rubbing with metal wool, hand sanding, dry sandblast, grit blast, or vapor honing may be used to prepare stainless steel. Care should be exercised to ensure that abrasive materials such as grit, sand, cloth, sandpaper, or brushes are free from contamination that may be spread or rubbed onto the steel. Dry grit or sand blasting tends to warp thin sheet materials; these methods are suited only for thick-section parts. Vapor blasting by water or steam and an abrasive is the most effective method and does not warp parts, if done carefully. Peening action of any blast process may work-harden the surface of some base metals. Hand sanding, rubbing with metal wool, and wire brushing are the least effective methods. Exercise care in using any mechanical method to prevent deep gouges or rough surfaces, which are not conducive to good bonding. Wash, vapor-degrease, and brush or air blast (dry oil-free air) surfaces to remove all traces of the abrasives. Apply the primer or adhesive immediately after the treatment.

7.14.4.2 Acid Etch Method 1

1. Etch for 10 minutes at 65°C–71°C in the following solution by weight:
   90 parts water
   37 parts sulfuric acid (specific gravity 1.84)
   0.2 parts sodium dodecylbenzene sulfonate such as Nacconol® 90G [9]
2. Rinse thoroughly and remove smut with a stiff brush if necessary.
3. Immerse for 10 minutes at room temperature in the bright dip solution detailed in Table 7.22.

TABLE 7.22 Composition of Etching Solution for Stainless Steel

Solution Component	Part by Weight
Nitric acid (specific gravity 1.42)	15
Hydrofluoric acid (35.35%, specific gravity 1.15)	2
Deionized water	88

TABLE 7.23 Composition of Etching Solution for Stainless Steel

Solution Component	Parts by Weight
Hydrochloric acid (specific gravity 1.2)	200
Orthophosphoric acid (specific gravity 1.8)	30
Hydrofluoric acid (35.35%, specific gravity 1.15)	10

TABLE 7.24 Composition of Etching Solution for Stainless Steel

Solution Component	Parts by weight
Sodium metasilicate	1
Sodium salt anionic surfactant, polyether sulfonate type (Triton® X200K)	1.8
Water	47.2

7.14.4.3 Acid Etch Method 2

Immerse for 2 minutes at approximately 93°C in the solution described in Table 7.23, heated by a boiling-water bath.

7.14.4.4 Acid Etch Method 3

Immerse for 15 minutes at 63 ± 3°C in a solution, by volume, of 100 parts sulfuric acid (specific gravity 1.84) to 30 parts saturated sodium dichromate solution.

7.14.4.5 Sodium Metasilicate

Immerse for 15 minutes at 63 ± 3°C in the solution described in Table 7.24.

7.14.4.6 Acid Etch Method 4

1. Immerse for 10 minutes at 60°C–65°C in the solution described in Table 7.25.
2. Rinse thoroughly.
3. Immerse for 5 minutes at 50°C–65°C in sulfuric acid–dichromate solution used for preparing aluminum (Table 7.2).
4. Rinse thoroughly.
5. Dry at a temperature below 93°C.

TABLE 7.25 Composition of Etching Solution for Stainless Steel

Solution Component	Parts by Weight
Hydrochloric acid (specific gravity 1.2)	50
Formalin solution (40%)	10
Hydrogen peroxide (concentration, 30% to 35%)	2
Water	45

7.14.4.7 Commercial Household Cleaner

Stainless steel surfaces may be prepared by vigorous scouring with a wet cloth and a commercial household cleaner. This method should be used only in cases where other methods are unavailable. A somewhat lower bond strength results from this method compared to the other techniques described.

7.15 TIN

Solvent cleaning with methylene chloride, chloroform, or trichloroethylene is recommended prior to abrading. Scraping, fine sanding, or scouring are suitable methods of abrading, which should be followed by solvent cleaning [2]. This is one of the few metals for which abrasion may be used without being followed by an acid etch [30].

7.16 TITANIUM

Titanium is a very important metal with uses in many industries. Titanium is a space-age metal used widely in aerospace applications requiring high strength-to-weight ratios at elevated temperatures above 300°C [42]. Adhesive-bonded helicopter rotor blades consisting of titanium skins in the form of sandwich panels have been in use since the 1950s. This section provides a somewhat chronological description of the surface treatment methods. The methods published by ASTM method D2651-01 are presented in Section 7.16.7.

The earliest surface treatment methods for these titanium surfaces were based on cleaning and etching in alkaline mixtures. These processes provided good joint strength. However, they were sensitive to the chemical composition of the titanium alloy and, within the same alloy, were sometimes affected by batch-to-batch variations. At a later date, a pickling etchant for stainless steel was used. This etchant was based on mixtures of nitric and hydrofluoric acids. However, when intermediate-curing-temperature adhesives were developed, these adhesives did not give strengths as high with titanium as they did with aluminum alloys. Consequently, the phosphate–fluoride process was developed, described later in this section. This process had been the most widely used surface preparation procedure for titanium in aerospace applications. It has been described in ASTM D2651-01 [8] and MIL-A-9067 [48].

In the late 1960s, army helicopters in Southeast Asia began to develop severe debonding problems in sandwich panels of titanium and glass-reinforced epoxy composite skins bonded to aluminum honeycomb core. These failures were attributed to the ingress of moisture into the interface. The combined effects of moisture and stress would have thus accelerated joint failures. As a result of this problem, a number of research programs were undertaken to explain the mechanism of failure in hot, humid environments and to improve the environmental resistance of titanium adhesive-bonded joints. One such program resulted in the development of a modified phosphate–fluoride process by Picatinny Arsenal workers (Wegman et al. [47]). At a later date, two commercial surface preparations, Pasa-Jell and VAST, became available as alternatives. These procedures are described in Section 7.20. Many surface preparation methods were developed in the 1970s and 1980s due to the need for superior adhesive bonds of titanium joints in aerospace and other applications [49].

One of the problems that requires close control in titanium processing is hydrogen embrittlement. The formation of hydrogen gas is inherent in the acid etching and anodizing processes. Hydrogen adsorption on titanium surfaces can lead to embrittlement. Extreme caution must be taken when treating titanium with acid etchants that evolve hydrogen. Immersion times must be closely controlled and minimized [2].

7.16.1 Stabilized Phosphate–Fluoride Treatment [39]

This method, developed by Wegman at Picatinny Arsenal, is an improvement over the basic phosphate–fluoride method described in MIL-A-9067. The improvement is achieved by the addition of sodium sulfate in the pickle. The method is reported to give good initial bond strength and excellent durability under adverse conditions, including high temperature, and high humidity (60°C and 95% RH) under load. In this method, the proper crystalline structure is established by the phosphate–fluoride process, then stabilized by the incorporation of sodium within the crystalline structure.

1. Vapor-degrease or clean with acetone.
2. Alkaline-clean by immersing parts in alkaline cleaner (non-silicated), 5%–10% by volume, for 5 minutes at 66°C. For suitable formulations (Oakite Products), please contact Chemetall Corporation in New Providence, NJ (www.chemetallamericas.com).
3. Rinse in running tap water at 40°C for 2 minutes.

TABLE 7.26 Composition of Etching Solution for Titanium

Solution Component	Amount
Hydrofluoric acid, 70% wt. concentration	15.7–23.6 g/l
Sodium sulfate, anhydrous	23.6 g/l
Nitric acid, 70% wt. concentration	315–394 g/l
Deionized water	to make 1 liter

TABLE 7.27 Composition of Phosphate Fluoride Etching Solution for Titanium

Solution Component	Amount
Trisodium phosphate	51.2–55.1 g/l
Potassium fluoride	19.7–23.6 g/l
Hydrofluoric acid, 70% wt. concentration	17.3–19.7 g/l
Deionized water	to make 1 liter

4. Modified HF acid-nitric acid pickle: immerse for 2 minutes at room temperature in the solution described in Table 7.26.
5. Rinse in running tap water at service temperature.
6. Phosphate–fluoride etching is done by soaking parts for 2 minutes at room temperature in the solution described in Table 7.27.
7. Rinse in running tap water at service temperature for 2 minutes.
8. Immerse in deionized water at 66°C for 15 minutes.
9. Rinse in water at room temperature at 71°C for 15 minutes.
10. Dry at 60°C for 30 minutes in a preheated air-circulating oven.
11. Wrap the parts in clean kraft paper until ready to bond.

This method yields excellent durability for both 6,4 titanium and chemically pure (CP) titanium. The former, however, exhibits a loss of lap-shear strength after 5 years outdoor weathering. The CP titanium does not show this effect.

7.16.2 Alkaline Cleaning [45]

Use steps 1, 2, 3, 10, and 11 only, under the procedure for stabilized phosphate–fluoride method in Sec. 7.16.1. The results, however, give poorer durability than do those by that method.

7.16.3 Alkaline Etch [45]

1. Vapor degrease or clean with acetone.
2. Alkaline clean: Immerse parts in non-silicated alkaline cleaner for 5 minutes at 66°C. Examples of commercial products include Oakite Pyrene 1038 in water (for a reliable recommendation for suitable formulations please contact Chemetall Corporation in New Providence, NJ (www.chemetallamericas.com).
3. Rinse in running tap water at 40°C for 2 minutes.
4. Rinse in running deionized water for 1 minute at service temperature.
5. Alkaline etch: Immerse for 5–10 minutes at <100°C in a cleaner (such as Cleaner AD-25, available from Turco Products Div. of Henkel Surface Technologies, Madison Heights, Michigan, www.henkelna.com).
6. Rinse under running tap water at service temperature for 2 minutes.
7. Rinse under running deionized water at service temperature for 1 minute.
8. Dry at 60°C for 30 minutes in a preheated air-circulating oven.
9. Wrap the parts in clean kraft paper until ready for bonding.

7.16.4 Pasa-Jell® Treatment [49]

Pasa-Jell® is a proprietary chemical marketed by Semco Division, Products Research and Chemical Corp., Glendale, CA. This formulation is available either as a thixotropic paste suitable for brush application, or as an immersion solution for tank treatment. Pasa-Jell® 107 is a blend of mineral acids, activators, and inhibitors, thickened using inorganic agents to permit application in localized areas. The approximate chemical constituents are 40% nitric acid, 10% combined fluorides, 10% chromic acid, 1% couplers, with the balance being water. The immersion process requires nonmetallic tanks made of PVC, polyethylene, or polypropylene. A recommended mixture uses 1:1 dilution for 12 minutes. The thixotropic Pasa-Jell® paste causes a reaction time of 10–15 minutes, which assures durable bonds.

7.16.5 VAST Process [49,50]

VAST is an acronym for Vought abrasive surface treatment. It is a development of Vought Systems of LTV Aerospace Corp., Dallas, now Vought Aircraft Division of Triumph Aerostructures (www.triumphgroup.com). In the VAST process, titanium is blasted in a specially designed chamber with slurry of fine abrasive containing fluorosilicic acid under high pressure. The particles are made of aluminum oxide at about 280 mesh in size and the acid concentration is maintained at 2%. The process produces a gray smut on the surface of 6A1-4V-Ti alloy, which must be removed by a rinse in 5% nitric acid. The joint strength resulting is superior to that provided by the unmodified phosphate fluoride process. The film produced is crystalline, having an anatase structure containing Ti, O, Si, F, Pb, and C. The oxide is stable up to 175°C, but starts converting to a rutile structure at higher temperatures.

The VAST process, because of its need for special equipment, has found limited use. The process details are as follows [39]:

1. Wipe the surface with methyl ethyl ketone.
2. Alkaline clean.
3. Rinse with deionized water at room temperature.
4. Use the VAST process for 5–10 minutes in a suitable chamber. The slurry consists of 2000 ml of 2% hydrofluorosilicic acid (H_2SiF_6) plus 500 ml of 240-grit purified aluminum oxide (Al_2O_3). The white aluminum oxide (available from the St. Gobain Abrasives Carborundum Division, Niagara Falls, NY 14302) is acceptable.
5. Rinse with tap water spray at room temperature.
6. Immerse for 1 minute in 5% nitric acid solution (optional, depending on titanium alloy).
7. Rinse in deionized water at room temperature.
8. Air-dry.
9. Bond within 4 days after treatment to be safe, although experiments have shown no changes over up to 9 days.

7.16.6 Alkaline-Peroxide Etch (RAE Etch) [49]

When titanium is immersed in alkaline hydrogen peroxide solutions, depending on the concentration of sodium hydroxide and hydrogen peroxide, the metal is either etched or oxidized. The concentrations that produce gray oxides produce adhesive-wettable surfaces. A recommended mixture is 2% caustic soda and 2.2% hydrogen peroxide. Exposure to the oxidizing solution may be at room temperature, but 10–36 hours are required under these conditions to produce high-strength durable joints. Good bonding surfaces are produced within 20 minutes at 50°C–70°C. Oxidized adherends are washed with plain and acidified water before rinsing with acetone and drying at 100°C.

High bond strengths (50–55 MPa) are produced if the titanium surface is subjected to alumina blasting prior to the treatment. An alkaline peroxide etch has many advantages over the acid-based treatments. The chemical constituents are less toxic, the treatment does not require acid-resistant containers, the process is free from hydrogen pick-up, and the waste material is easily disposed of. The process, however, is limited to batch production because of the high instability of hydrogen peroxide at high exposure temperatures. This process was developed jointly by British Aerospace and the Royal Aircraft Establishment (RAE). Durability is reported to be excellent with 91%–92% of initial joint strength (control) being retained after 4 years in a warm, wet environment.

7.16.7 ASTM Suggested Methods

ASTM D2651-01 [8] lists the following mechanical and chemical methods for surface preparation of stainless steel.

7.16.7.1 Mechanical Abrasion

See Section 7.14.4.1.

7.16.7.2 Chemical Methods

1. Acid etch (hydrochloric, orthophosphoric, and hydrofluoric).
2. Immerse for 2 minutes at room temperature in the following solution: 841 ml orthophosphoric acid (reagent grade, 85–87%), to 63 ml hydrofluoric acid (reagent grade, 60%).
3. Rinse.
4. Oven dry for 15 minutes at 88°C–93°C.

7.16.7.3 Acid Etch (Nitric Hydrofluoric)

1. Immerse for 15 minutes at 76°C in the following solution: 6 to 8 fl oz of a caustic cleaner, such as Turco® Vitro-Klene 10 or equivalent, to 3.6 l (1 gal) water.
2. Rinse in cold tap water.
3. Immerse for 5 minutes at room temperature in the following solution by weight:
   48% nitric acid (specific gravity 1.5)
   3% ammonium bifluoride (technical)
   49% water
4. Rinse in cold tap water.
5. Air-dry at room temperature.
6. Immerse for 2 minutes at room temperature in the following solution:
   50.0 g trisodium phosphate (technical)
   8.9 g sodium fluoride (technical)
   26.0 ml hydrofluoric acid (48%)
   water to make 3.6 liters of solution
7. Air-dry at room temperature.

7.16.7.4 Stainless Steel Methods

Processes used to prepare titanium alloys for adhesive bonding can be much the same as for stainless steel. Good bond strengths have resulted by bonding titanium alloys that have been anodized by proprietary processes.

7.16.7.5 Chromic Acid–Fluoride Anodizing (a variant of CAA Chromic Acid Anodizing)

Stainless steel surface processes, as described in Sections 7.14.4.4–7.14.4.6, have been found satisfactory.

7.16.8 Sol-gel Process

See Section 7.2.1.

7.17 TUNGSTEN AND ALLOYS

7.17.1 Hydrofluoric–Nitric–Sulfuric Acid Method [29]

1. Degrease in a vapor bath of trichloroethylene.
2. Abrade the surface with medium-grit emery paper.
3. Degrease again as in step 1. Using equipment constructed of fluoropolymer resins, polyethylene, or polypropylene, prepare the solution detailed in Table 7.28. Blend the hydrofluoric acid and the nitric acid with water. Then slowly add the sulfuric acid, stirring constantly with a polytetrafluoroethylene or polyethylene rod. Add a few drops of 20% hydrogen peroxide.
4. Immerse for 1–5 minutes in the above solution at room temperature.
5. Rinse under tap water.
6. Finish rinsing in distilled water.
7. Dry in an oven at 71–82°C for 10–15 minutes.

7.18 URANIUM

For any adhesive to bond well with this metal, the uranium must be freshly pickled and dried. The dark-colored, loosely adherent surface oxide layer must be removed. Cleaning is satisfactory when the metal surface becomes bright and shiny.

TABLE 7.28 Composition of Etching Solution for Tungsten

Solution Component	Parts by Weight
Nitric acid, specific gravity 1.41	30
Sulfuric acid, specific gravity 1.84	50
Hydrofluoric acid, 60% wt. concentration	5
Deionized water	15

7.18.1 Abrasive Method

This method, developed by Picatinny Arsenal workers, involves degreasing the uranium in a vapor bath of trichloroethylene, and then sanding the bonding surfaces in a pool of the adhesive to be used for bonding [39]. This is done to prevent further oxidation of the uranium block by exposure to air after sanding, and also to prevent contamination of the surrounding area with radioactive particles. Polyamide-epoxy and polyurethane adhesives are recommended. The problem of oxidation remains, however, when only the two bonding surfaces of the uranium block are cleaned and adhesive coated. Under these conditions, the oxide layer spreads from the uncleaned side of the uranium until the bonding surfaces themselves are completely oxidized. By sanding all six sides of each uranium block and then coating with a protective layer of adhesive, the problem is solved.

7.18.2 Acetic Acid–Hydrochloric Acid Method

This method is used with aluminum-filled adhesives when no primer or surface coating is to be applied [39].

1. Pickle uranium bars clean in 1:1 nitric acid and water.
2. Rinse briefly in distilled water.
3. Immerse in 9:1 acetic acid–hydrochloric acid bath for 3 minutes. Note that 200 ml of this bath can accommodate no more than 4 bars of 11.4 cm by 2.5 cm by 0.3 cm dimensions without a vigorous reaction taking place and a black film forming on the uranium surface.
4. Rinse briefly in distilled water.
5. Rinse in acetone.
6. Air-dry.

7.18.3 Nitric Acid Bath

This method is to be used when a primer or surface coating is applied [39].

1. Pickle-clean in 1:1 nitric acid and water.
2. Rinse in acetone.
3. Immerse for 10 minutes in a coating bath of 1.0 g purified stearic acid dissolved in 95–99 ml of acetone and 1–5 ml of nitric acid.
4. Air-dry.
5. Rinse with carbon tetrachloride spray.
6. Air-dry.
7. Store in distilled water in a polyethylene bottle at 60°C.

7.19 ZINC AND ALLOYS

The most common use of zinc is in galvanized metals. Zinc surfaces are usually prepared mechanically. One mechanical and two chemical methods follow.

7.19.1 Abrasion (for General-Purpose Bonding) [27]

1. Grit- or vapor-blast with 100-grit emery cloth.
2. Vapor degrease in trichloroethylene (TCE).
3. Dry at least 2 hours at room temperature, or 15 minutes at 93°C to remove all traces of TCE.

7.19.2 Acid Etch [29]

1. Vapor degrease in trichloroethylene.
2. Abrade with medium-grit emery paper.
3. Repeat the degreasing in step 1.
4. Etch for 2–4 minutes at room temperature in the solution specified in Table 7.29.

7.19.3 Sulfuric Acid–Dichromate Etch [27]

1. Vapor degrease in trichloroethylene.
2. Etch for 3–6 minutes at 38°C in the solution detailed in Table 7.29.
3. Rinse in running tap water.
4. Rinse in distilled water.
5. Dry in air at 40°C.

7.19.4 Conversion Coatings

Phosphate and chromate conversion coatings are available for zinc from commercial sources as proprietary materials.

TABLE 7.29 Composition of Etching Solution for Zinc

Solution Component	Parts by Weight
Sulfuric acid, conc., specific gravity 1.84	2
Sodium dichromate (crystalline)	1
Deionized water	8

7.20 WELD BONDING METALS

Chemical etching is essential to assure high-strength, weld-bonded joints. There are differences in the surface requirements after cleaning for welding and bonding. In order to achieve class A resistance welds, it is necessary to have a chemically active surface which may have a high surface resistance. Final selection of the surface preparation method should be based on the end use of the hardware and consideration of the relative importance of weld quality and adhesive joint strength [51].

Lockheed workers Kizer and Grosko [52] have studied weld bonding using four different surface preparation techniques on aluminum. A technique developed for aluminum by the Northrop Grumman Corporation for the Air Force follows. It was found to be superior to the previously recommended FPL etch plus 60-minute dichromate seal [53,54].

1. Vapor degrease in trichloroethane vapor for 60 seconds, followed by spray rinse of condensed trichloroethane fluid for an additional 60 seconds. All parts must be free of water prior to vapor degreasing.
2. Alkaline clean in Cleaner® 85 solution. (Cleaner® 85 is available from Turco Products Div., Henkel Surface Technologies, Madison Heights, Michigan, www.henkelna.com).
3. Immediately spray-rinse in cold deionized water for at least 5 minutes and inspect for water break-free condition. Should water breaks occur, repeat the above steps.
4. Immerse the alkaline-cleaned parts in a deoxidizer solution consisting of 8%–16% by volume of nitric acid and 20.1–24.6 g/l Amchem® 7 for 7–10 minutes at room temperature. When chemical addition is required to maintain the strength of the solution, use Amchem® 17 instead of Amchem® 7. (These products are available from Amchem® Products, Inc., Ambler, PA.)
5. Immediately spray-rinse in cold deionized water for at least 5 minutes.
6. Anodize at 1.5 ± 0.2 volts for 20–25 minutes in a solution of 10.1 ± 1.1 g/l phosphoric acid and 10.1 ± 1.1 g/l sodium dichromate. Anodizing should be conducted in a room-temperature solution using a "ripple-free" DC power supply.
7. Immediately spray-rinse in cold deionized water for at least 5 minutes.
8. Oven-dry 30–60 minutes at 60°C–66°C.
9. Cleaned parts must be handled only with clean white cotton gloves and may be stored for periods of up to 21 days prior to weld bonding by wrapping them in chemically neutral paper.

For titanium, Northrop Grumman Corporation has developed two surface treatments for the US Air Force. They are given below [55].

7.20.1 Vapor Honing/Pasa-Jell® 107-M Procedure

1. Remove organic contaminants using a methyl ethyl ketone (MEK) solvent rinse.
2. Remove inorganic contaminants by immersion in a non-etch alkali (such as Oakite® 61 B) at 60°C. (Oakite® 61 is available from Chemetall Corporation, in New Providence, NJ, www.chemetallamericas.com).
3. Immerse in Pasa-Jell® 107-M solution at ambient temperature for 10 minutes. (This is a proprietary solution containing HNO_3, $Na_2Cr_2O_7$, H_2SiF_6, and proprietary surfactants supplied by Semco Division, Products Research and Chemical Corp., Glendale, CA). Semco Pasa-Jell® 107 is inorganically thickened to permit application in localized areas. Semco Pasa-Jell® 107-M does not contain the inorganic thickener, permitting immersion of large surfaces. Semco Pasa-Jell® 107 is designed to provide a balanced ratio of components which will maintain the effectiveness of the etch rate and will inhibit embrittlement as long as the total acid content is maintained above 20% [56].
4. Use the VAST process [57] of surface impingement in a suitable chamber. The slurry consists of 2% hydrofluorosilicic acid (H_2SiF_6) and 500 ml of 240-grit purified aluminum oxide (Al_2O_3), such as white aluminum oxide available from the St. Gobain Abrasives Carborundum Division, Niagara Falls, NY 14302.
5. Rinse with tap-water spray at room temperature.
6. Immerse for 1 minute in 5% nitric acid solution (optional, depending on the titanium alloy).
7. Rinse in deionized water at room temperature.
8. Air-dry.
9. Bond within 4 days after treatment. (Experiments show, however, that no changes take place in up to 9 days.)

7.21 CONCLUSIONS

Metals require surface treatment to remove contamination and prepare their surfaces for adhesion bonding. Both initial bond strength and bond durability are affected by the surface preparation technique. Most metals require unique methods of treatment for optimal bond strength formation. Most surface preparation treatment methods use or generate chemicals that have various serious health effects and must be used with extreme caution.

REFERENCES

[1] Review of Current Trends in Surface Pretreatment prior to Structural Adhesive Bonding, John Bishopp Star Adhesion Limited, Waterbeach Cambridge, at NPL, Teddington, Middlesex November 20, 2003.

[2] Snogren RC. Handbook on surface preparation. New York: Palmerton Publishing Co.; 1974.
[3] Arkles B. Commercial applications of sol-gel-derived hybrid materials. MRS Bulletin 2001; 402.
[4] Osborne JH. Nonchromate conversion coatings in use at boeing, Boeing Phantom Works – Seattle, <www.asetsdefense.org>; 2006
[5] Arnott D, Rider A, Mazza J. Surface treatment and repair bonding. In: Baker AA, Rose LRF, Jones R, editors. Advances in the bonded composite repairs of metallic aircraft structure. Elsevier; 2002.
[6] Fiebig JW, Mazza JJ, McKray DB. An ALC consideration of simple solgel preparations for improved durability of field and depot-level bonded repairs. In: Proceedings of the aging aircraft conference. Galaxy Scientific; 2002.
[7] A&M Environmental Technical Notes, Volume 4, Number 3, Boeing Company, August 1999.
[8] American Society for Testing and Materials (ASTM). ASTM 3933-98, Standard Recommended Practice for Preparation of Metal Surfaces for Adhesive Bonding, vol. 15.06. Annual Book of ASTM Standards; 2001.
[9] Nacconol 90G, available from the Stephan Co., Northfield, IL 60093.
[10] Tim Meara. New Honing Options for Hypereutectic Aluminum Cylinder Bores, Modern Machine Shop, Sep 18, 2008, <www.mmsonline.com/articles/new-honing-options-for-hypereutectic-aluminum-cylinder-bores>.
[11] Davis GD, Venables JD. Surface treatment of metal adherends. In: Chaudhury M, Pocius AV, editors. Adhesion Science and Engineering – vol. 2: Surfaces, Chemistry and Applications. Elsevier; 2002. pp. 947–1008.
[12] Stepan Chemical Co., Northfield, IL, <www.stepan.com>.
[13] Russell WJ, Garnis EA. Chromate-free method of preparing aluminum surfaces for adhesive bonding: an etchant composition of low toxicity, U.S. Army Armament Research and Development Command, Large Caliber Weapon Systems Laboratory, Technical Report ARLCD-TR-78001, ARRADCOM, Dover, New Jersey (May 1978).
[14] Russell WJ. Chromate-free process for preparing aluminum for adhesive bonding. In: Bodnar MJ, editor. J. Appl. Polym. Sci. Applied Polymer Symposia No. 32. Interscience Publishers; 1977, pp. 105–17.
[15] Wegman RF, et al. The function of the P2 Etch in treating aluminum alloys for adhesive bonding, preprint book. 29th National SAMPE Symposium, vol. 29, Reno, Nevada (April 3–5, 1984). pp. 273–81.
[16] Rogers NL, Russell W. Evaluation of Nonchromated Etch for Aluminum Alloys (P2-Etch), U.S. Army Armament Research and Development Command, Large Caliber Weapon Systems Laboratory, Contractor Report (Bell Helicopter Textron), ARLCD-CR-80008, ARRADCOM, Dover, New Jersey (April 1980).
[17] Great Britain Patent, GB290903, assigned to Charles Hugh Roberts Gower, Stafford O Brien And Partners, May 24, 1928.
[18] American Anodizing Council, <www.anodizing.org>; 2013.
[19] <www.hmwire.com>; 2013.
[20] Davis GD, Venables JD. In: Chaudhury M, Pocius AV, editors. Chapter 21 – Surface Treatments of Metal Adherends in Adhesion Science and Engineering, vol. 1. Elsevier; 2002.
[21] American Society for Testing and Materials (ASTM), ASTM D3922-89 (1993). Standard Recommended Practice for Preparation of Aluminum Surfaces for Structural Adhesives Bonding, (Phosphoric Acid Anodizing), vol. 15.06, Annual Book of ASTM Standards; 1993.

[22] Brewis DM. In: Kinloch J, editor. Aluminum adherends, durability of structural adhesives. London and New York: Applied Science Publishers; 1983.

[23] Boerio FJ, Davis GD, deVries JE, Miller CE, Mittal KL, Opila RL, et al. Current status, research needs, surface treatment of metal adherends 1003 and opportunities. In: Czanderna AW, Landgrebe AR, editors. Applications of surface processing to transportation and utilities technologies. Golden, CO: National Renewable Energy Laboratory; 1992. pp. 23–1; and Crit. Rev. Surf Chem. 3: 81; 1993.

[24] Geiss PL, Gramsch-Kempkes S. Mechanical Adhesion in Adhesively Bonded Metal-Joints with Retentive Laser-Structured Adherends, 36th Annual Meeting of the Adhesion Society, 3–6 Mar 2013, Daytona Beach, FL (USA).

[25] Broad R, French J, Sauer J. New effective, ecological surface pretreatment for highly durable adhesively bonded metal joints. Int J Adhes Adhes 1999;19:2–3.

[26] Bishopp J. Handbook of adhesives and sealants, vol. 1. Elsevier; 2005. pp. 163–214.

[27] Shields J. Adhesives handbook. 2nd ed. London: Newnes-Butterworth; 1976 (The 3rd Edition was published by Butterworths, London, in 1984.)

[28] Rogers NL. Surface preparation of metals for adhesive bonding. In: Bodnar MJ, editor. Applied Polymer Symposia No. 3, structural adhesives bonding. Interscience Publishers; 1966. pp. 327–40.

[29] Guttman WH. Concise guide to structural adhesives. New York: Reinhold; 1961.

[30] Cagle CV. Surface preparation for bonding beryllium and other adherends. In: Cagle CV, editor. Handbook of adhesive bonding. New York: McGraw-Hill; 1973.

[31] Phoenix Brands, <www.phoenixbrands.com> August, 2013.

[32] American Society for Metals (ASM). Metals handbook. Surface engineering nonferrous metals, vol. V. ASM International; 1994.

[33 Military Specification, MIL-M-45202C, Anodic Treatment of Magnesium Alloy (April 1, 1981).

[34] Military Specification, MIL-M-3171C (superseded by SAE-AMS-M-3171, May 1998), Magnesium Alloy, Processes for Pretreatment and Prevention of Corrosion (March 14, 1974).

[35] Heating and pickling huntington alloys, Huntington Alloy Products Division, The International Nickel Co., Inc., Huntington, WV; 1968.

[36] CHEMLOK adhesives, A Guide to Handling and Application, Application Guide, published by LORD Corp.; 1999.

[37] Keith RE, et al. Adhesive Bonding of Nickel and Nickel-Base Alloys, NASA TMX-53428 (Oct. 1965).

[38] Petrie EM. Plastics and elastomers as adhesives. In: Harper CA, editor. Handbook of plastics, elastomers and composites. 4th ed. New York: McGraw-Hill; 2002.

[39] Landrock AH. Processing handbook on surface preparation for adhesive bonding, Picatinny Arsenal Technical Report 4883. Dover, NJ: Picatinny Arsenal (Dec. 1975).

[40] Devine AT. Adhesive bonded steel: bond durability as related to selected surface treatments, U.S. Army Armament Research and Development Command, Large Caliber Weapon Systems Laboratory, Technical Report ARLCD-TR-77027 (Dec. 1977).

[41] Brockmann W. In: Kinloch J, editor. Steel adherends, durability of structural adhesives. Applied Science Publishers; 1983.

[42] DeLollis NJ. Adhesives, adherends adhesion. Huntington, NY: Robert E. Krieger Publishing Co.; 1980. (This book is the 2nd edition of a 1970 publication under the name *Adhesives for Metals: Theory and Technology*).

[43] Vazirani HN. Surface preparation of steel for adhesive bonding. J Adhes 1969;1:222–32.

[44] Keith RE, et al. Adhesive Bonding of Stainless Including Precipitation Hardening Stainless, NASA TMX-53574 (April 1968). (Available from NTIS as AD 653 526.)
[45] Slota SA,Wegman RF. Durability of Adhesive Bonds to Various Adherends, Picatinny Arsenal Technical Report 4917 (June 1976).
[46] Atkins RW, et al. Explosives Research and Development Establishment, Ministry of Defence (UK), Waltham Abbey, Essex, England. An Investigation into the Influence of Surface Pre-Treatment by Particle Impact on the Strength of Adhesive Joints Between Like Steel Surfaces, ERDE-TR-120, January 1973. (Available from NTIS as AD 771 003.)
[47] Wegman RF, et al. Evaluation of the Adhesive Bonding Processes Used in Helicopter, Manufacture, Part I. Durability of Adhesive Bonds Obtained as a Result of Processes Used in the UH-1 Helicopter. Picatinny Arsenal Technical Report 4186, Dover, NJ: Picatinny Arsenal (Sept. 1971).
[48] Military Specification, MIL-A-9067C(1). Adhesive Bonding, Process and Inspection Requirements for Naval Air Systems Command (Feb 23, 1979). (This specification was cancelled for Air Force use by Notice 1, dated 9 December 1974.).
[49] Mahoon A. In: Kinloch AJ, editor. Titanium adherends, durability of structural adhesives. London and New York: Applied Science Publishers; 1983.
[50] Hohman AE, Lively GW. Surface Treatment of Titanium and Titanium Alloys, U.S. Patent 3,891,456 filed (Oct. 17, 1973).
[51] Beemer RD. Introduction to weld bonding. SAMPE Q 1973;5:37–41.
[52] Kizer JA, Grosko JJ. Development of the weldbond joining process for aircraft structures. In: Bodnar MJ, editor. Applied Polymer Symposia No. 19, processing for adhesives bonded structures. NY: Interscience Publishers; 1972. pp. 353–70.
[53] Bowen BB, et al. Improved Surface Treatments for Weldbonding Aluminum, AFML-TR-159 (Oct. 1976).
[54] Wu KC, Bowen BB. Advanced aluminum weldbonding manufacturing methods, Preprint Book. 22nd National SAMPE Symposium, 22:536–54, San Diego, CA (April 16–28, 1977).
[55] Mahon J, et al. Manufacturing Methods for Resistance Spotweld-Adhesive Bond Joining of Titanium, AFML-TR-76-21 (March 1976).
[56] Bergdahl Associates, Inc., <www.bergdahl.com/pasaJell_107.htm>; August 2013.
[57] Lively GW, Hohman AE. Development of a mechanical-chemical surface treatment for titanium alloys for adhesive bonding, proceedings, 5th National SAMPE Technical Conference, Kiamesha Lake, NY (Oct. 9–11, 1973). pp. 145–55.

Chapter 8

Surface Preparation of Thermoplastics, Thermosets, and Elastomers

8.1 INTRODUCTION

A reader who is unfamiliar with surface treatment of materials should begin by first reading Chapter 6 of this book. That chapter teaches the general considerations in the preparation of materials surfaces, including metals and plastics. The present chapter describes the common methods used for the preparation of specific plastic adherends. These procedures provide practical ways to obtain strong, reproducible adhesive bonds that readily fit in commercial processes. To add to the usefulness of this chapter, the authors have made liberal citation of commercial plastics when describing treatment techniques. It is difficult to describe surface preparation methods of specific plastics without identifying them; plastics manufacturers are the most frequent sources of basic preparation methods.

8.2 THERMOPLASTICS

Thermoplastic surfaces, unlike thermosetting materials, ordinarily require physical or chemical modification to achieve acceptable bonding. This is especially true in the case of crystalline thermoplastics such as polyolefins (primarily polyethylene and polypropylene), linear polyesters, and fluoropolymers. Methods used to improve the bonding characteristics of these surfaces are described in Chapter 6.

Surface preparation methods are usually suggested for conventional bonding where an adhesive is applied to the bonding surfaces. Solvent cementing and thermal welding do not require chemical treatment of the plastic surface. As with metallic substrates, the effectiveness of the treatment imparted to plastics surfaces decreases with time. It is, therefore, important to carry out the priming or bonding steps as soon as possible after surface preparation [1].

Surface Treatment of Materials for Adhesive Bonding. DOI: http://dx.doi.org/10.1016/B978-0-323-26435-8.00008-3

8.2.1 Acetal Copolymer (Celcon®)

This plastic (made by Ticona Corp.) is a highly crystalline copolymer with excellent solvent and chemical resistance. Two suggested surface-preparation methods are described below.

8.2.1.1 Chromic Acid Etch [2]

1. Wipe parts with acetone.
2. Air-dry.
3. Etch the surfaces for 10–15 seconds in a solution made up as summarized in Table 8.1.
4. Flush with tap water.
5. Rinse with deionized water.
6. Oven-dry at 60°C.

8.2.1.2 Hydrochloric Acid Etch [3]

1. Immerse parts for 5 minutes in concentrated HCl at room temperature. Use a glass rod to move fresh acid into contact with the acetal copolymer. About 8–13 μm of acetal copolymer is removed per minute of the etching process.
2. Rinse and dry at room temperature for 4 hours. Heat-formed or machined acetal copolymer should be stress-relieved before etching. A specially developed primer used in very thin coats gives excellent results.

8.2.2 Acetal Homopolymer (Delrin®)

This is a highly crystalline polymer (from DuPont Company) that has excellent solvent resistance. The chemical structure and solvent resistance of homopolymer acetal lead to relatively poor adhesion of solvent-based paints and cements. This is largely overcome by surface treatment prior to applying the coating. Originally, a chemical etching process, termed "sanitization," was developed where a mildly acidic solution produced uniformly distributed anchor points on the surface [4]. This process is no longer used due to environmental concerns with some of the chemicals in the etching bath. Two

TABLE 8.1 Composition of Etching Solution for Acetal Copolymer

Solution Component	Parts by Weight
Sulfuric acid	400
Potassium dichromate	11
Distilled water	44

methods for surface preparation of homopolymer acetal parts to form relatively strong bonds are described below.

8.2.2.1 Three-Step Procedure

A more recent 3-step etching procedure (Table 8.2), using chemicals more compatible with today's environmental requirements, has also been successfully tested [5].

8.2.2.2 Chromic Acid Etch [6]

1. Wipe with acetone or methyl ethyl ketone.
2. Immerse in a chromic acid solution (Table 8.3) for 10–20 seconds at room temperature (20°C–30°C). Dissolve the potassium dichromate in clear tap water; then add the sulfuric acid in increments of about 200 g, stirring after each addition.
3. Rinse in running tap water for at least 3 minutes.
4. Rinse in distilled or deionized water.
5. Dry in an air-circulating oven at 38°C for about 1 hour.

Adhesives are strongest in compression and shear modes, but perform comparatively weakly in tension, cleavage, and peel modes. Adhesion depends on the surface preparation of the parts. Cleaning and degreasing is an absolute minimum. Surface roughening, mechanically with sandpaper or chemically with an etching bath, strongly improves the results. Several types of adhesives are considered, including two-component epoxy, two-component polyurethane or cyano-acrylate [5].

8.2.3 Acrylonitrile–Butadiene–Styrene (ABS)

This terpolymer may be prepared for bonding by the following techniques.

8.2.3.1 Abrasion with Primer [7]

1. Sand with a medium grit sandpaper.
2. Wipe free of dust.
3. Dry in an oven at 71°C for 2 hours.
4. Prime with a silane primer such as Dow Corning A-4094 or General Electric SS-4101.

8.2.3.2 Warm Chromic Acid Etch [8]

1. Degrease in acetone.
2. Etch in the solution detailed in Table 8.4 for 20 minutes.
3. Rinse in tap water.
4. Rinse in distilled water.
5. Dry in warm air.

TABLE 8.2 Three-Step Etching Procedure for Homopolymer Acetal Polymer [5]

	Step 1	Alternative Step 1 A	Step 2	Step 3
Solution	71% o-H_3PO_4 (orthophosphoric acid). Mix 5 vol. parts of commercial concentrated (85%) o-H_3PO_4 with 1 vol. part of DI water (deionized water).	Preparation: Mix 5 vol. parts of (85%)o-H_3PO_4 with 1 vol. part glycerin DAB7H and 0.5–1.0 g/l sodium alkylarylsulfonate(1) (phenylsulfonate)	12% NaOH (sodium-hydroxide). Preparation: Dissolve 120 g NaOH in 1 l of DI water	1% acetic acid. Preparation: Add 20 ml of glacial acetic acid to 1 l of DI water.
Bath temperature	70°C	85°C	100°C	80°C
Dipping time	5 minutes	5 minutes	6 minutes	10 minutes
Rinse	Thoroughly, 1 min/23°C with water.	As for Step 1.	As for Step 1.	As for Step 1, followed by a supplementary rinse in DI water.
Heating elements	Quartz or heating elements coated with Teflon®.	As for Step 1.	Stainless steel, titanium, heating elements coated with Teflon® or quartz.	As for Step 2.
Material for dipping tanks recommended	PVC. Polypropylene (PP), high density polyethylene (HDPE) or glass.	PVC. PP. HDPE or glass.	Stainless steel, mild steel or glass.	PVC, PP HDPE or glass.
Remarks	If the surface of the parts is significantly contaminated (mold release agent, finger prints, traces of oils, etc.) the parts should be degreased prior to the dipping.	The solution of Step 1 A has some degreasing character but may give less good adhesion.		

(1) Common detergent. All the above information is subject to disclaimer.

TABLE 8.3 Composition of Etching Solution for Acetal Homopolymer

Solution Component	Parts by Weight
Sulfuric acid, concentrated	300
Potassium dichromate ($K_2Cr_2O_7$)	15
Water	24

TABLE 8.4 Composition of Etching Solution for Acrylonitrile–Butadiene–Styrene

Solution Component	Parts by Weight
Sulfuric acid, concentrated	26
Potassium dichromate	3
Distilled water	11

8.2.4 Allyl Diglycol Carbonate (CR-39® by PPG Industries)

This resin may be cast as a film. It was developed in 1940 and is sold in monomer form. It is used to produce highly transparent plastics for optical applications such as lenses for glasses. The following surface treatment method has been recommended [9].

1. Degrease in methanol, isopropanol, or detergent.
2. Abrade by grit- or vapor-blast, or use 100-grit emery cloth. Sand or steel shot are suitable abrasives.
3. Solvent-degrease in methanol or isopropanol.

8.2.5 Cellulosics [Cellulose Acetate, Cellulose Acetate Butyrate (CAB), Cellulose Nitrate, Cellulose Propionate, Ethyl Cellulose]

Cellulosics are ordinarily solvent-cemented, unless they are to be bonded to dissimilar adherends, using conventional adhesives. In such cases, the following surface preparation method may be used.

8.2.5.1 Abrasion and Cleaning [10,11]

1. Solvent-degrease in methyl alcohol or isopropyl alcohol.
2. Grit- or vapor-blast, or use 220-grit emery cloth.

3. Repeat step 1.
4. Heat 1 hour at 93°C and apply adhesive while still hot.

8.2.6 Ethylene–Chlorotrifluoroethylene Copolymer (ECTFE)

This fluoropolymer has excellent chemical resistance. See Chapter 6 for more information about adhesion treatment of ECTFE [12].

8.2.7 Ethylene–Tetrafluoroethylene Copolymer (ETFE)

This high-temperature-resistant polymer can be easily processed by conventional methods, including extrusion and injection molding. It can be joined by screw assemblies, snap-fit, press-fit, cold or hot heading, spin-welding, and heat-bonding. See Chapter 6 for more information about the adhesion treatment of ETFE [12].

8.2.8 Ethylene–Vinyl Acetate (EVA)

This copolymer is, in actuality, an elastoplast, with properties intermediate between that of a plastic and an elastomer. A suggested surface preparation method, the SIRA method [9], involves:

1. Degrease in methanol.
2. Prime with epoxy adhesive.
3. Fuse into the surface by heating for 30 minutes at 100°C.

8.2.9 Fluorinated Ethylene–Propylene Copolymer (FEP)

This fluoropolymer is a true thermoplastic and a copolymer of tetrafluoroethylene and hexafluoropropylene. FEP can be heat-sealed, heat-bonded, laminated, and combined with many materials. It can be used by itself as a hot-melt adhesive. FEP can be treated for adhesion bonding using techniques described in Chapter 6 [12].

8.2.10 Ionomer [10]

Ionomers are polymers in which ionized carboxyl groups create ionic crosslinks in the intermolecular structure. The properties are similar to crosslinked thermosets, even though these resins are processed at conventional temperatures as with other thermoplastic resins (for example, Surlyn® resins by the DuPont Company). These resins, like polyethylene, are unable to develop high-strength adhesive bonds. The following abrasive [9] surface preparation technique is suggested:

1. Solvent-degrease in acetone or methyl ethyl ketone.

2. Grit- or vapor-blast, or use 100-grit emery cloth. ISO-grit alumina is a suitable abrasive.
3. Repeat step 1.

8.2.11 Nylon (Polyamide) [10]

Both solvent cementing and adhesive bonding are used with nylon (nylon 6,6 is most commonly used). The solvents, however, are quite different from those used for cellulosics and polystyrene. Nylon molding resins are not ordinarily cemented by conventional solvent-type cements due to their lack of solubility in ordinary nontoxic solvents. Nylon is frequently bonded to metals, introducing problems not found in nylon-to-nylon bonds. Conventional adhesives must be used in nylon–metal bonds. Nylons are relatively crystalline thermoplastics that are not considered solvent-sensitive, requiring relatively strong solvents for surface preparation. Suitable solvents for cleaning/priming polyamide surfaces include:

- Acetone (B.P. 56.5°C).
- Methyl ethyl ketone (B.P. 79.6°C).
- Mixture of resorcinol, ethanol and *p*-sulfonacid [48].
- A nitrilphenol based solution.
- A resin based on resorcinol–formaldehyde.

Effective chlorinated solvents (except perchloroethylene,) such as 1,1,1-trichloroethylene and Freon® TMC (a 50/50 mixture of methylene chloride and 1,1,2-trichloro-1,2,2-triflouoroethane), have been banned and can no longer be used.

Detergents have also been used instead of organic solvents in cleaning nylon. Nylon, like polycarbonate, will readily absorb moisture from the air. For this reason, it is desirable to keep the humidity low before bonding.

Abrasive method [13]:

1. Wash with acetone.
2. Dry.
3. Hand-sand with 120-grit abrasive cloth until gloss is removed.
4. Remove sanding dust with a short hair stiff brush.

Bonding or priming should be carried out as soon as possible. Parts should be warmed slightly before bonding. Recommended primers include nitrile phenolic, resorcinol–formaldehyde, vinyl phenolic, silane, etc. Epoxies have been used on metallic surfaces, with the nylon bonded to the epoxy [14].

8.2.12 Perfluoroalkoxy Resins (PFA)

Perfluoroalkoxy resins combine high-temperature performance with a capability for thermoplastic processing. See Ch. 6 for surface preparation methods [12].

8.2.13 Phenylene-Oxide-Based Resins (Noryl® Polyaryl Ethers by GE Plastics)

This family of plastics is composed of polystyrene-modified polyphenylene oxide. These resins have outstanding dimensional stability at high temperatures, broad use-temperature range, and other desirable properties. Solvent cementing is the usual method of joining. Surface preparation for bonding consists of sanding or the following acid etch procedure [15,16].

1. Solvent-clean with isopropanol or aqueous solutions of most commercial detergents.
2. Sand or etch with chromic acid. If chromic acid etch is used, etch 1 minute at 80°C in the solution described in Table 8.5.
3. Rinse in distilled water.
4. Dry.

Note that studies made with the original unmodified polyphenylene oxide etched in the above solution for 3 minutes at 65°C have revealed that chromic acid solution became ineffective after its first use. The evaluation was made with AF-30 nitrile-phenolic adhesive and modified T-peel tests. To avoid this possible problem, vacuum blasting, followed by an acetone wipe, can be used [10].

8.2.14 Polyarylate (Ardel® by BP plc)

This new engineering thermoplastic announced in 1980 is aromatic polyester based on bisphenol A and phthalic acid. According to the manufacturer, its surface preparation is similar to that of polysulfone (see "Polysulfone," Sec. 8.2.28).

8.2.15 Polyaryl Sulfone (Astrel® 360 plastic) [10]

This engineering thermoplastic supplied by the 3M Company is structurally useful at temperatures up to 260°C. The manufacturer's literature mentions welding by thermal, ultrasonic, or solvent techniques. Air Force studies

TABLE 8.5 Composition of Etching Solution for Polyaryl Ethers

Solution Component	Amount (g)
Sulfuric acid, concentrated	375
Potassium dichromate	18.5
Distilled water	30

evaluated sandblasting, acid etch, and solvent wash methods, with the latter effective for surface preparation of steel-plastic-steel bonds.

Solvent wash is a triple wash (3 successive washes) in a 65/35 mixture by volume of a fluorocarbon and reagent-grade isopropyl alcohol.

8.2.15.1 Sandblast

1. Ultrasonic-clean with an alkaline-etching solution.
2. Cold-water rinse.
3. Alcohol-wash.
4. Sandblast with 150-mesh silica sand.
5. Alcohol-wash.
6. Dry with dry nitrogen.

8.2.15.2 Acid Etch

1. Ultrasonic clean in an alkaline cleaning solution.
2. Cold-water wash.
3. Bathe 15 minutes at 66°C–71°C in acid solution shown in Table 8.6.
4. Cold-water wash.
5. Dry at 66°C in air-circulating oven.

8.2.16 Polycarbonate

(Supplied as Lexan® by General Electric Co., Calibre® from Dow Chemical Co., and Tuffak® from Altumax UK, a Division of Atoglas, the Atofina Subsidiary.) [10]

Solvent cementing is the common method for bonding this resin to itself, or to plastics soluble in the same solvents. The latter include cellulose acetate butyrate, acrylics, and polyurethane. Adhesive bonding, however, must be used in bonding to metals, wood, rubber, and most other plastics. Polycarbonate is hygroscopic, like nylon. For this reason, it is important to keep humidity low before bonding.

In surface preparation of polycarbonate for adhesive bonding, the customary cleaning solvents include methyl alcohol, isopropyl alcohol, petroleum ether, heptane, VM&P naphtha, and white kerosene. Ketones, toluol, and benzene

TABLE 8.6 Composition of Etching Solution for Polyaryl Sulfone

Solution Component	Amount (%)
Sulfuric acid, concentrated	96.6
Sodium dichromate	3.4

should not be used since polycarbonate is incompatible with these solvents; they cause crazing or cracking. A number of other cleaning solvents, including paint thinners, are similar to ketones in that they cause grazing or cracking, so care should be used in the selection of solvents. Light solutions (1%) of detergents, such as alkanox or Joy® are also suggested [10,17].

8.2.16.1 *Flame Treatment* [18]

1. Wipe with ethyl alcohol to remove dirt and grease.
2. Pass the part through the oxidizing portions of the flame of a BernzOmatic propane torch (<www.bernzomatic.com>).
3. Treatment is complete when both sides are polished to a high gloss, and free of scratches and visible flaws. The process usually requires 5–6 passes on both sides.
4. Allow the part to cool for 5–10 minutes before bonding.

8.2.16.2 *Abrasion* [13]

1. Wipe with ethyl alcohol, methyl alcohol, or other acceptable solvent or cleaner mentioned above.
2. Air-dry.
3. Sand with fine-grit (120-grit, 400-grit maximum) abrasive cloth or sandpaper.
4. Remove sanding dust with a clean dry cloth or a short-hair stiff brush.
5. Repeat solvent wipe.

8.2.16.3 *General Electric Procedure for Use with RTV Silicone Adhesives* [19]

RTV (room-temperature-vulcanized) silicone adhesives are effective for applications requiring excellent bond strength, a high service temperature, and good thermal expansion. Silicone adhesives may be used with all standard polycarbonate resins, including glass-reinforced grades. For optimum performance, the following procedure should be applied:

1. Lightly abrade mating surfaces with fine emery.
2. Clean surfaces of grease or foreign matter with methyl alcohol or other compatible cleaning solvent.
3. Prime abraded surfaces with GE Silicone Primer SS-4004 and allow a minimum of one-hour drying time. Unprimed surfaces significantly reduce bond strength.
4. Apply silicone adhesive in desired thickness. The final bond thickness may range from 0.127 mm to 0.762 mm.
5. Assemble.

8.2.17 Polychlorotrifluoroethylene (PCTFE) [20]

This resin is a fluoropolymer with excellent chemical resistance. See Ch. 6 for more information about the treatment methods for PCTFE.

8.2.18 Polyester (Saturated) [10]

Thermoplastic polyesters are available as polyethylene terephthalate (PET), used primarily in films (Mylar® and Melinex® by DuPont Co.) or in fibers (Dacron® by Koch Industries, and Hostaphan® by Mitsubishi Polyester Film GmbH), and polybutylene terephthalate (PBT), used in molding (Valox® by GE Plastics Co., <www.Geplastics.com>, Celanex® from Ticona Corp., Florence, KY, and Tenite® by Eastman Chemicals, <www.eastman.com>). Glass-filled grades of these resins are also available. Thermoplastic polyesters are generally resistant to solvents and chemicals, including alcohol, ethers, and aliphatic hydrocarbons, most chlorinated hydrocarbons, aqueous salt solutions, and aqueous acid and basic solutions under specified conditions. Specific surface preparations for the above thermoplastic polyesters are described below.

8.2.18.1 Treatments for Valox® Thermoplastic Polyester [21,22]

1. Abrade-degrease:
   a. Sand lightly with 240-grit sandpaper.
   b. Degrease with toluene, isopropyl alcohol, petroleum naphtha, heavy Terpene, and toluene which is a common solvent for the unreinforced Valox®.
2. Plasma treatment [23]: Activated gas plasma, using oxygen, argon, or water vapor plasma, has produced bonds 3 to 4 times stronger than those obtained with the abrasion method.

8.2.18.2 Treatment (Abrade-Degrease) for Celanex® Thermoplastic Polyester [22]

1. Sandpaper.
2. Degrease by wiping with a solvent such as acetone.

8.2.18.3 Treatment for Tenite® Cellulosic Plastics [24]

Examples include Tenite® cellulose acetate, butyrate and propionate. Acetate based resin can be cleaned with a mixture of 70% acetone/30% ethyl lactate or 80% acetone/20% Eastman PM acetate. Butyrate and propionate resins may be cleaned with 80% acetone/20% Eastman PM acetate, 70% isopropyl alcohol/30% toluene, and 95% isopropyl alcohol/5% Eastman PM acetate [25].

8.2.19 Polyetheretherketone (PEEK) [26]

This material is available from Victrex plc. PEEK is an aromatic polyether, high-performance thermoplastic used in coating, insulation of high-performance wire and cable, and molded parts. Acetone has a slight crazing effect on PEEK, but neither isopropyl alcohol or toluene have any effect, and may be used in cleaning.

Surfaces to be bonded should be clean, dry, and free from grease. This may be achieved by wiping the joint surfaces with a degreasing solvent. Excellent resistance of PEEK to hydrocarbons provides a choice from a broad list of solvents. In addition, roughening of the surfaces with an abrasive will form a more reliable bond. Degreasing should be repeated after abrading. Bond strengths may be further increased by more vigorous surface activation procedures – in particular, flame treatment with a blue (oxidizing) flame, or treatment with a chromic acid etching solution. Table 8.7 shows results obtained with these various treatments when bonding PEEK to aluminum with Araldite® 2010 (Huntsman Corp.), a general-purpose epoxy adhesive.

8.2.20 Polyethersulfone (PES) (Victrex plc) [27]

This is another high-performance engineering thermoplastic. Methyl alcohol, ethyl alcohol, isopropanol, and low-boiling petroleum ether are solvents that may be used for cleaning. Acetone, methyl ethyl ketone, tetrahydrofuran, toluene, and methylene chloride should not be used since they have an adverse effect on the plastics. Polyethersulfone can be bonded to itself by solvent cementing adhesives or ultrasonic welding. Mating surfaces should be degreased with non-polar hydrocarbon solvent before adhesive bonding [28].

8.2.21 Polyetherimide (Ultem® from GE Plastics) [28]

Polyetherimide is an amorphous engineering thermoplastic offered by GE Plastics. It has a number of desirable properties including a high heat-distortion temperature. Unlike most amorphous resins, polyetherimide is

TABLE 8.7 Bond Strength* of PEEK vs. Treatment Method

Surface Treatment	Bond Strength (kg)
Abrading + degreasing	40
Abrading + degreasing + flame oxidation	48
Abrading + degreasing + chromic acid etch	37

**Araldite® epoxy adhesive, cured 17 hours at 40°C; lap joint 13 mm wide with 13 mm overlap. Total bond area 170 mm^2; substrate aluminum. Testing method: tested shear at 20 mm/min.*

resistant to a wide range of chemicals including most hydrocarbons, alcohols, and fully halogenated solvents. Only partially halogenated hydrocarbons such as methylene chloride and trichloroethane will dissolve polyetherimide, and these compounds should not be used for cleaning. (Both solvents have been banned.)

8.2.22 Polyethylene (PE) [10]

Polyethylene has a nonpolar, nonporous, and inert surface. For this reason, adhesives cannot link chemically or mechanically to untreated polyethylene surfaces. Since polyethylene is relatively inert to most solvents, solvent cementing cannot be used. For bonding to itself or to other materials, an adhesive with a suitable surface preparation method must be used. There are a number of suitable surface treatment techniques including chemical, electronic, flame, and primer methods. Oxidation treatments are the most successful. These include immersion in a chromic acid solution, exposure to corona discharge, flame oxidation, immersion in an aqueous solution of chlorine, or exposure to chlorine gas in the presence of ultraviolet light. The chromic acid oxidation method is probably the most convenient for use with molded plastic parts of diverse contour.

8.2.22.1 Chromic Acid Etch [10]

1. Wipe with acetone, MEK, or xylene.
2. Immerse in chromic acid solution (Table 8.8) for 60–90 minutes at room temperature, or 30–60 seconds at 71°C. (Dissolve the potassium dichromate in clean tap water, and then add the sulfuric acid in increments of about 200 g, stirring after each addition.)
3. Rinse in running tap water for at least 3 minutes.
4. Dry in an air-circulating oven at 38°C for approximately 1 hour.

8.2.22.2 Oxidizing Flame Method [17]

This method utilizes an oxyacetylene burner, which is passed over the faying surface until it appears glossy. To make certain that excessive oxide is not

TABLE 8.8 Composition of Etching Solution for Polyethylene

Solution Component	Parts by Weight
Sulfuric acid, concentrated	300
Potassium dichromate	15
Water	24

formed on the surface, it should be scoured lightly with soap and water. The surface should then be washed with distilled water and dried at room temperature. Flame treatment is fast and provides bond strength greater than the chromic acid treatment described above. However, it requires very careful control to prevent heat warpage. The procedure is safest with pieces of thick cross section. See Chapter 6 for details about flame treatment.

8.2.22.3 Gas Plasma Treatment [17]

This treatment is very effective and recommended for use with very complex geometric surfaces that cannot be sanded or flame-treated adequately, and where very strong bonds are required. See Chapter 6 for details about plasma treatment.

8.2.23 Polymethylmethacrylate (PMMA) (Plexiglas® – Arkema Corporation)

In bonding PMMA to itself, solvent cementing is ordinarily used. However, when bonding PMMA to a dissimilar substrate, conventional adhesive bonding must be used. Surface treatment is as follows [10]:

1. Wipe with methanol, acetone, MEK, isopropanol, or detergent.
2. Abrade with fine grit (180–400 grit) sandpaper, or use abrasive scouring with small amounts of water, dry grit blasting, or wet abrasive blasting.
3. Wipe with a clean, dry cloth to remove particles.
4. Repeat solvent wipe.

8.2.24 Polymethylpentene (TPX®) [29]

This material is now supplied by Mitsui Petrochemical Industries, Ltd. It is a copolymer of 4-methylpentene and useful up to 204°C. Chemical behavior is similar to that of other polyolefins. It is attacked by strong oxidizing agents. Some light hydrocarbons and chlorinated solvents can cause swelling and subsequent strength loss. Surface preparation methods include:

8.2.24.1 Degreasing-Abrasion (for General-Purpose Bonding) [9]

1. Solvent-clean in acetone.
2. Grit- or vapor-blast, or use 100-grit emery cloth.
3. Solvent-clean again in acetone.

8.2.24.2 Chromic Acid Etch [9]

1. Solvent-clean in acetone.
2. Immerse for 1 hour at 60°C (Table 8.9).
3. Rinse in tap water.
4. Rinse in distilled water.
5. Dry in warm air.

TABLE 8.9 Composition of Etching Solution for Polymethylpentene

Solution Component	Parts by Weight
Sulfuric acid, concentrated	26
Potassium dichromate	3
Water	11

8.2.24.3 Acid Permanganate Etch [9]

1. Solvent clean in acetone.
2. Immerse for 5–10 minutes at 90°C in a saturated solution of potassium permanganate acidified with sulfuric acid.
3. Rinse in tap water.
4. Rinse in distilled water.
5. Dry in warm air.

8.2.25 Polyphenylene Sulfide (PPS) (Ryton® – Chevron Phillips Chemical Co.)

This crystalline polymer has a high melting point (288°C), outstanding chemical resistance, thermal stability, and flammability resistance. It has no known solvents below 191°C–204°C. The surface may be prepared as follows:

8.2.25.1 Solvent-Sandblast-Solvent [30]

1. Solvent degrease in acetone.
2. Sandblast.
3. Solvent degrease again in acetone.

8.2.25.2 Picatinny Arsenal Method [31]

1. Wipe the faying surfaces with ethanol-soaked lint-free paper.
2. Sand with 120-grit sandpaper.
3. Clean off the dust with a stiff, bristled brush.

8.2.26 Polypropylene (PP)

This polyolefin is similar to polyethylene, and may be treated using the surface treatment method for polyethylene. Thermal treatment, however, is somewhat more rigorous than prescribed for polyethylene. While the treatment for polyethylene is 60–90 minutes at room temperature, or 30–60 seconds at 71°C, polypropylene should be treated for 1–2 minutes at 71°C.

8.2.26.1 Chromic Acid Etch [6,30]

1. Wipe with acetone, MEK, or xylene.
2. Immerse in the chromic acid solution (Table 8.10) for 1–2 minutes at 71°C. (Dissolve the potassium dichromate in clean tap water; then add the sulfuric acid in increments of about 200 g, stirring after each addition.)
3. Rinse in running tap water for at least 3 minutes.
4. Rinse in distilled or deionized water.
5. Dry in an air-circulating oven at 38°C for about 1 hour.

8.2.26.2 Oxidizing Flame Method [17]

Same as for polyethylene.

8.2.26.3 Gas Plasma Method [17]

Same as for polyethylene.

8.2.27 Polystyrene (PS) [17]

This material is ordinarily bonded to itself by solvent cementing, although conventional adhesive bonding, thermal welding, and electromagnetic bonding have been successfully tried. Conventional adhesive bonding may be necessary for bonding polystyrene to a dissimilar material. Surface preparation methods are as follows:

8.2.27.1 Abrading, Sanding [14]

1. Degrease with methyl or isopropyl alcohol.
2. Abrade with 200-grit sandpaper and remove dust particles.

8.2.27.2 Sodium Dichromate–Sulfuric Acid Process [14]

1. Degrease with isopropyl or methyl alcohol.
2. Immerse for 3–4 minutes in the solution (Table 8.11) maintained at 99–104°C.
3. Rinse thoroughly with distilled water.
4. Dry below 49°C.

TABLE 8.10 Composition of Etching Solution for Polypropylene

Solution Component	Parts by Weight
Sulfuric acid, concentrated	300
Potassium dichromate	24
Water	24

Note: This is a somewhat unusual modification of typical chromic-acid etching processes. Since no water is used to dilute the acid, a much higher immersion temperature may be used without boiling off the water. The usual high-temperature immersion in the conventional chromic acid solution is about 71°C, while 99–104°C is used here.

8.2.27.3 Non-Immersion Etch Process [14]

If polystyrene parts are to be used in high-frequency electrical applications, it may be desirable that only the faying surfaces be treated as follows:

1. Degrease with methyl or isopropyl alcohol.
2. Apply to the faying surfaces a thixotropic paste (Table 8.12).
3. Heat parts to 82°C and hold for a few minutes with the paste on the surface.
4. Rinse thoroughly with distilled water.
5. Dry below 66°C.

8.2.28 Polysulfone (Udel® – Solvay Advance Polymers) [17]

Polysulfone is a thermoplastic resin with very high strength and one of the highest service temperatures (171°C) of any melt-processible thermoplastic. It is highly resistant to mineral acids, alkalis, salts, detergent solutions, oils, and alcohol. On the other hand, it is attacked by polar organic solvents such as ketones, chlorinated hydrocarbons, and aromatic hydrocarbons. Polysulfone stress-cracks easily and is considered notch-sensitive. It can be solvent

TABLE 8.11 Composition of Etching Solution for Polystyrene

Solution Component	Parts by Weight
Sulfuric acid, concentrated	90
Sodium dichromate	10

TABLE 8.12 Composition of Etching Paste for Polystyrene*

Solution Component	Parts by Weight
Sulfuric acid, concentrated	3
Powdered Potassium	1

**Add Cab-O-Sil (Cabot) fused silica earth as required to obtain a thixotropic paste.*

cemented, conventionally adhesive bonded, or ultrasonically welded. Air Studies have evaluated three surface-preparation techniques: (1) sandblast; (2) acid-etch; and (3) solvent-wash. The acid-etch procedure was found to give the best results. (For details, see procedures for polyaryl sulfone in Sec. 8.2.15.)

Another method, using a grit-blast and ultrasonic cleaning, was used by General Electric. In this method, polysulfone parts were prepared for bonding to niobium with AF-42 epoxy-nylon film adhesive (3M Co). The procedure is as follows [17]:

1. Degrease in alcohol.
2. Grit-blast with 27–50 micron aluminum oxide (Al_2O_3). Use S.S. White Airbrasive® Equipment or equivalent (available from S.S. White Industrial Products, Piscataway, NJ).
3. Clean in an ultrasonic cleaner as follows:
   - Immerse parts in Neutra-Clean® 7 (a trademark of AMREP Inc., Marietta, GA, <www.amrep.com>) (90 g/l).
   - Rinse with tap water.
   - Rinse with distilled water.
   - Rinse with isopropyl alcohol for at least 30 seconds.
   - Flash with dry nitrogen.
   - Dry with warm air at 66°C maximum.

Other methods include vapor degreasing in methanol, sanding, or vapor blasting, and alcohol-wiping and/or light sanding, plus vinyl, urethane, or silicone primer.

8.2.29 Polytetrafluoroethylene (PTFE) [20]

PTFE does not melt like a true thermoplastic, but "sinters." Like other fluoroplastics, PTFE is relatively inert chemically. It is prepared for bonding by the techniques described in Ch. 6.

8.2.30 Polyvinyl Chloride (PVC)

The procedure given here is intended for use with the homopolymer rigid compounds. These compounds contain up to 5% plasticizer, making them difficult to bond with epoxy and other non-rubber types of adhesives. The homopolymer is not readily soluble, making it difficult to bond by solvent-cementing techniques. Thermal welding, especially hot-gas welding, is commonly used in joining rigid PVC. In general, a solvent is always used to remove plasticizer, grease, and dirt. Abrasion may or may not be used. If it is, sandpaper (various grits), vapor blasting, steel wool, or scouring powder is used. The following procedure is suggested [10].

1. Wipe with solvent, such as methanol, low-boiling petroleum ether, MEK, toluene.
2. Abrade with medium-grit (200-grit) sandpaper.
3. Blow off dust.
4. Repeat solvent-wipe.
5. Dry.
6. For maximum strength, prime with nitrile-phenolic adhesive, or bond immediately.

8.2.31 Polyvinyl Fluoride [12]

DuPont has three types of Tedlar® film: Type A, with one side treated for bonding, Type B, with both sides bondable, and Type S, which is untreated and used as a release film. Type B is used in laminating to metals, plastics, wood and other materials; it requires no further surface preparation for adhesive bonding. The methods for preparing the untreated film for adhesive bonding are similar to those of polyolefins and fluoropolymers (see Ch. 6).

8.2.32 Polyvinylidene Fluoride (PVDF) [12]

Polyvinylidene fluoride is a high-performance, high-molecular-weight homopolymer. It is resistant to most acids and bases, aliphatics, aromatics, alcohols, and chlorinated solvents. Strongly polar solvents such as ketones and esters cause partial solvation, especially at high temperatures. The methods for preparing the untreated film for adhesive bonding are similar to those of polyolefin and fluoropolymers (see Ch. 6).

8.2.33 Styrene–Acrylonitrile (SAN) (Tyril® – Dow Chemical Co.) [10]

This copolymer is usually solvent-cemented with solvents similar to those used for polystyrene, although effective solvents are more limited. In applications where solvent cements cannot be used, as in bonding to metals, the procedures suggested for polystyrene may be used [10]. Gasoline has also been used with no further treatment [23].

8.3 THERMOSETS

Most thermoset plastics are relatively easy to bond. Solvent cements obviously cannot be used because these materials are not soluble. In some cases, solvent solutions can be used to join thermoplastics to thermosets. In most cases, however, conventional adhesive bonding is the only practical non-mechanical way to join a thermoset to a thermoset or to non-plastic materials. Ultrasonic bonding can be used only with ultrasonic adhesives.

Frequently a mold-release agent is present on the surface of thermoset materials and must be removed before adhesive bonding. Mold-release agents are usually removed after a detergent wash, solvent-wash, or solvent-wipe, followed by light sanding to break the smooth surface glaze. A final solvent-wipe with clean solvent and clean lint-free cloth or paper tissue is usually effective. Solvents include acetone, methyl ethyl ketone (MEK), toluene, low-boiling petroleum ether, and isopropyl alcohol. Fine abrasive paper (sand, carborundum, or aluminum oxide-abrasive grit/metal, sand or oxide), metal wool (steel, aluminum, or brass), or steel shot are frequently used for abrasion of thermoset surfaces [17].

8.3.1 Diallyl Phthalate (DAP) [14]

This resin has exceptional electrical insulating properties, high-temperature stability, and strong resistance to most chemicals and moisture. Surfaces are hard and tough and pick up very little moisture. DAP parts are ordinarily molded or laminates with glass fabric and are not easy to bond with an adhesive. Sanding or buffing appear to be the best surface treatments. Only filled resins are available. Besides glass, fillers include minerals, Orlon® (acrylic fiber, originally developed by DuPont Co.), and asbestos. Surface preparation is as follows:

1. Wipe with acetone, MEK, toluene, low-boiling petroleum ether, or isopropyl alcohol.
2. Sand-, grit-, or vapor-blast or use steel wool.
3. Wipe with a clean dry cloth to remove grit and particles.
4. Repeat solvent-wipe in step 1.

8.3.2 Epoxies [14]

Epoxy resins are among the workhorses of thermosetting resins. They have a wide range of properties. They possess good dimensional stability and good electrical properties and mechanical strength. They have good creep resistance and are useful over a wide temperature range (−73°C to 427°C). These resins are easy to bond, requiring only a clean dry surface, usually solvent-cleaned and sanded. Both filled and unfilled grades are available. Fillers include minerals, glass, and glass microballoons. Epoxy laminates should be treated as covered in Sec. 8.5 on reinforced thermosets (laminates). A recommended surface treatment is as follows:

1. Wipe with acetone or other suitable solvent (see Sec. 8.3.1 diallyl phthalate) or vapor degrease.
2. Abrade by scouring or other treatment suggested in Section 8.3, "Thermosets."
3. Wipe with a clean dry cloth to remove grit.

4. Repeat solvent treatment in 1.
5. Apply a silane-type primer, such as Chemlok® 607 (Lord Corporation), diluted with 5–10 volumes of methanol or ethanol. This step is optional.

8.3.3 Melamine–Formaldehyde (Melamines) [14]

The amino resins like urea-formaldehyde are hard, rigid, and abrasion-resistant. They are self-extinguishing and have superior electrical properties as well as excellent dimensional stability and good creep resistance. They are noted for their high impact strength and resistance to water and solvents. Only filled resins are available. The fillers include cellulose, rag, and glass. A suggested surface preparation is as follows:

1. Scrub with an abrasive household detergent.
2. Rinse with tap water, then deionized water.
3. Dry.
4. Sand.
5. Wipe with isopropyl alcohol.
6. Dry.
7. Prime or bond.

The procedure for epoxies in Section 8.3.2 may also be used.

8.3.4 Phenol–Formaldehyde (Phenolics) [14]

Phenolics have an excellent combination of high physical strength and high-temperature resistance, good electrical properties, and good dimensional stability. Phenolics are widely bonded, not only on molded parts, but also as laminates and castings. Both filled and unfilled types are available. Fillers include asbestos, cellulose, wood, flour, and glass. Methods of surface preparation given above under "Epoxies" is suggested.

8.3.5 Polyester [14]

These unsaturated, nonlinear resins are similar to epoxies and phenolics in their surface-preparation requirements, basically requiring only sanding for good results. Polyesters have good resistance to oils and solvents. The method of surface preparation given for epoxies in Section 8.3.2 is suggested.

8.3.6 Polyimide [9,32,33]

This is a class of polymeric materials distinguished by exceptional thermal stability. They are available commercially in several forms, including fabricated parts and shapes, molding resins, films, and coatings for wire and fabrics. Some polyimides, condensation polymers, are essentially linear, with structures similar to those of thermoplastics. Others, addition polymers used as engineering resins in parts (moldings and laminates), are thermosetting. DuPont's Vespel® precision parts are available in graphite-PTFE- and MoS_2-filled bearing compositions, as well as unfilled form. These parts can withstand temperatures up to 482°C for short periods. A typical surface preparation method for Vespel® parts follows:

8.3.6.1 Vespel® Parts

1. Remove surface contamination such as dirt and oil with solvents. Clean by refluxing in perchloroethylene, or clean ultrasonically in chlorinated solvent.
2. Mechanically abrade with a wet or dry abrasive blast. Use light abrading (approx. 50–100 microinches) to maintain a uniform adhesive thickness. This type of abrading can be obtained with an air- or air-liquid abrasive blast.
3. Remove residual with the solvent in step 1.
4. Dry.

8.3.6.2 Sodium Hydroxide Etch [9]

1. Degrease in acetone.
2. Etch for 1 minute at 60°C–90°C (Table 8.13).
3. Rinse in cold water.
4. Dry in hot air.

8.3.7 Polyurethane [14]

This resin, like polyimides, is sometimes thermoplastic. It is formed from the reaction of a polyisocyanate with compounds containing a reactive hydrogen. The plastic may be either flexible or rigid, and can be obtained in different densities and forms (sheets, molding and casting resins, etc.). Polyurethanes are unsurpassed as cryogenic materials and have excellent adhesion and skid resistance, good chemical resistance, and superior impact resistance, but are limited at elevated temperatures (to 121°C). Polyurethanes, or urethanes as they are frequently called, also have good electrical properties. Cleaning usually

TABLE 8.13 Composition of Etching Solution for Polyimide

Solution Component	Parts by Weight
Sodium hydroxide	5
Water	95

involves light sanding and a dry bonding surface. A primer (urethane or silane) will usually improve adhesion. Polyurethanes are considered to have better cohesive than adhesive properties, which is why the interface may be the "weak link." A suggested surface preparation is as follows:

1. Wipe with acetone or methyl ethyl ketone.
2. Abrade with 100-grit emery cloth, sandpaper, or steel wool.
3. Wipe with a clean dry cloth to remove grit and particles.
4. Wipe again with acetone or methyl ethyl ketone.
5. Dry.
6. Use urethane or silicone primer to improve adhesion.

8.3.8 Silicone Resins [8,34]

Silicone resins are available in several forms. They are used as thermally stable electrical insulation resins, paint vehicles, molding compounds, laminates, impregnating varnishes, encapsulating materials, and as baked-on release agents. Recommended surface preparation is as follows:

1. Wipe with acetone or other solvent listed in the general discussion for thermosets in Section 8.3.
2. Sand-, grit-, or vapor-blast, or abrade with steel wool.
3. Wipe with a clean dry cloth to remove grit and particles.
4. Repeat solvent wipe in step 1.

Silicone laminates should be treated according to Section 8.4, "Reinforced Plastics/Thermosets."

8.3.9 Urea–Formaldehyde (U-F) [8,14]

Like melamine-formaldehyde resins, these amino resins are hard, rigid, abrasion-resistant materials. They have excellent dimensional stability and good solvent and creep resistance. They are self-extinguishing and have superior electrical properties. They are noted for high impact strength and resistance to water and solvents. Only cellulose-filled resins are available. The following surface preparation method is used:

1. Scrub with an abrasive household detergent.
2. Rinse with tap water, then deionized water.
3. Dry.
4. Sand.
5. Wipe with isopropyl alcohol.
6. Dry.
7. Prime or bond.

The procedure given in Section 8.3.2 for epoxies may be used.

8.4 REINFORCED PLASTICS/THERMOSETS [10]

The reinforced thermosets of most importance are the glass-reinforced plastics (GRP). The methods often used in preparing GPR laminates for bonding involve removing the original resin from the surface, yielding a roughened faying surface for bonding. Removal of surface gloss is usually sufficient to dispose of surface soils, exuded resin impurities, absorbed gases of vapors, and release agents used during the manufacture of the GPR. The methods of surface preparation most commonly used for laminated plastics are provided below.

8.4.1 Tear-Ply (Peel-Ply) Method [10]

During the manufacture of the GPR laminate, either one or both outer surfaces of the laminate are made of a layer of a weakly adhering material, the outer layer of which can later be stripped or peeled off easily. Nylon and Dacron® are frequently used for this purpose. This ply is cured as an integral part of the laminate. The resin used in the construction of the GPR adheres poorly to the nylon or Dacron cloth incorporated in the outer layer or layers, permitting the nylon or Dacron cloth to be peeled uniformly in one piece. The texture of the nylon or Dacron fabric layer is reproduced in the outer layer of the laminate. When the adhesive-bonding operation is to be performed, a thin knife is slid under a ply, and the desired distance peeled and trimmed. Loose particles are removed by brushing or blowing the surface with clean filtered air. There is very little chance of surface contamination with tear ply. The tear ply is applied only to the areas that are to be adhesive bonded. This technique is preferred over machine- or handsanding due to the reduced danger of surface contamination.

8.4.2 Sanding [10]

The surface of laminated plastics may be sanded lightly with medium-grit (80 to 120 grit) emery paper or other abrasive paper capable of roughening the surface without substantially damaging the reinforcing fibers. It is desirable to have the direction of bonding parallel to the surface fiber orientation in order to minimize damage to these fibers. The technique works best on large faying surfaces where the edges, rounded as a result of sanding, can be trimmed, or where the edge effects are negligible. Sanding dust may be removed with a short-hair stiff brush. If water-break-free surfaces are not obtained, the procedure should be repeated.

8.4.3 Grit-Blasting

Grit-blasting or sandblasting, if available, may also be used. A specially-trained operator is needed to produce uniformity within pieces by sandblasting. A clean, uniform-size grit is essential for proper surface preparation with sandblasting. This may be a problem when treating GPR surfaces, and for that reason this technique is rarely used.

After sanding or sandblasting, the surface is sometimes wiped with solvent such as MEK, acetone, or toluene, depending on the known mold lubricants. In some cases, a solvent is used before and after abrading the surface. If water-break-free surfaces are not obtained, the procedure should be repeated.

Glass-reinforced plastic laminates prepared for bonding by hand- and machine-sanding can be stored to 30 days at 23°C and 50% RH with no adverse effect on bond strength. Machine sanding has given slightly better results than hand sanding. Tear ply and sanding gave about the same results, but the tear-ply method has less risk of surface contamination. Variations in bond strength are more likely to occur as a result of changes in sanding techniques than by the difference in methods.

In general, bond quality diminished with increased surface exposure time (SET). The best overall adhesive evaluated was epoxy film adhesive, which was found to be the least sensitive to the method of surface preparation. In general, the best result was obtained when GPR laminates were bonded within four hours after sanding. If absolutely necessary, bonding can be carried out after periods of time up to 14 days SET with only moderate strength loss.

8.4.4 Manual Scouring [35]

In this method, the faying surface is scrubbed with tap water and an abrasive household-type cleaner to remove contaminants and release agents. Clean cloths or nonmetallic bristle brushes will facilitate the scouring. The surface is rinsed with clean running tap water followed by a rinse in distilled water, and then dried at 54–66°C. The parts should exhibit a water-break-free surface, otherwise the above procedure should be repeated.

8.4.5 Solvent-Soak and Abrading [35]

If water-break-free surfaces cannot be obtained by tear-ply, sandblasting, or manual scouring, the plastic laminates should be treated as follows:

1. Soak for 48 hours in reagent-grade acetone.
2. Dry for 3–4 hours at 88°C–104°C.
3. Sand lightly with 200-grit sandpaper.
4. Remove particles by air-blast or vacuum.
5. Check for water-break-free condition. If achieved, the parts may be bonded. If not, resoak the laminates in acetone for an additional 24 hours minimum and repeat the rest of the procedure. Parts that do not pass the water-break-free test should be rejected.

8.5 REINFORCED THERMOPLASTICS (GLASS-REINFORCED) [10]

Surface preparation of glass-reinforced thermoplastics is not necessarily similar to the technique prescribed for the base unreinforced resin. In general, there is a drop of up to 50% in the expected bond strength with glass reinforcement. The reason for this drop is not fully known, although it may be the result of the finishes applied to glass during its manufacture. In many cases, solvent cementing will be effective. Examples are aqueous phenol for glass-reinforced nylon 6 and acetone for polystyrene.

8.6 PLASTIC FOAMS [8]

Plastic foams are generally treated by the same surface-preparation methods used for the base polymer. This is particularly true for the thermoplastic "structural" foams such as polycarbonate and modified polyphenylene oxide. In some cases, these thermoplastic foams may be solvent cemented. The thermosetting foam most commonly encountered is polyurethane, which is available in rigid, semi-rigid, and flexible form. The rigid foams are obtainable in a wide range of densities. Obviously, the denser grades, with their increased surface area of actual material exposed, will be easier to prepare by the usual technique of sanding lightly and then removing the resulting dust by vacuum or by blowing-off. The cellular structure of plastic foams offers an advantage in that a physical interlocking of the two faying surfaces is possible in a foam-to-foam bond. With rigid polyurethane foams, a conventional adhesive bond may not be necessary, since foam "poured-in-place" or "foamed-in-place" will adhere tenaciously to most substrates during curing.

8.7 SURFACE PREPARATION OF RUBBERS

Almost all types of elastomers can be bonded to a large variety of substrates by the process known as vulcanization. The strength of similar bonds by adhesive bonding, or post-vulcanization (PV) bonding, as it is also called, was originally believed to be much weaker and less resistant to environmental exposure. Published work has discredited this theory, with some exceptions. The advantages of PV bonding are listed [36].

- A variety of vulcanized elastomers can be bonded.
- Common elastomer surface treatments can be used.
- Bonded systems possess good environmental resistance.
- Bonds are comparable to vulcanization bonding.
- The process is applicable to rubber-to-metal parts as well as to rubber-to-rubber assemblies.
- Large or complicated shapes that may (not be economically feasible by the "bond-in-the-press" technique) may be PV bonded.

- Vulcanization bonding is impractical for field repair where the required equipment is usually unavailable.

Table 8.14 shows bond values obtained with seven elastomers, comparing the bond values obtained with both vulcanization bonding and post-vulcanization (adhesive) bonding, using a 149°C cure. In most cases the results are comparable. EPDM was the only elastomer where a significant decrease in bond strength was obtained by post-vulcanization bonding.

Solvent washing and abrading are common surface treatment methods for most elastomers, but chemical treatment is often applied for maximum strength and other properties. Many synthetic and natural rubbers require "cyclizing" with concentrated sulfuric acid until hairline cracks appear on the surface. Some rubbers require primers for optimum bonding.

Mechanical abrasion is usually accompanied by sanding or buffing the surface with 80–140 grit sandpaper or a buffing wheel. The dust from the buffing process, or other contamination, is usually removed by means of a clean cloth dipped in a suitable solvent. The solvent must be reasonably compatible with the chemical type or rubber being cleaned. If the solvent is strongly incompatible with the rubber involved, or if too much is used, the rubber may swell excessively, curl unacceptably, or may be degraded. Particular care must be taken not to trap solvent in a system that is totally closed, or else the rubber may be damaged. On the other hand, a mild wipe of a somewhat aggressive solvent may help tackify the rubber surface. Methyl ethyl ketone (MEK) and toluene are solvents commonly used for cleaning elastomers. If such curling occurs, a more compatible solvent, such as toluene, may be substituted. Cyclization, mentioned above, and chlorination are

TABLE 8.14 Comparison of Vulcanization and Post-Vulcanization Bonding*[34]

Material	Tear Strength-Peel (N/m)	
	Vulcanization	Post-Vulcanization
Natural rubber	7880	7529
Natural rubber (B)	11,132	9806
SBR	33,094	17,513
Neoprene	16,284	20,312
Butyl rubber	17,685	11,623
Nitrile rubber	16,635	2814
EPDM	22,763	4903

**Adhesive used was Chemlok® 234B (LORD Corporation).*

other common surface-preparation methods and will be described in the next section on neoprene (Sec. 8.7.1).

Research has shown that the bondability of elastomers is a function of polarity. The less polar EPDM and butyl elastomers are more difficult to bond than the more polar nitrile, neoprene, SBR, and natural rubber [36].

8.7.1 Neoprene (Polychloroprene) (CR) [10]

Neoprene is the generic name for polymers of chloroprene (2-chloro-1,3-butadiene) first manufactured by DuPont in 1931. There are many types of solid and latex neoprene.

8.7.1.1 Abrasive Treatment [8,10]

1. Scrape the surface with a sharp blade to remove layers of wax, sulfur, and other compounding ingredients which may have floated to the surface.
2. Solvent-wipe with ethyl, isopropyl, or methyl alcohol, MEK, or toluene.
3. Uniformly abrade surfaces with 80–120-grit abrasive paper. Machine sanding with a "jitterbug" oscillating sander is preferred over hand-sanding because the machine sanding produces a more uniform surface with less effort. A buffing wheel may also be used.
4. Solvent-wipe again as in step 2 to remove particles.

8.7.1.2 Cyclization [8,37]

This process is often applied to prepare for bonding with flexiblized epoxy adhesives. It has been used successfully in many rubbers, in addition to neoprene. A commonly used cyclization procedure is as follows:

1. Scrape surface with a sharp blade to remove layers of wax, sulfur, and other compounding ingredients which may have floated to the surface.
2. Solvent-wipe with ethyl, isopropyl, or methyl alcohol, MEK, or toluene.
3. Immerse rubber surface in concentrated sulfuric acid (specific gravity 1.84) for 5–45 minutes. The optimum time must be determined experimentally.
4. Rinse thoroughly with tap water, preferably hot.
5. Rinse thoroughly with distilled water.
6. Neutralize by immersing for 5–10 minutes in a 10%–20% solution of ammonium hydroxide (sodium hydroxide is also used).
7. Rinse thoroughly with tap water.
8. Rinse thoroughly with distilled water.
9. Dry at temperatures up to 150°C.
10. Flex the resultant brittle surface of the rubber with clean rubber or plastic gloves until a finely-cracked appearance is produced.

Cyclizing hardens and slightly oxidizes the surface, thereby permitting the necessary wetting of the adhesive bond. The finely cracked surface indicates that the rubber is ready for bonding. Light lacy lines on the surface indicate insufficient immersion time. Deep, coarse cracks and a thick crusty surface indicate excessive immersion. If immersion is not feasible, the acid may be made up into a thick paste by the addition of barium sulfate (barytes) or Cab-O-Sil® (Cabot Corp., Boston, MA) or use 100 part by weight (pbw) concentrated sulfuric acid: 5 pbw Santocel® C (Monsanto Co.), or equivalent. A stainless steel or other acid-resistant spatula should be used to apply the paste. After the paste is applied, the remaining steps of the treatment (4 through 8) should be carried out.

8.7.1.3 Cyclization (Boeing-Vertol Modification) [38]

Boeing-Vertol has developed a much milder method, as follows:

1. Clean in toluene.
2. Force-dry 1 hour at 60°C.
3. Immerse 2 minutes in concentrated sulfuric acid.
4. Rinse in tap water.
5. Force-dry 1 hour at 60°C.

8.7.1.4 Chlorination [10]

1. Scrub neoprene in 49°C–60°C nonionic detergent solution (2%–3% by weight).
2. Rinse thoroughly in tap water.
3. Rinse thoroughly in distilled water.
4. Air-dry.
5. Immerse 1½ to 3 minutes in the solution detailed in Table 8.15 at room temperature (prepare solution just prior to use, adding ingredients in the order listed).
6. Rinse thoroughly with distilled water and dry at temperatures up to 66°C.

TABLE 8.15 Composition of Chlorination Solution for Neoprene

Solution Component	Parts by Weight
Hydrochloric acid	0.3
Sodium hypochlorite	0.3
Water	97

8.7.1.5 Activated Gas Plasma

This method, described in Chapter 6, has been used successfully with neoprene. The best results have been obtained with ammonia, air, and nitrogen plasmas at 30 minutes' exposure [39].

8.7.2 Ethylene–Propylene–Diene Terpolymer (EPDM)

This rubber, along with the copolymer of ethylene and propylene (EPM), forms a classification called ethylene–propylene rubbers. However, since EPDM, the terpolymer, is in much greater use than EPM, the former is frequently called ethylene–propylene rubber. Its outstanding property is a high resistance to ozone and weathering. EPDM is relatively difficult to bond due to its lower polarity [17].

8.7.2.1 Abrasive Treatment [49]

As for neoprene (Sec. 8.7.1.1). Acetone and MEK are among the effective solvents.

8.7.2.2 Cyclization [10]

As for neoprene (see Sec. 8.7.1.2).

8.7.3 Silicone Rubber (Polydimethylsiloxane)

These rubbers are of completely synthetic materials with many applications. They comprise two types: heat-vulcanizing and room-temperature vulcanizing (RTV). Both types have unique properties unobtainable with organic rubbers, particularly where superior endurance and extended life expectancy are required. Silicone elastomers maintain their usefulness from −101°C to 316°C. Resistance to oxidation, oils, and chemicals is high, and stability against weathering is good [17].

8.7.3.1 Solvent Cleaning [8]

1. Sand with a medium-grit sandpaper.
2. Solvent-wipe surfaces with acetone, MEK, ethyl, methyl, or isopropyl alcohol, or toluene.

8.7.3.2 Soap-and-Water Wash [18]

A simple washing with a mild (Ivory®) soap and water, followed by thorough rinsing, has been found to give satisfactory results.

8.7.3.3 Primers [40,41]

Priming the silicone rubber with Chemlok® 607 adhesive (LORD Corporation) in methanol solvent has been found to give good results after cleaning with acetone or methanol. The primer should be dried at 200°C for 10–15 minutes. The Chemlok® 607 can also be used as the adhesive following the priming. Chemlok® silanol-type primers have also been developed. Other satisfactory silane-based primers can be obtained from Henkel Corp., Dow Corning, GE Silicones, and other suppliers. The primer should be air-dried at room temperature for 15–40 minutes before bonding.

8.7.3.4 Activated Gas Plasma [42]

This method, described generally above under "Activated Gas Surface Treatment of Polymers," has been used successfully with RTV vinyl-addition-type silicone rubber, using excited oxygen plasma. This treatment permitted the use of an epoxy adhesive for bonding. Usually, cured RTV silicones are considered unbondable by conventional adhesives such as epoxies.

8.7.4 Butyl Rubber (BR)

Butyl rubber has been commercially produced since 1942 and is a well-established specialty elastomer with a wide range of applications. It is a copolymer of isobutylene and isoprene. Its special properties are low gas permeability, thermal stability, ozone and weathering resistance, vibration damping, and relatively high coefficient of friction and chemical and moisture resistance [17]. Being less polar, butyl rubber is relatively difficult to bond.

8.7.4.1 Abrasive Treatment [2]

Similar to neoprene (Sec. 8.7.1.1); toluene is used as a solvent.

8.7.4.2 Cyclization [8]

Similar to neoprene (Sec. 8.7.1.2). Cyclization is not always recommended for butyl rubber [37,43].

8.7.4.3 Chlorination [8]

Similar to neoprene (Sec. 8.7.1.3). Sharpe [37] does not recommend chlorination for butyl rubber.

8.7.4.4 Primers

Prime with butyl rubber adhesive in an aliphatic solvent [44], or use silicone primers.

8.7.4.5 Activated Gas Plasma

This method has been used successfully with butyl rubber; good results were obtained with air and nitrogen plasmas at 30 minutes in Picatinny Arsenal studies.

8.7.5 Chlorobutyl Rubber (CIIR)

This modified butyl rubber contains about 1.2 wt% of chlorine and has greater vulcanization flexibility and enhanced cure compatibility than other general-purpose elastomers [10].

8.7.5.1 Abrasive Treatment [8]

As for neoprene (Sec. 8.7.1.1).

8.7.5.2 Other Methods [8]

See Table 8.16 for a comparison of results obtained by a number of different surface-preparation methods.

8.7.6 Chlorosulfonated Polyethylene (Hypalon® – DuPont Co.)

This synthetic rubber, also known as chlorosulfonyl polyethylene, introduced in 1952, is characterized by ozone resistance, light stability, heat resistance, weathering, resistance to deterioration by corrosive chemicals, and good oil resistance. Presently available types contain from 25% to 43% chlorine and 1.0% to 1.4% sulfur [10].

TABLE 8.16 Peel Strength of Epoxy–Polyamide Resin Adhesive to Chlorobutyl Rubber Pretreated by Various Methods [7]

Surface Treatment	N/m
Cyclization	3152
Solvent-clean and chlorinate	2452
Solvent-clean, abrade and chlorinate	2802
Detergent-clean and chlorinate	5254
Ethanol-wipe	1226
Abrade and ethanol-wipe	1226
MEK-wipe	1051
Abrade and MEK-wipe	1138

8.7.6.1 Abrasive Treatment [8]

As for neoprene (Sec. 8.7.1.1); acetone and MEK are used as solvents.

8.7.6.2 Primers [38]

1. Wipe with toluene.
2. Force-dry 1 hour at 60°C.
3. Prime with an adhesive primer such as Scotchcast® epoxy powder coating resins (by 3M).

8.7.7 Nitrile Rubber (Butadiene–Acrylonitrile) (NBR)

Nitrile rubbers are copolymers of butadiene and acrylonitrile, frequently referred to as Buna N. Their properties vary with the acrylonitrile content. Nitrile rubbers exhibit a high degree of resistance to attack by oils at both normal and elevated temperatures [10].

8.7.7.1 Abrasive Treatment [8]

As for neoprene (Sec. 8.7.1.1); methanol and toluene work as solvents.

8.7.7.2 Cyclization [8]

As for neoprene (Sec. 8.7.1.2), except that the exposure should be for 10–15 minutes.

8.7.7.3 Chlorination [8]

As for neoprene (Sec. 8.7.1.4).

8.7.7.4 Primers [9,37,44]

Primers have been used successfully with Buna N rubber. Those that have been particularly effective include:

- Chemlok® AP-131 Primer (LORD Corporation) for bonding to steel.
- Chemlok® 205 adhesive (LORD Corporation) for bonding to aluminum.
- Chemlok® 220 adhesive (LORD Corporation) for bonding to aluminum.
- Scotchcast® epoxy powder coating resins (3M) for bonding to aluminum.

Depending on the adhesive to be used, better results may be obtained by using a toluene wipe, with or without sanding, before priming. One possible procedure is as follows [10]:

1. Wipe with toluene.
2. Force-dry at 60°C.
3. Prime with an adhesive primer such as from those listed above.

8.7.7.5 *Activated Gas Plasma*

This method has been used successfully with Buna N rubber. The best results were obtained with ammonia, air, and nitrogen plasmas at 30 minutes [39].

8.7.8 Polyurethane Elastomers

This category covers thermosetting materials derived from the reaction of an isocyanate with a hydroxy compound. Their unique elastomeric properties include exceptionally high abrasion resistance, very high tear and tensile strength, and high hardness with good mechanical strength. Both polyester (AU) and polyether (EU) types are available, as are thermoplastic polyester urethane. They are usually completely soluble in strong polar solvents such as tetrahydrofuran and dimethyl formamide, and at high temperatures they can be molded, extruded, or calendered [17]. Treatments are described below.

8.7.8.1 *Abrasive Treatment*

As for neoprene (Sec. 8.7.1.1); methanol is a solvent.

8.7.8.2 *Primers* [45]

A number of primers can be used, such as Chemlok® primer AP-134, and AP-131 (LORD Corporation) or a chlorosilane supplied by GE Silicones, Rohm & Haas, or Dow Corning Corp. Methanol can be used as a degreasing agent prior to priming.

8.7.9 Synthetic Natural Rubber (Polyisoprene) (IR)

This rubber approximates the chemical composition of natural rubber (NR). Its predominant structure is *cis*-1,4-polyisoprene, the same as that of natural rubber. In general, the synthetic polyisoprenes are lower in modulus (more flexible) and higher in elongation than the natural rubber. Synthetic polyisoprene was first introduced commercially in 1960 [17]. Suggested surface treatments are as follows:

8.7.9.1 *Abrasive Treatment* [8]

As for neoprene (Sec. 8.7.1.1); methanol and isopropanol are suggested as solvents.

8.7.9.2 *Cyclization* [8]

As for neoprene (Sec. 8.7.1.2), except that exposure should be for only 5–10 minutes since this material is not acid-resistant. Some workers suggest much shorter periods (10–45 seconds).

8.7.9.3 *Chlorination* [8]

As for neoprene (Sec. 8.7.1.1).

8.7.10 Styrene–Butadiene Rubber (SBR) (Buna S)

This type of rubber, once called GR-S, is one of the most important synthetic rubbers. It is a copolymer of styrene and butadiene [17]. Treatments are as follows:

8.7.10.1 *Abrasive Treatment* [8]

As for neoprene (Sec. 8.7.1.1). Toluene is recommended as a solvent. Excessive toluene, however, causes the rubber to swell. A 20-minute drying will restore that part to its original dimensions.

8.7.10.2 *Cyclization* [8]

As for neoprene (Sec. 8.7.1.2), except that the exposure time should be 10–15 minutes.

8.7.10.3 *Chlorination* [8]

As for neoprene (Sec. 8.7.1.4).

8.7.10.4 *Primers* [8,45]

Recommended primers include butadiene-styrene adhesive in aliphatic solvent and Chemlok® 205 adhesive (LORD Corporation) in methyl isobutyl ketone (MIBK) diluent.

8.7.10.5 *Activated Gas Plasma* [39]

This method has been used somewhat successfully with Buna S rubber; best results were obtained using helium, air, and nitrogen plasmas.

8.7.11 Polybutadiene (Butadiene Rubber)

Although this type of rubber was first produced in Europe in the early 1930s, it was almost unknown in the USA until 1960 [17]. Treatments are as follows:

8.7.11.1 *Abrasive Treatment* [8]

As for neoprene (Sec. 8.7.1.1); methanol is recommended as a solvent.

8.7.11.2 *Cyclization* [8]

As for neoprene (Sec. 8.7.1.2), except that the exposure time should be 10–15 minutes since this is a non-acid-resistant rubber.

8.7.11.3 Chlorination [8]

As in neoprene (Sec. 8.7.1.4).

8.7.11.4 Solvent Wipe [9]

A simple solvent wipe with methanol may be used for general-purpose bonding.

8.7.12 Fluorosilicone Elastomers

These rubbers have become the second largest in volume of the fluoroelastomer types. The non-silicone-containing fluoroelastomers are discussed below. Fluorosilicone rubbers retain most of the useful qualities of the regular silicone rubbers and have improved resistance to most fluids, with the exception of ketones and phosphate esters. They are most useful when low-temperature resistance is required in addition to fluid resistance [17]. An effective treatment is a solvent wipe with methyl, ethyl, or isopropyl alcohol, or with toluene.

8.7.13 Fluorocarbon Elastomers

These materials, also called fluoroelastomers, include Kel-F® and Fluorel® elastomers (both 3M). Viton® (DuPont Company) and Fluorel® are copolymers of vinylidene fluoride and hexafluoropropylene (FPM). Kel-F® is a copolymer of chlorotrifluoroethylene and vinylidene fluoride and is called polychlorotrifluoroethylene (CFM). Viton® and Fluorel® are the most important members of this group today. These elastomers have excellent resistance to ozone, oxidation, weathering, heat, aliphatic and aromatic hydrocarbons, and alcohols. They are also highly impermeable to gases, and have good strength, electrical resistivity, and resistance to abrasion, water, acids, and halogenated hydrocarbons. This type usually includes the fluorosilicone elastomers which are discussed above [16]. Treatments suggested are as follows:

8.7.13.1 Sodium Etch [46]

For optimum bond strength use this procedure, described above under "Ethylene-Chlorotrifluoroethylene Copolymer" (Sec. 8.2.6).

8.7.13.2 Dry Abrasion [46]

For relatively low adhesion with low environmental resistance:

1. Wipe, spray with, or immerse in acetone, MEK, toluene, or ethyl or isopropyl alcohol
2. Abrade lightly and uniformly with 180–320-grit abrasive paper
3. Repeat step 1.

8.7.13.3 Epichlorohydrin Elastomer

This elastomer homopolymer and the copolymer combine exceptional resistance to aromatic and halogenated hydrocarbons with high resistance to ozone, weathering, gas permeability, compression set, compacting and tearing. These compounds have good tensile strength, resilience, and resistance to abrasion, water, acids, and alkalis. They also have good low-temperature and heat resistance. Epichlorohydrin elastomers of both types require a simple cleaning with acetone (not MEK), alcohols, or aromatics, such as toluene. No abrasion is necessary [17].

8.7.14 Polysulfide Rubber (PTR)

The first introduction of a polysulfide rubber for commercial application was in 1930 when Thiokol Chemical Corp. developed Thiokol® (acquired by PolySpec Corporation). Since then a number of other types of Thiokol® have been developed (<www.polyspec.com>). Thiokol® resins are highly impermeable to gases, and have excellent resistance to ozone, oxidation, weathering, and aliphatic and aromatic hydrocarbons [17]. Treatments are as follows:

8.7.14.1 Abrasive Treatment [8]

As for neoprene (Sec. 8.7.1.1); methanol is used as a solvent.

8.7.14.2 Chlorination [9]

1. Degrease in methanol.
2. Immerse overnight in strong chlorine water.
3. Wash.
4. Dry.

Snogren [8], however, does not recommend chlorination for treating polysulfide.

8.7.14.3 Primers

An example of a silicone primers for use with polysulfide sealants is Chemlok® AP-131 (LORD Corporation) [45].

8.7.15 Propylene Oxide Rubber

This elastomer is a sulfur-vulcanizable copolymer of propylene oxide and allyl glycidyl ether. Its vulcanizates are particularly attractive for dynamic uses which require high resilience, excellent flex life, and flexibility at extremely low temperatures. It performs similarly to natural rubber in these applications, but has added advantages including:

1. Good resistance to aging at high temperatures.
2. Good ozone resistance.
3. Moderate resistance to loss of properties in contact with fuels, and/or some solvents.

The surfaces of this elastomer should be prepared by solvent wiping with xylene, toluene, or other appropriate solvent [47].

8.7.16 Polyacrylate (Polyacrylic Rubber)

Polyacrylic elastomers are noted primarily for their high resistance to heat, ozone, oxidation, weathering, aliphatic hydrocarbons, and sulfur-bearing oils. They also have good resilience, gas impermeability, and moderate strength. Abrasion resistance is not as strong as with nitrile rubber or SBR [17]. Surface preparation is as follows:

8.7.16.1 Dry Abrasion [46]

As for fluorocarbon elastomers (Sec. 8.7.13.2); methanol has also been used as a solvent.

8.8 THERMOPLASTIC ELASTOMER

This class of polymers combines the end-use properties of vulcanized elastomers with the processing advantages of thermoplastics. Because of the unique molecular configurations of these polymers, they may be processed with the same techniques utilized with other thermoplastics, but the mechanical properties of the final articles are essentially indistinguishable from those of similar articles fabricated from conventional vulcanized elastomers.

There are several types, including polyester (Hytrel® by DuPont Co.), polystyrene–butadiene–polystyrene block copolymers (Kraton® by Kraton Polymers U.S. LLC), polystyrene–isoprene–polystyrene block copolymers (Solprene® by Dynasol Elastomers), polyolefin (TPR® thermoplastics rubber by Uniroyal, Inc.) and polyurethane.

Since these materials are thermoplastic and may therefore be soluble in a wide range of organic solvents, it is quite possible that solvent cementing would be the most appropriate method of bonding the thermoplastic rubbers to themselves or mutually-compatible plastics. Solvent cementing might work with Kraton® (Kraton Polymers U.S. LLC), but not with TPR® and Hytrel® (DuPont), which do not have reactive bonding sites. In cases where the thermoplastic rubber must be bonded to metals or other non-plastic or non-rubber materials, neither solvent cementing nor fusion techniques should be considered. In such cases, conventional adhesives must be used. In the case of Hytrel® Polyester Elastomer, cleaning with MEK and then using commercial primers is effective. Abrasion is not recommended. Polyurethane rubber, on the other hand, has reactive sites and is cementable with solvents.

8.9 PAINTED SURFACES [8,47]

To obtain maximum-strength bonds, paint must be removed. Even if the bond between the adhesive, the paint, and the other adherend is good, the strength of the joint will be no better than the strength of the bond between the surface that has been painted and the paint itself. For temporary or such bonds that do not require maximum strength, clean the painted surface with a detergent solution, abrade with a medium-grit emery cloth, and wash again with detergent. After loose paint has been removed, alternate procedures may be used:

1. Solvent cleaning [2]. Immerse, spray, or wipe in either of the solutions in Table 8.17 or Table 8.18, which will degrease the surface and leave a dull finish to the paint.

TABLE 8.17 Composition of Solvent Cleaning Solution for Painted Surfaces

Solution Component	Volume (%)
Methylene chloride	15
n-Butanol	35
Mineral spirits	48
Methyl ethyl ketone	2

TABLE 8.18 Composition of Solvent Cleaning Solution for Painted Surfaces

Solution Component	Volume (%)
Methylene chloride	20
n-Butanol	20
Isopropanol	20
Methyl acetate	20
Toluene	20

2. Scrub surfaces with an alkaline cleaner, such as a solution of 2–4% by weight of trisodium phosphate (TSP) in hot water. Rinse with clean water.

8.10 CONCLUSIONS

Thermoplastics, thermosets, and elastomers require surface treatment to remove contamination and prepare their surfaces for adhesion bonding. Both initial bond strength and bond durability are affected by the surface preparation technique. Most materials require unique methods of treatment for optimal bond strength formation. Surface preparation treatment methods use or generate chemicals that may have various serious health effects and must be used with extreme caution.

REFERENCES

[1] Petrie EM. Plastics and elastomers as adhesives. In: Harper CA, editor. Handbook of plastics and elastomers. New York: McGraw-Hill; 1975.

[2] Ross MC, et al. Effects of varying processing parameters on the fabrication of adhesives-bonded structure, part X, adhesive bonding structural plastics. Picatinny Arsenal technical report 4318; July 1972.

[3] Smith DR. How to prepare the surface of metals and non-metals for adhesive bonding. Adhes Age 1967;10(3):25–31.

[4] U.S. Patent 3,235,426, W.M. Bruner, assigned to DuPont Company (Feb 15, 1966).

[5] Delrin® Design Information, Module II, No. L-10464, published by DuPont Co. (March 2002).

[6] Bodnar MJ, Powers WJ. Adhesive bonding of the never plastics. Plast Technol 1958;4(8):721–5.

[7] Wegman RF. Adhesive bonding of silicone rubber to kralastic for the launcher. Picatinny Arsenal technical memorandum 1865. Dover, NJ: Picatinny Arsenal; 1969.

[8] Snogren RC. Handbook on surface preparation. New York: Palmerton Publishing Co.; 1974.

[9] Shields J. Adhesives handbook. 3rd ed. London: Newness-Butterworth; 1984.

[10] Landrock AH. Processing handbook on surface preparation for adhesive bonding, Picatinny Arsenal technical report 4883. Dover, NJ: Picatinny Arsenal; 1975.

[11] Salomon G. In: Houwink R, Salomon G, editors. Introduction, adhesion and adhesives, vol. 2: applications. 2nd ed. Elsevier Publishing Co.; 1967.

[12] Ebnesajjad S. Fluoroplastics, vol. 2: melt processible fluoropolymers. Norwich, NY: William Andrew Inc.; 2002.

[13] Ross MC, et al. Effects of varying processing parameters in the fabrication of adhesive-bonded structures, part X. Adhesive bonding of structural plastics, Picatinny Arsenal technical report 4318. Dover, NJ: Picatinny Arsenal; 1972.

[14] Cagle CV. Bonding plastic materials. In: Cagle CV, editor. Handbook of adhesive bonding. NY: McGraw-Hill; 1973.

[15] Abolins V, Euckert J. Adhesive bonding and solvent cementing of polyphenylene oxide. Adhes Age 1967;10(7):22–7.

[16] <www.geplastics.com/resins/techsolution/technifacts.html>, Reinhard D. Adhesive bonding of Noryl resins, Noryl processing guide, NPG 12, undated. Selkirk, NY: General Electric Co., Plastics Department.
[17] Landrock AH. Effects of varying processing parameters in the fabrication of adhesive-bonded structures, part XVIII. Adhesive bonding and related joining methods for structural plastics–literature survey, Picatinny Arsenal technical report 4424. Dover, NJ: Picatinny Arsenal; 1972.
[18] Devine AT, et al. Bonding Lexan polycarbonate to silicone rubber for gas masks, Picatinny Arsenal technical report 3930. Dover, NJ: Picatinny Arsenal; 1970.
[19] General Electric Co., Plastics Department, Pittsfield, MA. Lexan Polycarbonate Resin Design, Brochure CDC 536; undated.
[20] Ebnesajjad S. Fluoroplastics, vol. 1: non-melt processible fluoroplastics. Norwich, NY: William Andrew Inc.; 2000.
[21] General Electric Co., Plastics Department, Pittsfield, MA. Valox Thermoplastic Polyester, Brochure, VAL-5A; undated.
[22] Ticona Corp., Florence, KY. Celanex® thermoplastic polyester–properties and processing, <www.Ticona.com>; 2005.
[23] Mr. Donald Croft, General Electric Co., Products Section, Pittsfield, MA (3 January 1975).
[24] Paul Rich R. Personal communication. Eastman Chemical Products, Inc.; 1972.
[25] <www.eastman.com/Literature_Center/P/PP110.pdf>; June 2013.
[26] ICI Petrochemicals and Plastics Division. PEEK aromatic polymer, provisional data sheet PK 11, adhesive bonding; 1979.
[27] Victrex PES aromatic polymer–data for design–unfilled grades, technical service note VX101. 3rd ed. 1978.
[28] Victrex PES aromatic polymer–adhesive for use with Victrex polyethersulfone, No. VX TS 1.78; August 1978.
[29] Ohi H, Mitsui Petrochemical Industries, Ltd., Tokyo, Japan. TPX® Polymethylpentene. Modern plastics encyclopedia. New York: McGraw-Hill; 1995.
[30] Anderson MD, Bodnar MJ. Surface preparation of plastics for adhesive bonding. Adhes Age 1964;7(11):26–32.
[31] Ross MC. Preliminary study of adhesive bonding of newer structural plastics, Picatinny Arsenal technical memorandum 2204; April 1976.
[32] VESPEL® S Line Design Handbook, Rep No. H-53668, DuPont Engineering Polymers, Geneva, Switzerland; May 2002.
[33] Properties of DuPont VESPEL® Parts, Rep No. H15724-1, DuPont Engineering Polymers, Geneva, Switzerland; Aug 1993.
[34] Trimineur RJ. Silicone, modern plastics encyclopedia. New York: McGraw-Hill; 1995.
[35] Cagle CV. Handbook of adhesive bonding. New York: McGraw-Hill; 1973.
[36] Spearman BP, Hutchinson JD. Post vulcanization bonding techniques. Adhes Age 1974;7(4):30–3.
[37] Sharpe LH. Adhesive bonding, fastening and joining reference issue. Mach Des 1969;14(21):119–20.
[38] McIntyre RT, et al. Effect of varying processing parameters in the fabrication of adhesive bonded structures, part VI, production methods, Picatinny Arsenal technical report 4162. Dover, NJ: Picatinny Arsenal; 1971.

[39] Westerdahl CA, et al. Activated gas plasma surface treatment of polymers for adhesive bonding-part III, Picatinny Arsenal technical report 4279. Dover, NJ: Picatinny Arsenal; 1972.
[40] Hall JR, et al. Activated gas plasma surface treatment of polymers for adhesive bonding. J Appl Polym Sci 1969;13:2085–96.
[41] International Plasma Corporation (IPC), Product Bulletin 2402.
[42] DeLollis NJ, Montoya O. Bondability of RTF silicone rubber. J Adhes 1967;3(1):57–67.
[43] Gaughan JE. Bonding elastomeric compounds. In: Cagle CV, editor. Handbook of adhesive bonding. New York: McGraw-Hill; 1973.
[44] Guttman WH. Concise guide to structural adhesives. New York: Reinhold; 1961.
[45] Various Hughson Chemicals (Division of LORD Corporation, Erie, PA) publications, 1972 and 1974 on Chemlok® primers and adhesives.
[46] Snogren RC. Selecting surface preparation processes for adhesive bonding, sealing and coating. Paper presented at Design Engineering Conference and Show, American Society of Mechanical Engineers, held at Chicago, IL, April 22–25, 1968. Paper 68-DE-45. Also published as Selection of Surface Preparation Processes, Parts 1 and 2, Adhesives Age, 7 and 8; July and August 1969.
[47] Vazirani HN. Surface preparation of steel for adhesive bonding. J Adhes 1969;1:222–32.
[48] Secondary Operations Guide, Assembly and Finishing of Engineering Plastics, DSM Engineering Plastics Corp, The Netherlands, 2006.
[49] Snogren RC. Handbook on surface preparation. New York: Palmerton Publishing Co; 1974.

Chapter 9

Plasma Treatment of Polymeric Materials

9.1 INTRODUCTION

Plasma (glow discharge) sometimes refers to the *fourth* state of matter. It is produced by exciting a gas with electrical energy. It is a collection of charged particles containing positive and negative ions (Fig. 9.1). Other types of fragments, such as free radicals, atoms, and molecules, may also be present. Plasma is electrically conductive and influenced by a magnetic field (Fig. 9.2). Plasma is also intensely reactive, which is precisely the reason that it can modify surfaces of plastics [3]. It can be used to treat parts to make their surfaces harder, rougher, more or less wettable, and more conducive to adhesion.

Plasma is comprised of a partially ionized gas containing an equal number of positive and negative charges, as well as some non-ionized gas particles. A *glow discharge* is defined as plasma formed by the passage of electric current through a low-pressure gas like a fluorescent lamp. Glow discharge is globally neutral, but contains regions of net positive and negative charge. Most thin polymeric film processes utilize glow discharges, but "plasmas" and "glow discharges" are often used interchangeably [4]. *Plasma density* is the number of species/cm^3, which is in the range of 10^7-10^{20}. Typical glow discharges and arcs have an electron and ion density $\sim 10^8-10^{14}$.

Plasma treatment has its origins in the 1857 invention of dielectric barrier discharge (DBD). Werner von Siemens subjected oxygen or air to an alternating electric field of sufficient amplitude in a narrow annular gap, which resulted in dielectric barrier discharge. The process generated ozone, which became the first application of DBD [5]. The electrodes were placed outside the plasma chamber; thus they did not come into contact with the plasma front. DBD can take place in air, or in vacuum as long as trace amounts of one or more gases is injected into the plasma chamber.

Plasma treatment is probably the most versatile surface treatment technique [6,7]. Different types of gases such as argon, oxygen, nitrogen, fluorine, carbon dioxide, and water can produce the unique surface properties

Surface Treatment of Materials for Adhesive Bonding. DOI: http://dx.doi.org/10.1016/B978-0-323-26435-8.00009-5

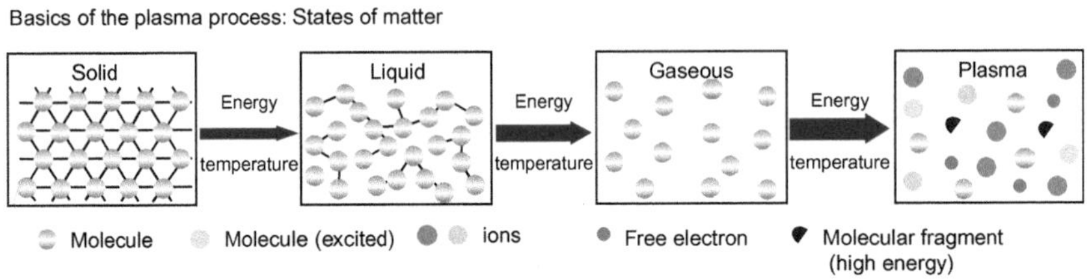

FIGURE 9.1 Plasma is the fourth state of matter.

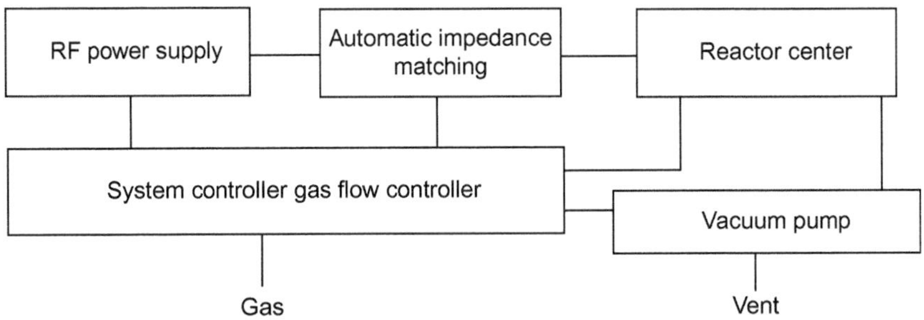

FIGURE 9.2 Schematic diagram of a plasma system [1,2].

required by various applications. For example, oxygen–plasma treatment can increase the surface energy of polymers, whereas fluorine–plasma treatment can decrease the surface energy and improve the chemical inertness. Crosslinking at a polymer surface can be introduced by inert gas plasma. Modification by plasma treatment is usually confined to the top several ten nanometers and does not affect the bulk properties. The main disadvantage of the low pressure (vacuum) plasma process is that it requires a vacuum system, which increases the cost of operation.

Corona treatment and low pressure and atmospheric pressure plasma treatment methods are all based on the dielectric barrier discharge phenomenon. All three processes have commercial applications which extend far beyond enhancement of adherability of surfaces. The relative cost and complexity of operating the three DBD-based treatment processes are shown in Fig. 9.3. The requirement of trace gases increases energy consumption, and special electrodes in atmospheric pressure plasma treatment (APT) render this technique more costly and complex than the standard corona treatment. The vacuum (low pressure) plasma method is more expensive than APT not only because of the higher capital cost of equipment but also by the requirement of extremely low pressure. The latter renders vacuum plasma treatment less productive and somewhat limited to higher value parts. More recently the industry's efforts have been focused on reducing costs and simplifying

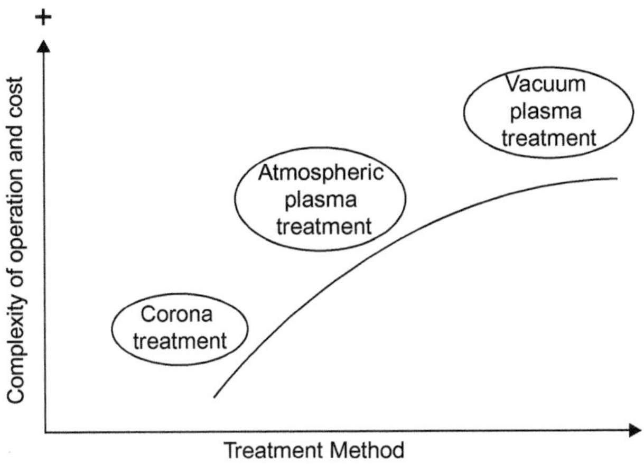

FIGURE 9.3 Complexity of operations and cost of treatment methods based on dielectric barrier discharge.

the low pressure plasma technique [8–10]. Good examples of cost analysis presented by the industry are available to readers as guidance for economic comparison of different plasma processes [11].

9.2 PLASMA REACTION WITH POLYMER SURFACES

Reactions of gas plasmas with surfaces polymers can be subdivided into three groups as per the following [12,13]:

1. Surface reactions include (a) reactions between the species in the gas phase and surface atoms and chemical groups and (b) reactions among the surface species that produce functional groups and crosslinking at the polymer surface. For example these reactions occur when a polymer surface is treated with plasma using gases such as ammonia, carbon monoxide, carbon dioxide, fluorine, hydrogen, nitrogen, nitrogen dioxide, oxygen, and water.
2. If the gases can act as monomers then plasma polymerization occurs at the surface forming a thin film on the surface of a polymer via polymerization of an organic monomer such as CH_4, C_2H_6, C_2F_4, and C_3F_6. The polymerization reaction scheme is rather complex because it involves reactions between gas-phase species, reactions between gas-phase species and surface species, and reactions between surface species.
3. Plasma treatment removes material from the polymer surfaces by chemical reactions and physical etching at the surface to form volatile products, and thus it can clean and etch polymer surfaces. Oxygen-containing plasmas are used to remove organic contaminants from polymer surfaces, e.g., oligomers, anti-oxidants, anti-block agents or mold-release agents.

Etching differs from cleaning only in the amounts of materials that are removed from the surface. Oxygen plasmas and oxygen- and fluorine-containing plasmas are frequently used for the etching of polymers [7].

There are different types of gas plasma depending on the properties of the gas; they include:

1. **Inert gas plasmas** – Helium, neon, and argon are the three inert gases used in plasma technology, although argon is the most common because of its low cost.
2. **Oxygen-containing plasmas** – Oxygen and oxygen-containing plasmas are most common for modifying polymer surfaces.
3. **Nitrogen-containing plasmas** – Nitrogen-containing plasmas are widely used to improve wettability, printability, bondability, and biocompatibility of polymer surfaces.
4. **Fluorine-containing plasmas** – When a plasma gas contains fluorine gas, surface reactions, etching, and plasma polymerization may take place simultaneously.

In general, plasma treatment increases the surface energy of polymers thus reducing wetting contact angle. The *depth of surface modification* mainly depends on the power level and treatment time. For plasma-treated polymer samples, the depth of the surface modification could be several tens of nanometers based on studies carried out using variation of angle-resolution in X-ray photoelectron spectroscopy (XPS) [14]. Concentration of functional groups created on a polymer surface by plasma treatment may vary depending on the *environment* and *temperature* over time. The underlying reason for surface chemical and physical changes of polymer surfaces is the extra mobility surface chains have over those in the bulk. The added mobility at the polymer surface lets the surface polymer molecules reorient in response to different environments [15]. Surface studies have demonstrated that the cause of decrease in the surface energy when the treated surface is placed in a low-energy medium such as air or vacuum is the rotation of the polar groups in the bulk or the migration of low-molecular-weight fragments to the surface to reduce the interfacial energy. Increasing the size of polar groups by grafting, an increase in crystallinity and crosslinking [16] have been found to be effective remedies for the prevention of physico-chemical surface changes of plasma treated polymers.

Another important factor that affects the speed of *aging* a plasma-treated polymer surface is temperature. A lower storage temperature reduces the rate of aging. Figure 9.4 shows the change of the water advancing contact angle on an oxygen–plasma-treated polypropylene surface as a function of aging time at different temperatures [17]. The rapid change of the contact angle at high temperatures supports the hypothesis that the changes in the surface are caused

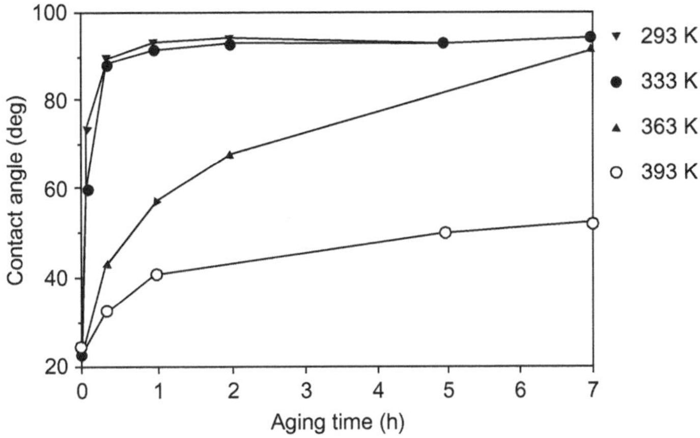

FIGURE 9.4 Advancing contact angle of water on oxygen–plasma-treated polypropylene as a function of aging time at different temperatures [17].

by polymer chain motion, reorienting the polar groups into the bulk. XPS studies of the chemical composition of the surface have shown little change after exposure to elevated temperatures. This is clearly an indication that any changes occur in layers thinner than the sampling depth of the XPS method [7].

9.3 LOW PRESSURE PLASMA TREATMENT (LPT)

In 1966, Schonhorn and Hansen [18] reported on a highly effective treatment for the surface preparation of low-surface-energy polymers for adhesive bonding. The techniques consisted largely of exposing the polymer surface to inert gas plasma at reduced pressure generated by electrodeless glow discharge (i.e., radio-frequency field). For polyethylene, only very short treatment times were necessary (approximately 9 seconds), while larger contact times were required for other polymers such as PTFE [19]. Plasma gases used (O_2, He, N_2) can be selected to include a wide assortment of chemical reactions. In the process, atoms are expelled from the surface of the polymer and new chemical groups are formed thus can produce a strong, wettable, and crosslinked skin [20].

Low-pressure plasma treatment can be applied to a variety of plastic parts, even powder additives like pigments and fillers. In practice, this type of plasma treatment is not widely used in the industry because it is conducted in a vacuum process (10^{-2}–10^{-3} mbar pressure), and is thus both expensive and inconvenient. Nevertheless, plasma treatment has been well researched and the academic literature is replete with papers.

The plasma used for treating material surfaces is called *cold plasma*, which means its temperature is room temperature. Cold plasma is created by introducing the desired gas into a vacuum chamber (Fig. 9.5), followed by

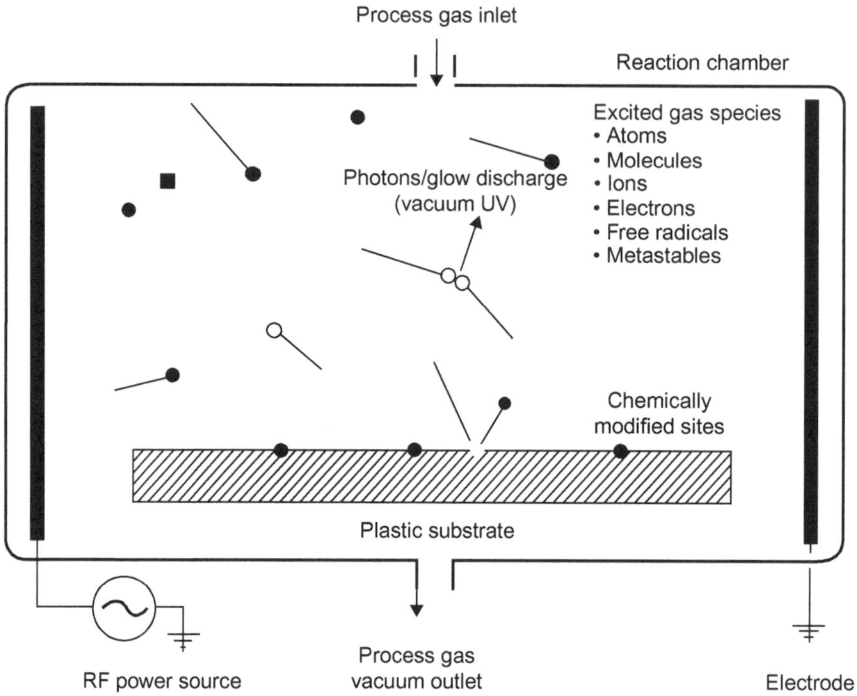

FIGURE 9.5 Schematic of the surface modification of plastic in gas plasma reactor.

radio frequency (13.56 MHz) or microwave (2450 MHz) excitation of the gas. The energy dissociates the gas into electrons, ions, free radicals, and metastable products. Practically any gas may be used for plasma treatment although oxygen is most common. The electrons and free radicals created in the plasma collide with the polymer surface and rupture covalent bonds, thus creating free radicals on the surface of the polymer as shown in Figure 9.6. The free radicals in the plasma may then recombine to generate a more stable product. After a predetermined time or temperature is reached, the radio frequency is turned off. The gas particles recombine rapidly and the plasma is extinguished. The most important components of a system are the vacuum chamber, the vacuum pump, and a high-frequency generator for plasma creation. The principal process of a low-pressure system can be most simply explained by Figures 9.7 and 9.8.

9.3.1 Applications of Low Pressure Plasma Treatment to Polymers

Plasma-treated polymers typically form adhesive bonds from two to four times stronger than bonds formed by traditional chemical or mechanical

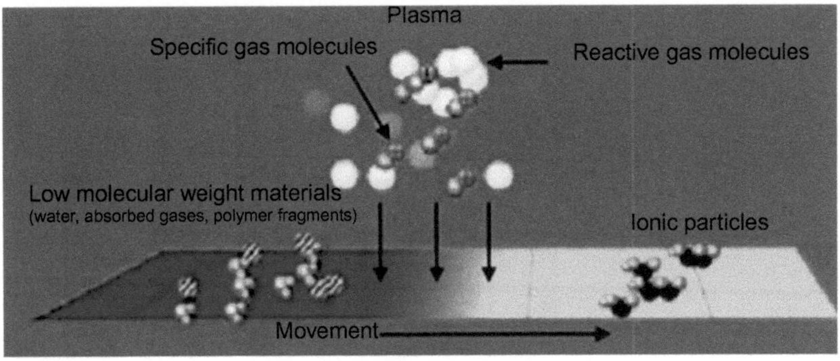

FIGURE 9.6 Schematic of surface modification of plastics in gas plasma [21].

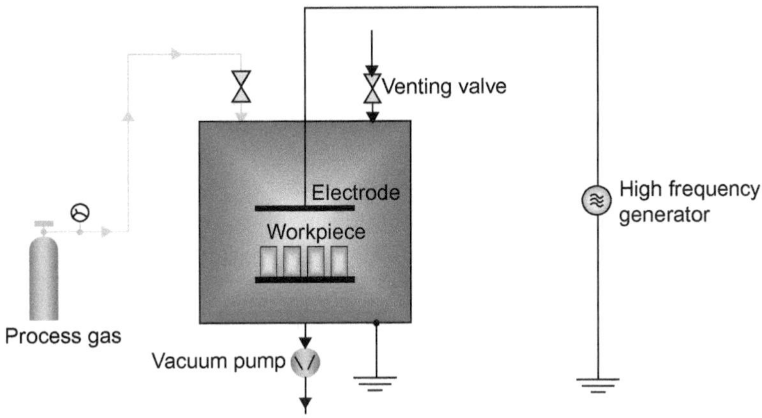

FIGURE 9.7 Basic schematic diagram of kHz and MHz low pressure plasma systems. (*Courtesy of Diener Electronic*, www.plasma.de/en/index.html) [22].

preparation. Researchers at Picatinny Arsenal conducted and reported on a study of a number of polymers, using activated gas plasma treatment [23]. A different study covered other plastics (Valox® polybutyl terephthalate polyester, polyethersulfone, polyarylsulfone, polyphenylene sulfide, ECTFE fluoropolymer, nylon 11, nylon 6/12, and nylon 12) [24].

Table 9.1 presents the lap shear joint bond strength of high density polyethylene as a function of gas used in the vacuum (low pressure) treatment and the width of the plastic film (Figures 9.9 and 9.10). The narrower bond area yields higher bond strength than the wider one (13 mm). Tables 9.1, 9.2, 9.3, and 9.4 show the effects of varying the process conditions of vacuum plasma treatment on the lap shear joint bond strength for a number of plastics. Clearly, plasma treatment is not equally efficient in rendering different plastics adhereable.

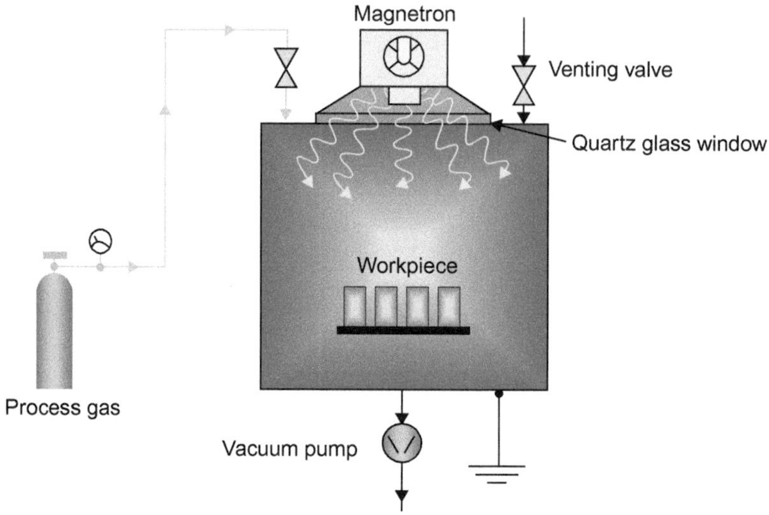

FIGURE 9.8 Basic schematic diagram of microwave plasma systems with metal chamber. (*Courtesy of Diener Electronic*, www.plasma.de/en/index.html) [22].

TABLE 9.1 Effect of Plasma Treatment on Lap Shear Joint Strength (kg/cm) of High-Density Polyethylene Sheet (1.6 mm Thick) [23]

Treatment Type (Duration)	13-mm Wide Lap Shear	6-mm Wide Lap Shear
Oxygen (30 min)	80	245
Oxygen (1 min)	84	248
Helium (1 min)	84	242
None – solvent wiped	9.3	–

FIGURE 9.9 Schematic of adherends (blue) and Adhesive layer (black) in Single-lap Joint Shear Test (per ASTM methods D1002 for metal/metal, D3164 for plastic/metal and D3163 for plastic/plastic adherends).

A significant amount of investigation has been conducted on the surface modification of fluoropolymers. Plasma treatment of polytetrafluoroethylene has been the subject of numerous and extensive studies and developments both in academia and in the industry because of the difficulty of obtaining

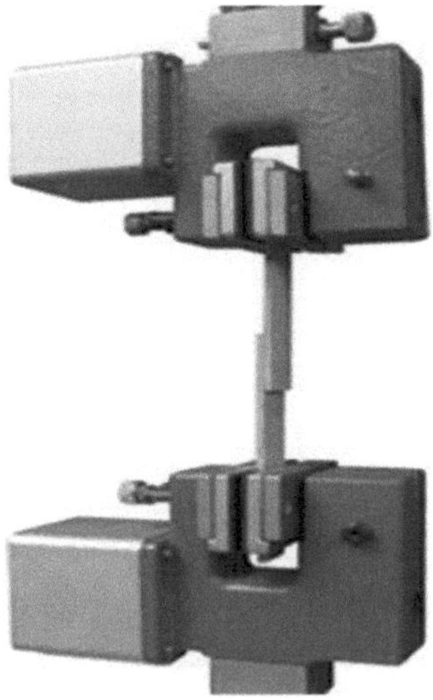

FIGURE 9.10 Lap joint shear test specimen in a tester device.

TABLE 9.2 Effect of Plasma Treatment Duration on Lap Shear Joint* Strength (kg/cm) and Failure Mode of Polyethylene (3.2 mm Thick) [23]

Treatment Type	Duration (min)	Bond Strength (MPa)	
		High Density	Low Density
None – solvent wiped	–	2.2 (A, PE)	2.6 (A, PE)
Oxygen	0.5	13.7 (A, PE)	–
Oxygen	1	9.2 (A, Al)	10 (C, PE)
Oxygen	30	16.8 (A, Al)	10.1 (C, PE)
Helium	0.5	6.4 (A, PE)	9.6 (C, PE)
Helium	1	9.4 (A, PE)	9.5 (C, PE)
Helium	39	21.6 (C, PE)	9.1 (C, PE)

*Notes:
Bonded to etched aluminum coupons using a crosslinked epoxy (a blend of Epon® 828 and Versamid® 140, 70/30 parts by weight).
A = adhesive failure, C = cohesive failure (see Ch. 5 for definitions of bond failure).
PE = polyethylene, Al = aluminum, PC = polycarbonate, PP = polypropylene.

acceptable adhesive bondability in addition to the strong desire to eliminate sodium etching.

A number of studies have been made on the vacuum plasma treatment of polymers [7], [25–27]. Chan et al. [7] have reviewed the low pressure (vacuum) plasma surface technology in a comprehensive study published in 1996.

TABLE 9.3 Effect of Plasma Treatment Duration on Lap Shear Joint* Strength (kg/cm) of Polycarbonate and Polypropylene Sheet [23]

Treatment Type	Duration (min)	Bond Strength (MPa)	
		Polycarbonate (1.6 mm Thick Sheet)	Polypropylene (0.12 mm Film)
None – solvent wiped	–	2.8 (A, PC)	2.6 (A, PP)
Oxygen	0.5	5.5 (A, PC)	12.9 (A, PP)
Oxygen	30	6.4 (A, PC)	21.2 (A, PP)
Helium	0.5	4.6 (A, PC)	3.1 (A, PP)
Helium	30	5.8 (A, PC)	1.4 (A, PP)

**Notes:*
Bonded to etched aluminum coupons using a crosslinked epoxy (a blend of Epon® 828 and Versamid® 140, 70/30 parts by weight).
A = adhesive failure, C = cohesive failure (see Ch. 5 for definitions of bond failure).
PE = polyethylene, Al = aluminum, PC = polycarbonate, PP = polypropylene.

Most of the development efforts since the early 1990s have, however, been focused on atmospheric pressure plasma treatment because of its advantageous economics and relative ease of operation.

In a study in 2006 low density polyethylene was treated by low pressure plasma using oxygen. Increasing the exposure time increased the surface energy (γ_s) and its polar component (γ_s^p) while the dispersed component (γ_s^d) decreased (Table 9.5) rather unsurprisingly. The majority of the increase in surface energy was gained during the first minute of exposure.

In a 1995 study [28], films of low density polyethylene, polypropylene, and polyethylene terephthalate were treated by low pressure microwave plasma using oxygen, hydrogen, and hexamethyldisiloxane mixed with oxygen and ethylene. The latter gases were monomers intended to plasma-polymerize on the film surfaces. The treated films were bonded to aluminum block with an epoxy adhesive on both sides and pulled, as seen in Figure 9.11, to measure bond strength.

Figure 9.12 shows that the bond strength of polyethylene treated by O_2 plasma increases strongly from the low value of 0.4 N mm^{-2} for the untreated polymer by a factor of 20.5, up to 8.2 N mm^{-2} after a plasma treatment of only 6 seconds [28]. The value of 8.2 N mm^{-2} is nearly the maximum bond strength reached for polyethylene and is not significantly influenced by a longer treatment than 900 seconds. At treatment time of 1800 seconds, a bond strength declination to the value of 6.4 N mm^{-2} is observed. In the case of polypropylene the improvement in the bond strength with oxygen plasma took place over a longer time and a slight decline of the bond strength is obtained for polypropylene for longer treatment times. Polyethylene terephthalate's

TABLE 9.4 Effect of Plasma Treatment Duration on Lap Shear Joint* Strength (kg/cm) of Various Plastics [23]

Treatment Type	Duration (min)	Bond Strength (MPa)				
		Polystyrene (0.25 mm Thick Film)	Polyethylene Terephthalate (0.25 mm Transparent Film)	Polyvinyl Fluoride (25 μm Transparent Film)	Cellulose Acetate Butyrate (15.2 mm Thick Sheet)	Nylon 6 (76 μm Thick Film)
None – solvent wiped	–	3.9 (A, PS)	3.6 (A, PET)	1.9 (A, PVF)	7.5, 4.5 (A, CAB)	5.8 (A, N)
Oxygen	0.5	–	–	9.4 (A, Al/PVF)	7.8 (A, CAB)	11.2 (A, Al)
Oxygen	1	–	–	–	3.1 (A, CAB)	10.5 (A, Al)
Oxygen	30	21.5 (C, PS)	9.4 (A, PET)	9.8 (A, Al/PVF)	9.5 (A, CAB)	24.1 (A, Al)
Helium	0.5	–	–	9.9 (A, Al/PVF)	9.6 (A, CAB)	9.4 (A, Al)
Helium	1	–	–	–	4.1 (A, CAB)	9.5 (A, Al)
Helium	30	27.7 (C, PS)	11.4 (A, PET)	9.3 (A, Al/PVF)	–	27.3 (A, Al)

**Notes:*
Bonded to etched aluminum coupons using a crosslinked epoxy (a blend of Epon® 828 and Versamid® 140, 70/30 parts by weight).
A = adhesive failure, C = cohesive failure (see Ch. 5 for definitions of bond failure).
Al = aluminum, PS = polystyrene, PET = polyethylene terephthalate, PVF = polyvinyl fluoride, CAB = cellulose acetate butyrate, N = nylon 6.

TABLE 9.5 Surface Energies (Polar and Dispersive Component) for Low Pressure O_2 Plasma-Treated LDPE Surfaces as a Function of the Exposure Time [15]

Exposure Time (min)	γ_s^p (mJ m^{-2})	γ_s^d (mJ m^{-2})	γ_s (mJ m^{-2})
0	3.7	24.2	27.9
1	59.8	2.6	62.4
2	59.6	3.3	62.9
5	63.3	2.6	65.9
10	62.9	2.2	65.1
15	66.7	2.0	68.7
20	67.0	1.6	68.6
25	68.1	1.3	69.4
30	67.0	1.8	68.8

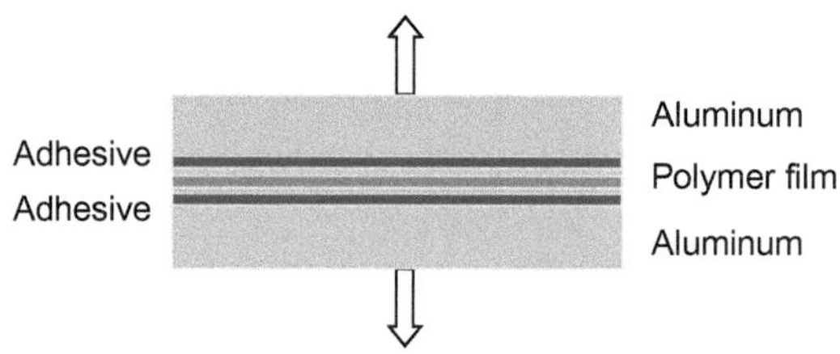

FIGURE 9.11 Schematic of test device for bond strength measurement of plasma treated polymer films.

bond strength of 5.1 N mm^{-2} is measured and is increased by only 20% to 6.2 N mm^{-2} by plasma treatments between 2 and 900 seconds. At a treatment time of 1800 seconds the bond strength declined to 4.3 N mm^{-2}. The improvements in adhesive bond strengths of polyethylene and polypropylene by oxygen plasma is due to the formation of hydroxyl, carboxyl, carbonyl, and ester groups in the surface layer of those polymers.

Surface modification of polyethylene terephthalate (PET) was investigated [3] by using low-pressure CF_4 plasma in a roll-to-roll double-sided treated system. The surface properties of the modified polymers are characterized by X-ray photoelectron spectroscopy (XPS) and water contact angle. Treated for a longer time and higher RF power (12 min, 600 W), the wettability of the PET surface shows two opposite extremes. The contact angle on one side of the PET film is super-hydrophilic, 7.56°, and the other side is hydrophobic, 108.63°. The XPS analysis measurement results also show different functional groups on the two sides of PET surfaces. The CF_3 in C (1 s) spectrum can enhance the hydrophobic surface, in spite of the fact that

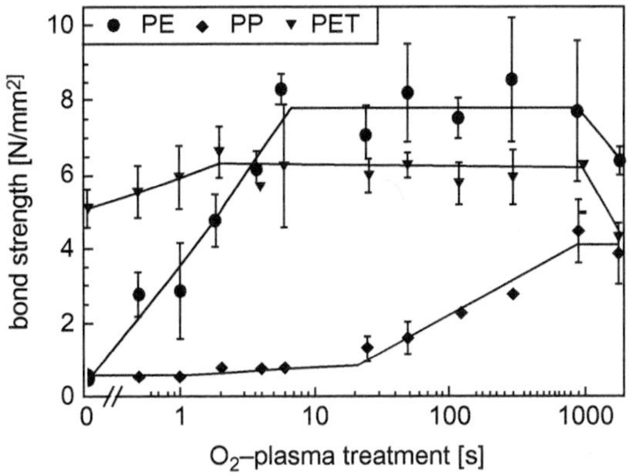

FIGURE 9.12 Peel strength of polymer films treated by O_2 plasma as a function of exposure time [28].

the relative intensity of CF_2 increases to 23.45%. Moreover, the surface dynamics of PET denote the concentration-dependent characteristics; that is, the F/C ratio decreases upon the concentration of hydroxyl ion increasing after dipping the sample into different $[OH^-]$ solutions.

Medical-grade polydimethylsiloxane (PDMS) elastomer was treated by low-powered plasma using four different gases: O_2, Ar, N_2, and NH_3. Changes to the surface chemistry after the treatment and the stability of those changes during aging in phosphate buffered saline (PBS) or in air for up to 1 month were investigated by Williams et al. [29]. Chemical analysis of the surface was conducted by X-ray photoelectron spectroscopy and dynamic contact angle and surface morphology were analyzed by optical microscopy and atomic force microscopy. Surface topography of untreated and plasma-treated specimens was examined using AFM operated in "tapping" mode. A typical AFM micrograph of Ar-treated PDMS is shown in Fig. 9.13. It can be seen that mild etching of the surface has occurred, producing surfaces that were more dimpled or pitted than the virgin surface.

All plasma treatments resulted in an increase in wettability, because of changes in chemistry combined with modest etching [29]. Furthermore, the primary site of attack of the plasma species appeared to be dependent upon the type of gas. The two main chemical changes observed after aging were due to reactions with the storage media and relaxation processes, resulting in further changes in wettability. The influence of the surface modifications on the blood compatibility of the materials was investigated by assessing contact phase activation using a partial thromboplastin time assay. It was demonstrated that the O_2 and Ar plasma treatments diminished the performance of the silicone but the N_2 and NH_3 treatments enhanced the activation of the coagulation cascade.

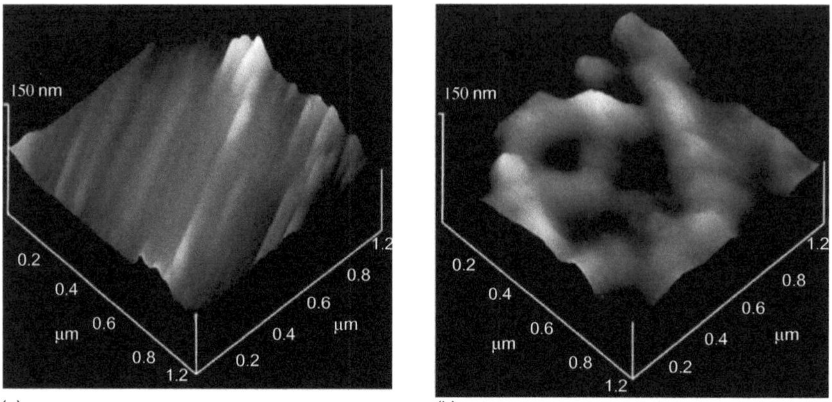

(a) (b)

FIGURE 9.13 AFM micrographs of (a) untreated; and (b) Ar plasma-treated PDMS, scan angle ϕ 30° [29].

Low-pressure glow discharge plasma with O_2 and N_2 was used to treat the surface of LDPE films to improve wettability to adhesives and to increase bond strength in a LDPE to form laminates with polyolefin foam. The main application of film/foam laminate is in the automotive industry. Typically a reactive polyurethane adhesive is used to bond the LDPE film to the polyolefin foam. Treatment by oxygen–plasma generated polar groups directly, compared to nitrogen plasma for which surface functionalization occurred after the plasma treatment upon exposure of the samples to wet air. That is when reactions with oxygen and water vapor take place producing polar chemical groups. Increase in surface roughness by plasma generated abrasion (etching) increases the surface area of bonding thus improves adhesive bond strength. The laminates exhibited good durability under aging conditions required for automotive testing (Figs 9.14 and 9.15) at different temperatures and relative humidities.

9.3.2 Application of Low Pressure Plasma Treatment of Fluoropolymers

Table 9.6 contains data that compares the results of plasma treatment and sodium etching for four fluoropolymers. Peel strengths of untreated and treated samples were measured by bonding them into T-peel specimens using the flexibilized epoxy adhesive Scotch-Weld® 3553 (available from 3M Corp.). The laminates were cured for several hours at 70°C and peel tested at a 12.5 cm/min pull rate. Polytetrafluoroethylene does not accept plasma treatment as well as PFA and FEP, as indicated by its relatively low peel strength. Plasma treatment does not impart sufficiently strong adhesive bonds

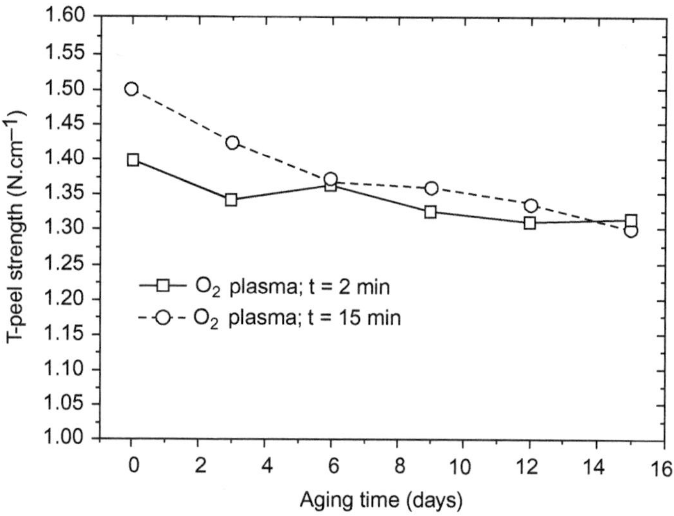

FIGURE 9.14 Aging effect on T-peel strength of oxygen plasma-treated LDPE films (conditions: 38°C and 95% relative humidity) [30].

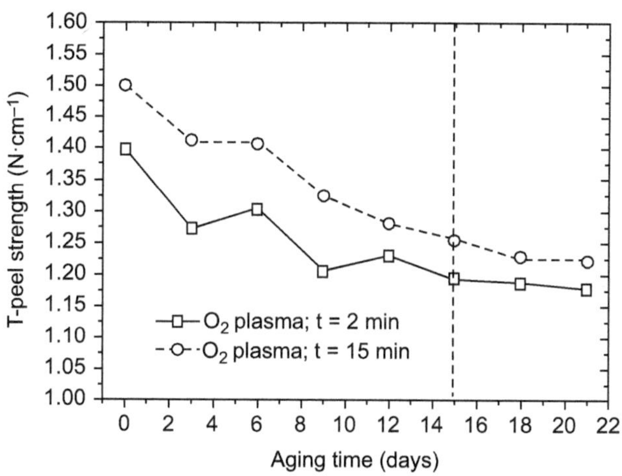

FIGURE 9.15 Aging effect on T-peel strength of oxygen plasma-treated LDPE films (conditions: 70°C and 55% relative humidity) [30].

in polytetrafluoroethylene, the most common fluoropolymer. Bond strength of plasma treated PTFE is at best less than half that obtained by sodium etching [33]. Sodium etching was the *only* effective method of modifying the surface of PTFE.

TABLE 9.6 Peel Strength of Adhesive Bonded Fluoropolymers [31,32]

Treatment	Material			
	PTFE	FEP	ETFE	PFA
Untreated	Negligible	0.1	Negligible	0.04
Sodium etched	5	8.2	–	6.4
Plasma treated	2.2	10.4	15.8	8.3

PTFE = Polytetrafluoroethylene, FEP = Fluorinated ethylene propylene copolymer.
ETFE = Ethylene tetrafluoroethylene copolymer, PFA = Perfluoroalkoxy copolymer.

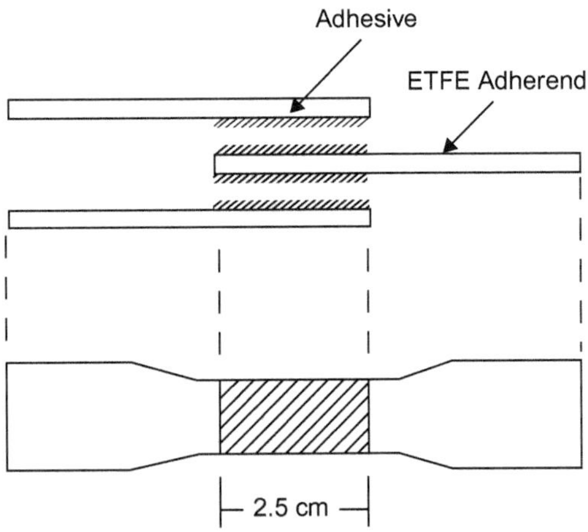

FIGURE 9.16 Schematic of the double-lap shear test specimen.

A study of adhesion improvement of ETFE by plasma treatment using oxygen, ammonia, and oxygen-plus-SF_6 gases has been reported [34]. Joints were made using commercial epoxy adhesives by a double-lap shear configuration (Fig. 9.16). The bond strength of the plasma-treated ETFE significantly exceeded the bond strength of the sodium-etched polymer specimen (Table 9.7). A range of values was obtained for bond strength with different adhesives.

An alternative plasma treatment technique is the glow discharge method that can be done at atmospheric pressure. Fluoropolymer films were treated by glow discharge in helium atmosphere [36]. Strips of the treated and untreated films were bonded to 0.2 mm thick aluminum foils using a urethane adhesive cured at 100°C for 15 minutes. The treatment conditions and bond strengths of the samples are given in Table 9.8. Glow discharge was effective for the

TABLE 9.7 Bond Strength of Plasma-Treated ETFE Using a Double-Lap Shear Test and Epoxy Adhesives [35]

Treatment Type	Bond Strength (MPa)
None	0.07
Tetra-Etch®*	0.78
Plasma ($O_2 + SF_6$)	1.49–2.31
Plasma (O_2)	1.47–1.83
Plasma (NH_3)	1.40–1.72

**Tetra-Etch® by WL Gore & Associates, Inc.*
ETFE = Ethylene tetrafluoroethylene copolymer.

TABLE 9.8 Effect of Atmospheric Glow Discharge Treatment in Helium Gas on Bond Strength [36]

Fluoroplastic	Untreated (g/cm)	Treated (g/cm)
PFA	0	600
FEP	0	640
ETFE	0	430
PVDF	100	830

Treatment conditions: gas flow rate = 5 1/min, discharge frequency = 5 kHz. Treatment time = 60 sec.
PFA = Perfluoroalkoxy copolymer, FEP = Fluorinated ethylene propylene copolymer.
ETFE = Ethylene tetrafluoroethylene copolymer, PVDF = Polyvinylidene fluoride.

treatment of a number of fluoropolymers. This technology has been commercialized.

A 1990 study [37] of plasma treatments of PTFE with air, oxygen, argon, and water vapor process gases created modified surfaces that were more polar and had a higher surface energy. A plasma treatment duration of 15 seconds (Fig. 9.17) was sufficient; this is considerably shorter than the treatment times in other reports. X-ray photoelectron spectroscopy (XPS) data showed the plasmas treatment with the gases used, mainly led to the loss of a considerable fraction of the fluorine atoms. In all cases the amount of oxygen uptake was much less than the amount of fluorine loss, indicating that the radicals generated by fluorine abstraction reacted with other radicals rather than oxygen-containing species from the plasma. The authors concluded the absence of an optically measurable level of conjugated double bonds suggested that double bond formation was less important than crosslinking of the surface for radical dissipation. The main effect of plasma

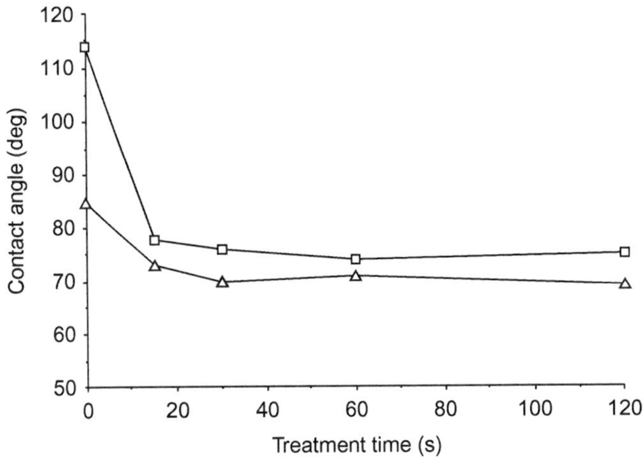

FIGURE 9.17 Contact angles of PTFE surfaces modified by treatment in a water vapor plasma, as a function of treatment duration. Distilled water (□) and methylene iodide (△) [37].

treatments of PTFE is not oxidative as is the case for most other polymers, and depends little on the plasma process gas used.

In a study [38], low-pressure plasma treatments of PTFE films using sulfur dioxide (SO_2) as the process gas were performed to impart functional groups containing sulfur and oxygen to the polymer surface. Plasma process parameters including gas flow rate, power, and treatment time were varied to determine the optimal plasma conditions for the desired surface modification. Wettability of the plasma-treated PTFE surface increased as demonstrated by a decrease in contact angle. Surface analysis by XPS showed decrease in fluorine content of the surface accompanied by the presence of oxygen and sulfur on the surface. The insertion of sulfur groups on the PTFE surface was not completely successful. The investigators followed up with another study.

In this study [39], PTFE films (100 μm thick) were first treated with low pressure plasma application using hydrogen as process gas. The second step was either a plasma treatment with SO_2 or by the immersion of the H_2 treated films in fuming sulfuric acid, containing SO_3. The contact angle of the treated samples decreased from the original 110° to 53° for H_2/SO_2 plasma and to 74° (H_2 plasma/SO_3), respectively. The sample "H_2/SO_2 plasma" showed only at the outer surface layer oxygen and sulfur traces as shown by XPS, while no evidence for the presence of functional groups in inner layers could be observed by attenuated total reflectance-Fourier transform infrared spectroscopy. Otherwise, the sample "H_2 plasma/SO_3" yielded exactly the contrary results: the observed bands in the infrared spectra assigned to associated OH and to SO_3 (at 3000–3700 cm^{-1} and 1,056 cm^{-1} respectively) are very strong but the O/C and S/C ratios given by XPS were not high enough. It was concluded from the different depths of surface penetration of both XPS and infrared spectroscopy that these two modification methods are able to insert functional groups into the polymer surface at different surface layers.

In another study, short-time plasma pre-treatment of polytetrafluoroethylene PTFE [40] with a low pressure microwave plasma was investigated. The source was Plasmodul® (Fig. 9.18), which allows the pre-treatment of PTFE substrates with ammonia plasma (NH_3). The newly developed plasma source Planartron®, which is derived from Duo-Plasmaline, was used for improving the adhesion properties of PTFE by generating oxygen and nitrogen plasmas. PTFE foils were modified on both sides by plasma treatment for typically 30 seconds. After bonding the films to an aluminum substrate using an epoxy adhesive, the bonding strength was measured directly by pull off tests. In Fig. 9.19 the bond strength is plotted vs. the exposure time in ammonia plasma. The figure shows that a short-time plasma treatment in the range of 15 seconds results in maximum bond strength ($3.5\ N\ mm^{-2}$) that is more than one order of magnitude higher compared to the untreated PTFE bond strength of $0.25\ N\ mm^{-2}$.

Next, PTFE films, pretreated with NH_3 plasma, were treated with N_2 plasma using the Planartron® device (Fig. 9.20) by placing the substrates at

FIGURE 9.18 Plasmodul® device with gas inlet system, plasma source, reaction chamber, and diagnostic ports [40].

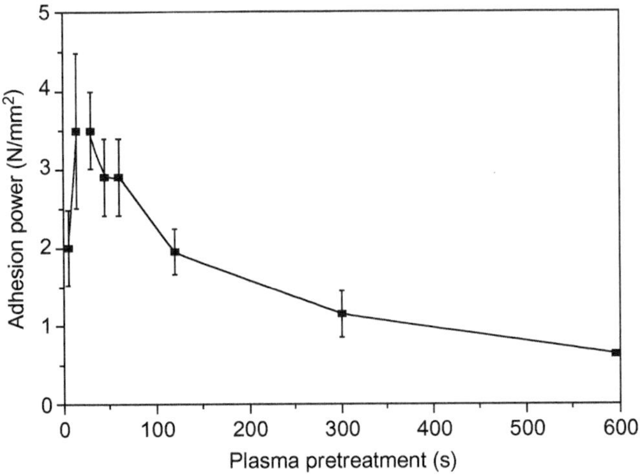

FIGURE 9.19 Bond strength versus plasma exposure time of PTFE treatment in Plasmodul® using NH_3 gas [40].

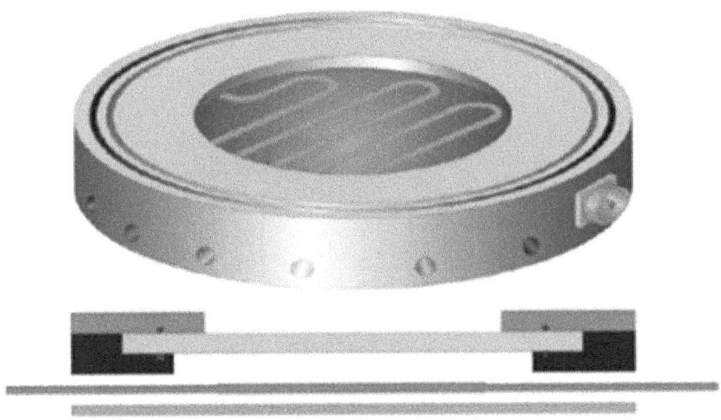

FIGURE 9.20 Schematic view of the Planartron® source. The antenna structure is placed under a quartz plate. This arrangement is especially designed for use in Plasmodul® [40].

distances of 2 and 0 cm from the plasma source. Figures 9.21 and 9.22 show the bond strength versus the exposure time in nitrogen plasma for 2 and 0 cm distances. Figure 9.21 shows that the bond strength of PTFE treated with nitrogen plasma at a 2 cm distance in a Planartron® source increases from 0.25 N mm^{-2} for the untreated material to 5.9 N mm^{-2} after 50 seconds of plasma treatment. At longer treatment times the bond strength decreases to 2.6 N mm^{-2} (after 200 seconds). The maximum bond strength of PTFE film, treated with N_2 at a distance of 0 cm, was reached in 5 seconds at 5 N mm^{-2}. This represents a factor of 20 compared to the untreated

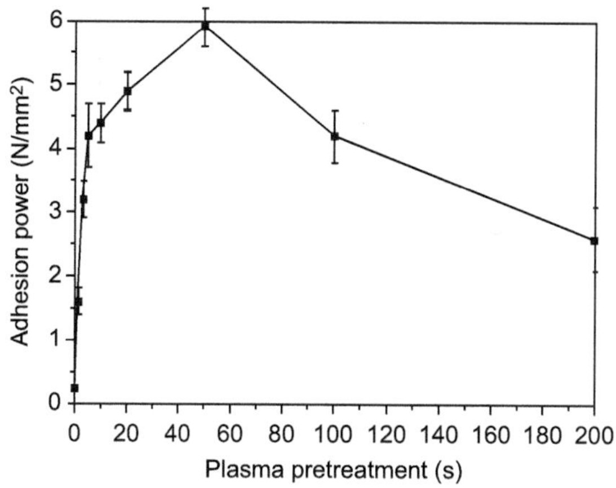

FIGURE 9.21 Bond strength versus plasma exposure time for PTFE treated with N_2 plasma at a distance of 2 cm [40].

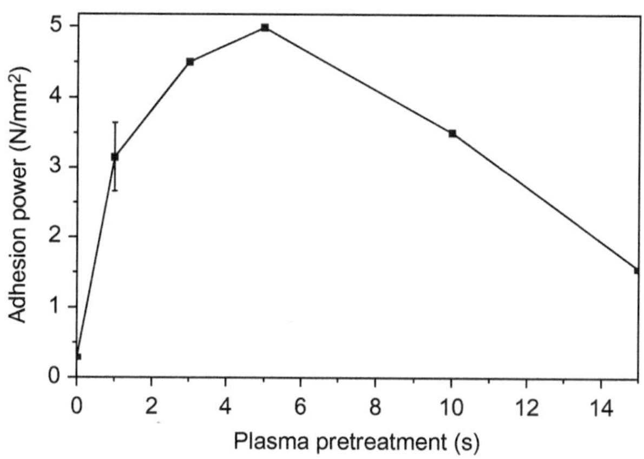

FIGURE 9.22 Bond strength versus plasma exposure time for PTFE treated with N_2 plasma at a distance of 0 cm [40].

material. Increasing the plasma exposition time again leads to a strong decrease in bond strength. It is remarkable that, already after a treatment of 5 seconds, the same bond strength is achieved as treating the PTFE for 20 seconds at a distance of 2 cm from the source. This type of development has reduced the cost of low pressure plasma treatment. A number of companies including Diener Electronic (www.plasma.de/en/index.html, Ebhausen, Germany), Henniker Plasma (www.plasmatreatment.co.uk, Warrington, UK), and Acton Technologies (www.actontech.com), Pittston, PA, USA offer treatment services.

A 2003 article [41] described the etching of polymer surfaces by low pressure plasma treatment. The topography of the surface is measured by atomic force microscopy (AFM). The as-received PTFE, Fig. 9.23(a), consisted of a dense assembly of 200–300 nm thick and 0.5–1.5 μm long cigars. These cigars arranged themselves on a micrometric rough surface. The PTFE films used had been produced by a cast process, where in a number of very thin layers were successively deposited and then fused all together. Figures 9.23b and 9.23c show the PTFE surface after a soft and a pronounced He plasma treatment, respectively. If the micrometric structure remains unchanged for both ion energies, an etching of the fine structure is clearly visible. The cigars become thinner by plasma treatments.

Effective PTFE surface treatment by low pressure plasma treatment (LPT) has been achieved thanks to combined academic and industrial research and development efforts. The new treatment processes have lower cost though are still more expensive than other methods. Companies such as Diener Electronic (www.plasma.de/en/index.html, Ebhausen, Germany), Henniker Plasma (www.plasmatreatment.co.uk, Warrington, UK), and Acton Technologies (www.actontech.com, Pittston, PA, USA) offer LPT for PTFE treatment.

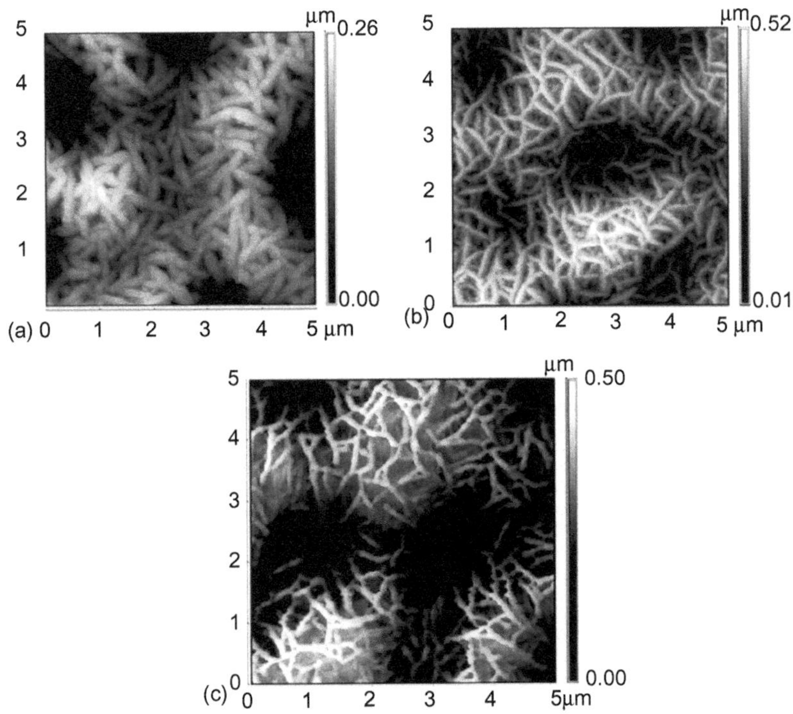

FIGURE 9.23 AFM profile of PTFE surfaces (a) untreated, (b) helium-plasma treated at a DC bias of 100 V for 2 min, and (c) helium-plasma treated at a DC bias of 200 V for 5 min [41].

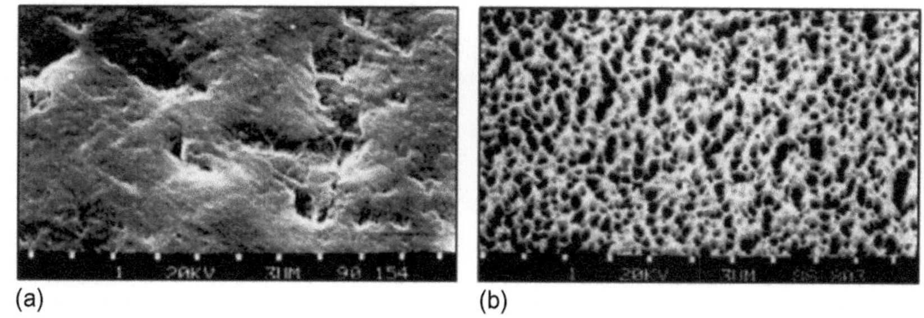

FIGURE 9.24 Scanning electron micrograph of PTFE: (a) before plasma treating and (b) after plasma treating.

Figure 9.24 shows the topical view of PTFE surface as-is and after low pressure plasma etching. The benefit of etching is increased surface roughness, which leads to an increase in available adhesive bonding surface area.

Many academic research papers on PTFE surface modification [42–49] have focused on the effect of LPT using different gases and compounds, grafting, plasma process variables, surface wettability, and analysis of surface composition and topography. These studies are useful in shedding light on the state of the modified PTFE surfaces versus the as-is sample. Indeed some of the articles report on good quality research work [48]. The studies seem to stop short of completing the rational research cycle because little has been reported about the effectiveness modification with respect to bondability of the treated surfaces or in other applications. Wettability improvement and impartation of polar functional groups on PTFE surface are prerequisite requirements of bonding using adhesives. In the end, however, the actual demonstration of successful bonding to the treated surfaces is the true measure of success of surface modification. If the surface treatment is not aimed at adhesive bonding and at other targets such as biocompatibilization, then the effectiveness of the modified surfaces to that end must be elucidated.

A study of low pressure plasma treatment of PTFE surface was made using scanning probe microscopy (SPM) [48]. The gases used were air, helium, and acrylic acid (AAc). The SPM image of the untreated PTFE film [Fig. 9.25(a)] shows a smooth surface, without any particular features on the surface. The PTFE films treated for 60 s in He plasma showed sphere-like aggregates which are formed on the surfaces [Fig. 9.25(d) and (e)]. Furthermore, the topography of the He-plasma-treated surface at 100 W [Fig. 9.25(e)] had higher sphere features than He-plasma-treated surface at 70 W [Fig. 9.25(d)]. However, less sphere-like structures were formed on surfaces treated with air plasma [Fig. 9.25(b) and (c)]. The morphologies of AAc-plasma-treated surface at 100 W [Fig. 9.25(f) and (g)] were similar to those for the He-treated surface [Fig. 9.25(e)], and only the surface treated with AAc-plasma for 60 s [Fig. 9.25(f)] was with lower sphere features and

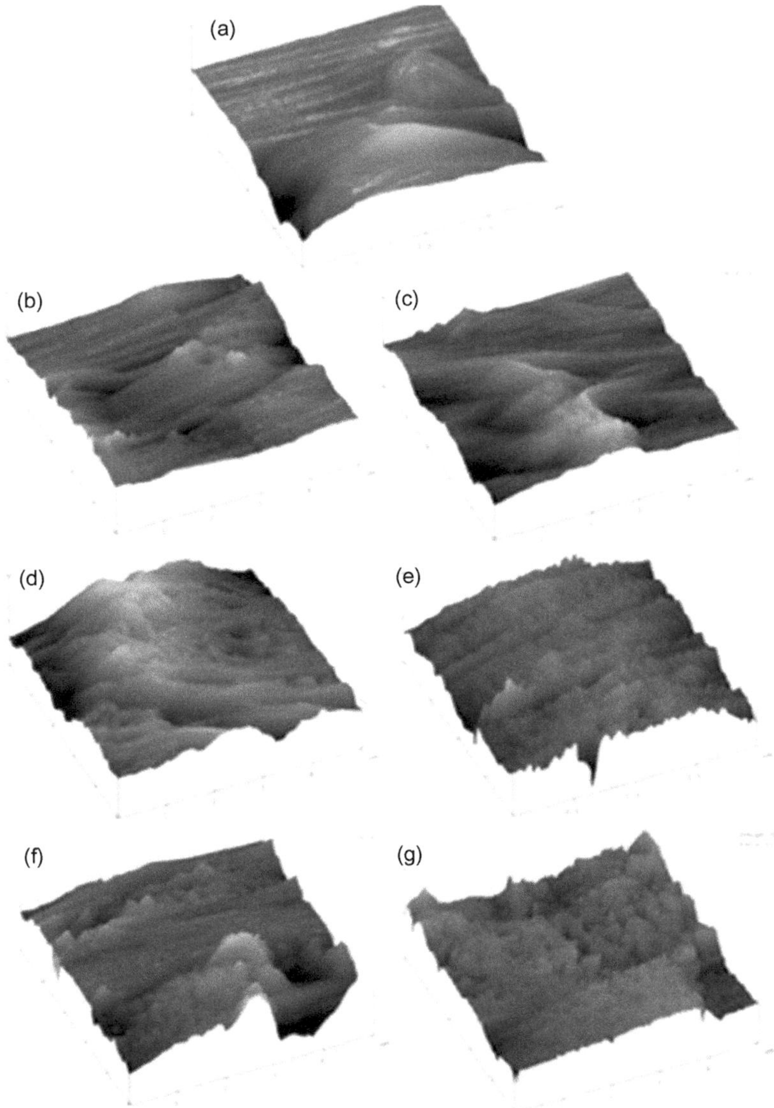

FIGURE 9.25 SPM images of untreated and treated PTFE surfaces: (a) untreated; (b) treated with air plasma at 70 W for 60 s; (c) treated with air plasma at 100 W for 60 s; (d) treated with He plasma at 70 W for 60 s; (e) treated with He plasma at 100 W for 60 s; (f) treated with AAc plasma at 100 W for 60 s; (g) treated with AAc plasma at 100 W for 120 s [48].

the surface treated for 120 s [Fig. 9.25(g)] had more intensive sphere-like peaks. As can be seen from Figure 9.25, for the same conditions, the PTFE films treated in air, He, or AAc plasma showed that the average roughness increased with a even higher power (100 W) or an even longer plasma exposure time (120 s). Also, the difference among the samples treated with air, He, or AAc plasma is noticeable: the sample treated with He plasma showed much rougher surface for the same treatment time and power, which was probably due to a different etching rate that was higher in the case of He plasma than in the case of air or AAc plasma. The formation of the rough surface after plasma treatment was mostly caused by the chemical erosion by atoms and physical erosion by ions in the plasma.

9.4 ATMOSPHERIC PRESSURE PLASMA TREATMENT

Atmospheric pressure plasma treatment (APT) systems have been developed out of necessity and for a number of reasons:

1. To take advantage of the capability of plasma technology.
2. In response to a number of shortcomings of 3D corona treatment systems discussed in Chapter 6 (see a comparison of atmospheric pressure plasma and a 3D corona treater given in Table 9.9).
3. To offer an alternative process to low pressure plasma methods which are costly and complex due to the vacuum requirement (see Table 9.10 for a comparison).

APT techniques are versatile and have found a number of important applications based on the properties of the plasma [51,52]. Some of these uses include:

1. Spectroscopic analysis.
2. Gas treatment: gas cleaning and synthesis.
3. Surface treatment: surface cleaning and etching, surface activation, surface coating.
4. Powder treatment.
5. Toxic waste treatment.
6. Plasma enhanced chemical vapor deposition.

Although the same physical phenomenon of electrical breakdown of gases at atmospheric pressure (electrical discharge) is used, the methods of creation and application of this discharge are completely different for 3D corona and atmospheric pressure plasma treatment techniques [50].

The plasma system is designed to create (Fig. 9.26) a uniform plasma cloud that completely surrounds small objects or spreads into the boundary layer of the surface. It can also be placed in the internal cavities, channels, etc. The same plasma system may treat the internal surfaces of a 50 μm capillary and cover a 50 mm diameter (and higher) surface area. One design of such a machine is based on the well known physical phenomenon [53]. The strength of the electrical field considerably increases in the vicinity of small radius objects. Applying a high voltage signal to a sharp edge body (e.g., a needle) causes electron leakage from the edge to the gaseous

TABLE 9.9 Comparison of an Atmospheric and a 3D Corona-Treater [50]

	Atmospheric Plasma	3D Corona
Average power	100 W	1000 W
Plasma carrier	Argon	Air
Plasma currents	Low (mA)	High (>10 mA)
Main direction of the energy transfer	From the electrode to the substrate surface	Between the electrodes parallel to the substrate surface
Plasma frequency	High (20,000 Hz)	Low (60 Hz)
Noise level	Low	High
Ozone generation	Low	High
Plasma flow temperature	Low (room)	High (>250°F)
Substrate exposure	Unlimited	Limited to thermal damage
Coverage from single head on the flat surface	Up to 3″ diameter circle or up to 5″ × 1″ strip	Up to 2″ × ¼″ strip
Ability to treat patterned surfaces	Unlimited	Limited
Ability to treat inner surfaces	Unlimited	Very limited
Ability to introduce special additives into the plasma for chemical surface modification	Limited	Very limited
Overall flexibility	High	Low
Overall efficiency	High	Low

environment (e.g., air). These free electrons, accelerated by the strong electrical field, have enough energy to ionize neutral gas molecules and produce other free electrons and ions.

These electron avalanches do not develop into an arc, but gradually decay as they move away from the edge, creating a uniform glowing cloud near the electrode. Since there is no well-defined second electrode, the currents in the plasma cloud are extremely low (100 μA) and plasma occupies a finite volume near to room temperature. The overall power to initiate and maintain this glow discharge usually does not exceed 100 watts. The low-temperature plasma cloud may be applied directly onto the surface of the treated material, or be considerably extended with the use of inert gases such as helium, argon, etc.

This kind of plasma is very effective since most of the micro-discharge trajectories end up on the surface of the treated material. A specially designed high frequency, low current power supply significantly increases

TABLE 9.10 A Comparison* of Advantages and Disadvantages of Low Pressure and Atmospheric Pressure Plasma Treatment Methods [8]

Applications properties	Low pressure plasma		Atmospheric plasma	
	Advantages	Disadvantages	Advantages	Disadvantages
Generation of plasma – generally –	Plasma is evenly distributed inside the plasma chamber. Chamber volume can vary from 2 to 12,000 liters	Complex vacuum technology. In-line plasma applications are limited	Plasma treatment is possible directly at the conveyor belt. In-line suitable. No vacuum necessary	The treatable area is limited to approx 8–12 mm (plasma generation principle). More nozzles are required to treat larger surfaces
Treatment of metal	Oxidation-sensitive objects can be treated with plasma. (e.g., H_2 process gas)	Microwave plasma can transfer the energy to the object, which then overheats. kHz do not cause overheating	When aluminum is treated with plasma, very thin oxide layers (passivation) can be created	Plasma treatment of oxidation-sensitive objects is limited
Treatment of polymers and elastomers	PTFE can be activated with plasma. (Etching process.) Advanced plasma processes for elastomer- and PTFE gaskets have been developed and are in use	For several materials (e.g., silicon) a bigger pump is required to reach the necessary process pressure	"Endless objects" (e.g., tubes or cables) can be treated with plasma. Very short process time	The plasma jet has a high temperature of 200–300°C. Process parameters have to be well-aligned to the surface to avoid burning the material. (thin materials)
3D Objects	All items in the plasma chamber are treated uniformly. Also cavities can be treated from inside. (e.g., water tanks, ignition coil)	None known	Local surface treatment is possible (e.g., gluing groove)	Complex robotic technology is necessary. Treatment of surfaces with deep grooves is limited

(*Continued*)

TABLE 9.10 (Continued)

Applications properties	Low pressure plasma		Atmospheric plasma	
	Advantages	Disadvantages	Advantages	Disadvantages
Bulk material	The rotary drum procedure enables uniform plasma treatment of bulk material. The quantity and volume can vary	Only one third of the rotary drum volume can be used	The objects can be treated directly in-line	The objects have to be positioned very accurately on the conveyor belt
Electronic/ semiconductors	Plasma treatment of electronic devices, printed circuit boards, and semiconductors is state of the art	None known	Plasma treatment of metal or indium tin oxide contacts is possibly directly before the bonding process. (e.g., LCD–TFT, chip production)	The high temperature of the plasma jet and the reduced ability to treat surfaces with deep grooves may limit the usage of atmospheric plasma in the electronics industry
Coating process	The coating layers are identical und uniform. Many PECVD and PVD processes have been developed and are in use	Plasma chamber can be contaminated by coating material	No industrial uses known yet	No industrial uses known yet

**Diener Electronic Company offers both low pressure and atmospheric pressure plasma treatment systems.*

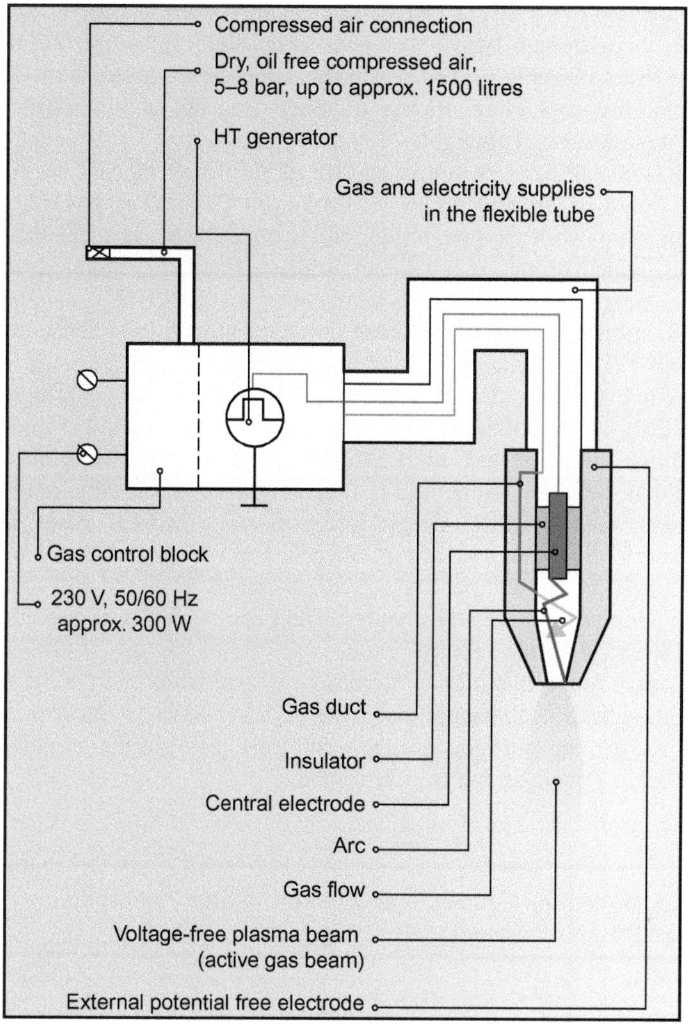

FIGURE 9.26 Schematic diagram of functional principles of an atmospheric pressure plasma system. (*Courtesy of Diener Electronic*, www.plasma.de/en/index.html) [8].

the efficiency by multiplying the number of micro-discharges in the cycle. The system is practically noiseless, produces very little ozone when operating in the open air, and generates no ozone when inert gases are used. A small amount of reactive gases may be added into the flow to obtain plasma with unique properties; this is frequently required for chemical surface modification. Thin film deposition by plasma polymerization may also be accomplished on the treated surface by adding monomers (C_2H_4, C_2H_2, etc.) into the plasma cloud. Some of the applications of APT for treating polymer surface are described in the rest of this section. Typical contact angle and adhesive bond strength before and after atmospheric pressure plasma treatment are listed in Table 9.11. The values in the table are subject to appreciable variability depending on the treatment process and measurement of contact angle and bond strength.

Early studies [55,56] demonstrated the effectiveness of APT for the treatment of polymethylmethacrylate (PMMA) and PTFE. For PMMA, helium and its mixtures with oxygen, argon, and nitrogen were used for the surface treatment, which produced bondable surfaces (Fig. 9.27) in relatively short exposure times. Similar results were obtained for PTFE as shown in Fig. 9.28, although a longer treatment time was needed to enhance the bondability of PTFE significantly.

Polyvinyl fluoride (PVF), poly[tetrafluoroethylene-co-hexafluoropropylene] (FEP), and poly[tetrafluoroethylene-co-perfluoroalkyl vinyl ether] (PFA) films were treated by atmospheric plasma treatment method to increase their adhesive strength [57]. The adhesive was Araldite, which is an epoxy resin sold by Ciba Geigy Company. The process gases included helium and its mixtures with argon, oxygen, and nitrogen. Adhesive strength was measured by a 180° peel test. Sodium-etched PFA and FEP film were tested in addition to untreated samples. The bond strengths of sodium-etched samples were called the "control".

This study found that the O_2/He plasma-treated films had the highest peel forces although their contact angles were highest and their amounts of incorporated oxygen or nitrogen atoms were least. He plasma treatment also showed better effects on the overall results.

TABLE 9.11 Wetting Contact Angle Before and After Atmospheric Pressure Plasma Treatment [54]

Plastic Type	Wetting Contact Angle (°)		Lap shear Adhesive Bond Strength (MPa)	
	Before	After	Before	After
Polypropylene	87	22	2.55	9.51
Low Density Polyethylene	87	22	2.55	10.00
High Density Polyethylene	–	–	2.17	21.55
Polyamide (nylon)	73	15	5.86	27.58
Polystyrene	–	–	3.93	27.58
Polyimide	79	10	–	–
Polycarbonate	75	33	2.83	6.40
Ethylene-tetrafluoroethylene copolymer	92	53	2.83	22.06

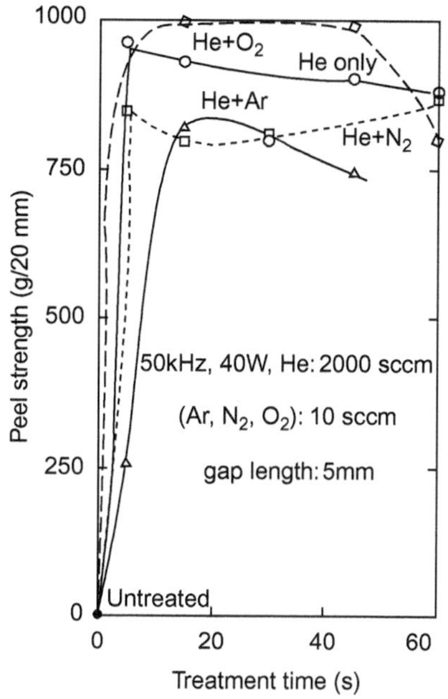

FIGURE 9.27 Peel strength of PMMA as a function of atmospheric pressure plasma treatment time [55].

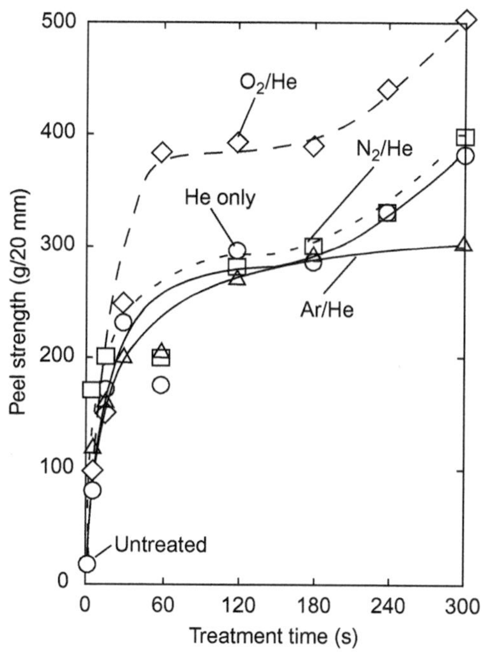

FIGURE 9.28 Peel strength of PTFE as a function of atmospheric plasma treatment time [57].

PVF treatment yielded the highest peel strength at >2 kg/mm (exceeded limits of test apparatus) compared to untreated film at 1.06 kg/mm. Surface roughness of treated PVF films increased significantly whilst treatment virtually did not affect FEP and PFA surfaces. Peel strength of treated FEP and PFA surfaces as a function of treatment time can be seen in Figs 9.29 and 9.30. Note: "control" in these figures refers to the sodium-etched film. Peel strength of APT-treated FEP film was close to or higher than the peel strength of sodium-etched film in contrast to FEP which had lower values. The maximum peel strength of FEP and PFA treated by plasma were approximately 180 (FEP) and 80 (PFA) times stronger than the forces of untreated films, whose values were 0.004 and 0.006 kg/mm. The best peel strength results for PFA were close to that of sodium-etched film.

Softal and Air Liquide companies have developed an APT technology marketed under the trade name Aldyne®. Air is substituted with a nitrogen-based gaseous mixture in the discharge area. In this process nitrogen-based chemical functional groups are grafted onto the surface, which generates both high level improvement in wettability and adhesive bond strength. For example, surface energy of biaxially oriented polypropylene is raised to 60 dynes/cm and up 72 dynes/cm for polyethylene terephthalate [58].

Plasma treatment has been applied [59,60] to textiles with the objectives of increasing or decreasing hydrophilicity, and improving dyability and printability of cotton, wool, and synthetic polymers. Atmospheric pressure plasma

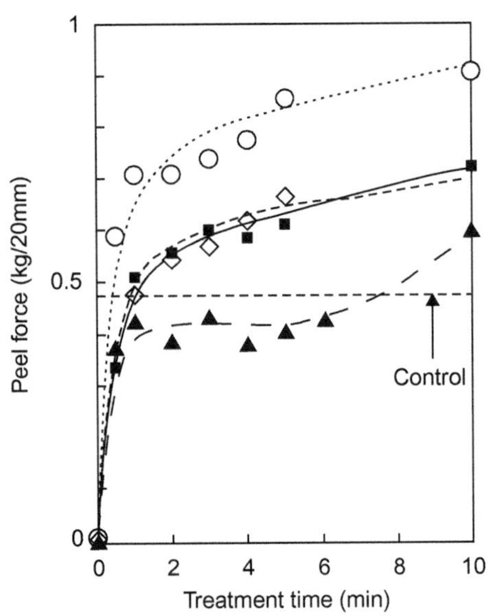

FIGURE 9.29 Variation of the peel strength of FEP film. Treatment gases were: ■ = He, ○ = O_2/He, ◇ = N_2/He, and ▲ = Ar/He [57].

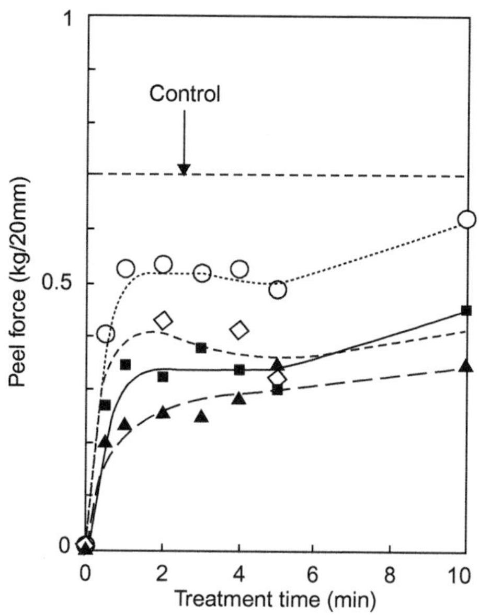

FIGURE 9.30 Variation of the peel strength of PFA film. Treatment gases were: ■ = He, ○ = O_2/He, ◇ = N_2/He, and ▲ = Ar/He [57].

treatment has been investigated as a possible means to treat textile fabrics because it represents the lowest cost plasma technique. For example, wool responded well to O_2−, air−, NH_3− gases which converted its surface from hydrophobic to completely hydrophilic. Another example was successful conversion of hydrophilic cotton surface to hydrophobic.

In another study [61] atmospheric pressure plasma, using a process gas of N_2/O_2 mixture, was applied to a polypropylene spun-bonded fabric. Chemical composition of the fabric before and after plasma treatment was analyzed by Fourier transform infrared spectroscopy. The spectra showed that oxygen- and nitrogen-containing groups were generated on the surface of the plasma-treated fabric. The scanning electron microscope was used to observe the surface morphology of the substrate. Physical properties like moisture and vapor transport, pore size distribution, and tensile strength were evaluated to understand the effect of the plasma treatment on spun-bonded polypropylene. It was evident from the capillary flow porometer results that pore size increased after plasma treatment, resulting in enhanced moisture vapor transport rate. No significant decrease in breaking strength of fabric was observed after the plasma treatment.

Commercialization of APT in the textiles industry has encountered difficulties [62]. The porous and uneven surface characteristics of fabrics make obtaining satisfactory results more difficult than with polymer films. The difficulty of inclusion of the treatment process in existing textile factories, and capital and operating costs of APT, remain barriers to its adoption in the industry.

Okubo et al. [63] modified PFA, PTFE, and polychlorotrifluoroethylene (PCTFE) by using atmospheric pressure argon and acrylic acid vapor plasma. The results of the T-type peeling test show that the peeling strength of the treated PTFE film is approximately over 100 times greater than that of the untreated film. It was confirmed from X-ray photoelectron spectroscopy and scanning electron microscope analyses that no chemical connections with F atoms existed on the surface and a hydrophilic layer was formed due to the plasma graft polymerization process.

9.4.1 In-line Plasma Treatment of Wire and Cable

Some atmospheric pressure plasma treatment methods allow inline surface modification of almost any wire insulation material [64,65]. Examples include polyamides, polyesters, and the most difficult fluoropolymers to treat. Following treatment, a surface can accept inkjet printing for marking purposes. Table 9.12 shows a comparison of the marking durability on ethylene tetrafluoroethylene copolymer (ETFE) and fluorinated ethylene propylene copolymer (FEP) for untreated (U), flame treated (Fl), and plasma treated (Pl) conditions. Some wires have been crosslinked by irradiation (IR) to enhance cut-through resistance of the insulation. Wire suppliers have been identified in each case and all ink was UV curable.

Table 9.13 shows the results of scrape abrasion testing on a Tefzel® ETFE insulation, both with and without plasma treatment. Clearly, plasma has a positive effect, enhancing the marking's resistance to removal by scrape abrasion. Contact angle data and calculated surface energy data can be found in Tables 9.14 and 9.15.

Enercon Industries has researched the in-line treatment effectiveness of different atmospheric surface activation systems on various polymer substrates. The study focused on the in-line effectiveness of the treatment methods because the processing of jacketed wires and cables is typically in-line (continuous) and printing of inks and coatings takes place immediately after the treatment. Table 9.16 shows a comparison for atmospheric pressure surface modification aimed at raising the surface energies of non-polar jacketing substrates. The data indicate that the APT process was effective in increasing the surface energy of PTFE and FEP, which are quite difficult to treat.

9.5 SURFACE TREATMENT OF BIOPOLYMERS

In the early 1980s, the plasma polymerization and plasma treatment of polymer surfaces of plasma polymers and/or plasma treated polymer surfaces in biomedical applications were contemplated seriously [66]. Some of the applications envisioned included improving the biocompatibility of polymer surfaces such as blood and tissue compatibility and controlled drug release by selective surface modification. Others were aimed at the improvement of

TABLE 9.12 Marking Durability of Flame and Plasma Treated Fluoropolymer Insulated Wires [64]

Insulation Type	ETFE			ETFE			IR ETFE						IR ETFE			FEP		
Wire Spec	**M17 5000-14TE1U00**			**M37 500 10TO2T14**			**M17 500 24102T23**			**M13-101364**			**WS191991-20-1**			**MIL-C 17G**		
Wire Vendor				**Thermax CDT, Belden Company**			**Teledyne Thermatics**			**Judd Wire**			**Delta Wire**			**Thermax CDT, Belden Company**		
Treatment Testing	U	Fl	Pl	U	Fl	Pl	U	Fl	Pl	U	Fl	Pl	U	Fl	Pl	U	Fl	Pl
Rub (finger)	F	P	P	F	F	P	F	P	P	F	P	P	F	P	P	F	F	F
Rub (eraser)	F	P	P	F	F	P	F	P	P	F	P	P	F	F	P	F	F	P
Alcohol	F	P	P	F	P	P	F	P	P	F	P	P	F	F	P	F	F	P
1,1,1-Trichlor	F	P	P	F	P	P	F	P	P	F	P	P	F	P	P	F	F	P
Freon® TMC	F	P	P	F	P	P	F	P	P	F	P	P	F	P	P	F	F	P
Water +	F	P	P	F	P	P	F	P	P	F	P	P	F	P	P	F	F	P
Acetone	F	F	P	F	P	F	F	P	P	F	F	P	F	F	P	F	F	P
Episolve 301	F	P	P	F	P	P	F	P	P	F	P	P	F	F	P	F	F	P
Masking tape	F	P	P	F	F	P	F	P	P	F	P	P	F	P	P	F	F	P
Thermal shock	F	P	P	F	F	P	F	P	P	F	P	P	F	F	P	F	F	P
ISO oil	F	P	P	F	P	P	F	P	P	F	P	P	F	P	P	F	F	P
Pen (40 dyne/cm)	F	P	P	F	P	P	F	P	P	F	P	P	F	P	P	F	F	F
Cold shock	F	P	P	F	P	P	F	P	P	F	P	P	F	F	P	F	F	P

Notes: U: untreated, Fl: flame-treated, Pl: plasma-treated. F: fail, P: pass.
All wires were marked with UV-curable ink ● Rub (finger): 20 strokes of moderate pressure ● Rub (eraser) per MIL-M-91531, 4.62 ● Alcohol, trichlor, Freon® TMC (50% methylene chloride, 50% 1,1,2-trichlorotrifluoroethane), water (plus surfactant to promote wetting), acetone, and Episolve per MIL-STD 202 ● Masking tape: press onto marking, wait 30 sec, peel off and read marking ● Thermal shock per WS 19185 ● ISO oil: soak for 48 hr and rub once ● Pen: check the spreading on the surface of the wire ● Cold shock: 5 min at −40°F, bend and rub.

TABLE 9.13 Results of a Scrape Abrasion Test Performed at Raychem, Menlo Park, CA, on Tefzel® Spec 55 Insulated Wire for Marking Identification Made by M-100J Ink-Jet Wire Marker [64]

Line Speed	30 ft/min	60 ft/min	120 ft/min
UV exposure	1 s	0.5 s	0.25 s
Average # of cycles Plasma off	7	11	12
Average # of cycles Plasma on	58	>125	48

Testing Method: Motor driven, reciprocating cam mechanism, equipped with a reset counter and capable of a 3/8″ stroke at 60 cycles per minute with a clamp for holding specimens. The contact surface is a smooth cylindrical steel mandrel, 0.025″ in diameter. Applied weight is 500 g. One cycle corresponds to two strokes. Note: The tests were stopped after 125 cycles.

TABLE 9.14 Contact Angles (Degrees) Measurement for Plasma Treated FEP Insulated Wire (22759/11-20) [64]

Plasma Treatment Conditions	Water Surface Tension (72 dyne/cm)	Methylene Iodide Surface Tension (50 dyne/cm)	Xylene Surface Tension (30 dyne/cm)
No treatment	101 ± 5	75 ± 4	46 ± 4
Air plasma	94 ± 8	70 ± 3	32 ± 3
Argon plasma	90 ± 4	66 ± 5	23 ± 5
Argon and ammonia	82 ± 4	6 ± 6	20 ± 7

adhesion in medical devices and applying protective coating to their surfaces. Since those early days, much progress has been made in the use of both technologies in modification of biopolymer surfaces. The extensive research and development and progress made in the area of compatibilization of biomaterials warrants coverage in a book devoted to the subject matter.

In the 1990s, extensive research and development on a number of biomedical devices took place. Some of the results were applied to products in orthopedic and medical devices such as knee, teeth, bone, pedicel screw, surgical tools, and others [67].

Two types of applications have been pursued for the use of plasma treatment [68]. First, modification of synthetic polymers intended for use with biological materials such as implants. Some examples include vascular prostheses, heart valves, and blood-bags. Second, plasma treatment can be used for biomolecule-immobilization of heparin and heparin-like molecules, collagen, albumin, and

TABLE 9.15 Surface Energy Calculation Based on Contact Angle Measurement for Plasma-Treated FEP Insulated Wire (22759/11-20) [64]

Plasma Treatment Conditions	Harmonic-Mean Approximation			Geometric-Mean Approximation		
	Dispersive Component (dyne/cm)	Polar Component (dyne/cm)	Total (dyne/cm)	Dispersive Component (dyne/cm)	Polar Component (dyne/cm)	Total (dyne/cm)
No treatment	17.3	5.4	22.7	14.3	2.6	16.9
Air plasma	20.5	7.2	27.9	17.7	3.9	21.6
Argon plasma	19.9	9.1	29.0	16.4	6.0	22.4
Argon and ammonia	23.0	11.5	34.5	19.3	8.4	27.7

TABLE 9.16 Effect of Surface Treatment Methods on Surface Energy of Polymers [65]

Polymer Name	In-line Surface Treatment Method	Energy Flux, $W/m^2/min$	Surface Energy (dynes/cm)	
			Before Treatment	After Treatment
PTFE	Atmospheric Gas-Phase Plasma	109	18.5	39
FEP	Atmospheric Gas-Phase Plasma	21.8	16–18	52
PVC	Flame Treatment	116.5	35	46

other molecules of biological origin to confer anti-thrombotic properties to polymer surfaces. Important research areas and specific examples of the use of plasma treatment with biomaterials are listed in Table 9.17.

Surface properties of poly (D,L-lactide) (PDLLA) were modified [69] by combining plasma treatment and collagen modification. A mouse model was used as model cells to evaluate the cell affinity of PDLLA before and after modification. Effects of different modification methods, including low pressure plasma treatment using ammonia and oxygen, collagen coating, and a combination of plasma treatment with collagen anchorage, were investigated and compared. The results showed that the hydrophilicity and surface-free energy were improved and reduced, respectively, after each modification. Plasma pre-treatment could improve the roughness as it incorporated the polar groups and positively charged groups on to the sample surface; so the plasma pre-treated surface would benefit in anchoring more collagen tightly.

TABLE 9.17 Common Research Areas and Applications of Plasma Treatment in Biomaterials Engineering [67]

Blood-compatible surface	Vascular grafts, catheters, stents, heart-valves, membranes (e.g., for hemodialysis), filters (e.g., for blood cell separation), biomolecules immobilized on surface
Non-fouling surfaces	Intraocular lenses (IOLs), contact lenses, wound healing, catheters, biosensors
Tissue engineering and cell culture	Cell growth, antibody production, essays, vascular grafts
Sterilization of surgical tools and devices	Cutting tools of surgeon, tweezers
Biosensors	Biomolecules immobilized on surfaces
Barriers coatings	Drug-release, gas-exchange membranes, device protection, corrosion protection, reduction of leaches (e.g., additives, catalysts, plasticizers)

As a result, cell affinity of PDLLA modified by combining plasma treatment with collagen anchorage was greatly improved. The modified materials could endure rinsing by PBS, which would facilitate further application when the modified materials were used as cell scaffolding in tissue engineering.

A medical grade segmented polyetherurethane (PEU) was treated with a low-powered low pressure gas plasma using O_2, Ar, N_2, and NH_3 as the treatment gases [70]. The influence of the surface modification to the polyurethane on the blood response to the polyetherurethane was investigated by measuring changes in the activation of the contact phase activation of the intrinsic coagulation cascade. The data demonstrated that the plasma treatment process caused surface modifications to the PEU that in all cases increased the polar nature of the surfaces. O_2 and Ar plasmas resulted in the incorporation of oxygen-containing groups that remained present following storage in an aqueous environment. N_2 and NH_3 plasmas resulted in the incorporation of nitrogen-containing groups but these were replaced with oxygen-containing groups following storage in the aqueous environment. In all plasma treatments there was a lowering of contact phase activation compared to the untreated surface, the N_2 and NH_3 treatments.

In this study [71] poly (lactide-co-glycolide) (PLGA) films were treated by oxygen plasma. The surface structure, topography, and surface chemistry of treated PLGA films were characterized by contact angle measurement, scanning electron microscope observation, atomic force microscopy, and X-ray photoelectron spectrum analysis. The cell (a mouse model) affinity of the oxygen–plasma treated films was evaluated under dynamic conditions by use of the parallel plate flow chamber (PPFC). The results showed that the hydrophilicity increased greatly after oxygen plasma treatment. High quantities of –C–O groups, such as hydroxyl and peroxyl groups, could be incorporated into the surface of PLGA (70/30) by controlling appropriate plasma treatment conditions. Moreover, the oxygen plasma treatment resulted in formation of peaks and valleys on the sample surfaces, and the roughness increased with treatment time. Cells stretched very well and the ability to endure the shear stress was improved greatly after the PLGA (70/30) was modified by appropriate plasma treatment, i.e., under 50 W for 2 or 10 min. However, when the treatment time was increased to 20 min, the percentage of adherent cells on the roughest surface decreased because the content of polar groups incorporated onto the surface decreased. The results showed that improved cell adhesion was attributed to the combination of surface chemistry and surface morphology of PLGA during plasma etching.

Biopolymers have continued to become increasingly important in the industry because of sustainability and environmental considerations. Hydrophobicity of poly(L-lactide) scaffolds is a drawback in applications such as tissue engineering. A study of plasma treatment is a useful technique to enhance the hydrophilicity of this polymer. The effect of this technique on the modifying depth and degradation of poly(L-lactide) has been studied. In this program the influence of NH_3 plasma treatment on the modifying depth and degradation of tissue scaffolds was investigated [72].

Cell attachment and proliferation are particularly important for the materials used for tissue-engineering purposes [73]. Polymethyl methacrylate (PMMA) and its copolymers have biomedical applications in vascular grafts, drug-releasing systems, and intraocular or contact lenses. In a study, polymethyl methacrylate films were prepared by solvent casting and their surfaces were modified by low pressure oxygen–plasma treatment by applying power of 20, 100, and 300 W. The effects of surface chemistry alterations on hydophilicity, work of adhesion, surface free energy, and cell adhesion were examined. The results demonstrated that there was an optimum value for hydrophilicity and surface free energy which enhanced cell attachment.

Radio frequency (RF) plasma treatment in O_2 was applied [74] to modify the surface of poly (L-lactic acid) (PLLA) and poly (D,L-lactic acid-co-glycolic acid) (PLGA) as biodegradable polymers. The surface structure, morphology, wettability, and surface chemistry of treated films were characterized by water drop contact angle measurement, scanning electron microscopy (SEM), optical invert microscope, differential scanning calorimetry (DSC), and ATIR–FTIR spectroscopy. The cell affinity of the oxygen–plasma treated film was evaluated by nerve tissue cell culture in stationary conditions. The results showed that the hydrophilicity increased greatly after O_2–plasma treatment. The results showed that improved cell adhesion was attributed to the combination of surface chemistry and surface wettability during plasma treatment. Cell culture results showed that B65 nerve cell attachment and growth on the plasma treated PLLA was much higher than for an unmodified sample and PLGA. Surface hydrophilicity and chemical functional groups with high polar component play an important role in enhancing cell attachment and growth.

The interaction of proteins and cells with polymers is critical to their use in scientific and medical applications [75]. In this study, plasma immersion ion implantation (PIII) was used to modify the surface of polytetrafluoroethylene (PTFE), enabling the covalent binding of a cell adhesive protein, tropoelastin, without employing chemical linking molecules. Tropoelastin coating of untreated or PIII treated PTFE simultaneously promoted and blocked cell interactions respectively, i.e., PIII treatment of the PTFE surface completely inverts the cell interactive properties of bound tropoelastin. This activity persisted over long-term storage of the PIII treated surfaces. The integrin binding C-terminus of tropoelastin was markedly less solvent exposed when bound to PIII treated PTFE than untreated PTFE, accounting for the modulation of cell adhesive activity. This presents a new methodology to specifically modulate cell behavior on a polymer surface using a simple one step treatment process, by adjusting the adhesive activity of a single extracellular matrix protein.

REFERENCES

[1] Liston EM. Plasma treatment for improved bonding: a review. J Adhes 1989;30:199–218.
[2] In-line plasma treatment of plastics. Adapted version from the German original: GAK, Dr. Gupta Verlag, Postfach 10 41 25, 40852 Ratingen, November Issue; 2008.
[3] Cheng TS, Lin HT, Chuang MJ. Surface fluorination of polyethylene terephthalate films with RF plasma. Mater Lett 2004;58:650–3.
[4] Plasma physics. Dr. Philip D. Rack, Professor, Department of Material Science and Engineering, University of Tennessee, Knoxville, Tennessee, USA, <http://web.utk.edu/~prack/Thin%20films/plasma>; 2013.
[5] Kogelschatz U. Dielectric-barrier discharges: their history, discharge physics, and industrial applications. Plasma Chem Plasma Process 2003;23(1):.
[6] Kolluri OS. Application of plasma technology for improved adhesion of materials. In: Pizzi A, Mittal KL, editors. Handbook of adhesive technology. 2nd ed. Marcel Dekker, Taylor & Francis; 2003.
[7] Chan CM, Ko TM, Hiraoka H. Surf Sci Rep 1996;24:1–54.
[8] Diener Electronic, <www.plasma.de/en/index.html>, Ebhausen, Germany; 2013.
[9] Henniker Plasma, <www.plasmatreatment.co.uk>, Warrington, UK; 2013.
[10] Acton Technologies, <www.actontech.com>; 2013.
[11] Plasma Technology, pub in 2007 by Diener Electronic, <www.plasma.de/en/index.html>.
[12] Yasuda HK, Yeh YS, Fusselman S. Pure Appl Chem 1990;63:1689.
[13] d'Agostino R, Cramarossa F, Fracassi F. In: d'Agostino R, editor. Plasma deposition, treatment, and etching of polymers. New York: Academic Press; 1990. pp. 95–162.
[14] Chan CM. Polymer surface modification and characterization. New York: Hanser; 1994.

[15] Sanchis MR, Blanes V, Blanes M, Garcia D, Balart R. Surface modification of low density polyethylene (LDPE) film by low pressure O_2 plasma treatment. Eur Polym J 2006;42:1558–68.

[16] Brennan WJ, Feast WJ, Munro HS, Walker SA. Polymer 1991;32:1527.

[17] Morra M, Occhiello E, Garbassi F. J Coll Interf Sci 1989;132:504.

[18] Schonhorn M, Hansen RH. A new technique for preparing low surface energy polymers for adhesive bonding, polymer letters. J Polym Sci B 1966;4:203–9.

[19] Dukes WA, Kinloch AJ. Preparing low-energy surfaces for bonding. In: Wake WC, editor. Developments in adhesives, vol. 1. London: Applied Science Publishers; 1977.

[20] Petrie EM. Plastics and elastomers as adhesives ISBN:0070266867 In: Harper CA, editor. Handbook of plastics, elastomers and composites. 4th ed. New York: McGraw-Hill; 2002

[21] Taylor W. Technical synopsis of plasma surface treatments, advisor: Dr. B. Welt, University of Florida, Gainesville, Florida, USA, <www.iopp.org>, December, 2009.

[22] Diener Electronic, <www.plasma.de/en/index.html>.

[23] Hall JR, et al. Activated gas plasma surface treatment of polymers for adhesive bonding. J Appl Polym Sci 1969;13:2085–96.

[24] Wegman RF, et al. Effects of varying processing parameters in the fabrication of adhesive-bonded structures-part 2, Important considerations for the bonding process, Picatinny Arsenal technical report 3999. Dover, NJ: Picatinny Arsenal; 1970.

[25] Egitto FD, Matienzo LJ. IBM J Res Deû 1994;38:423.

[26] Wertheimer MR, Martinu L, Klemberg-Sapieha JE. Plasma treatment of polymers to improve adhesion. In: Mittal KL, Pizzi A, editors. Adhesion promotion techniques: technological application. New York, Basel: Marcel Dekker; 1999. pp. 191–204.

[27] Greenwood OD, Boyd RD, Hopkins J, Badyal JPS. J Adhes Sci Technol 1995;9:311.

[28] Petasch W, Rauchle E, Walker M, Eisner P. Improvement of the adhesion of low-energy polymers by a short-time plasma treatment. Surf Coat Technol 1995;74–75:682–8.

[29] Williams RL, Wilson DJ, Rhodes NP. Stability of plasma-treated silicone rubber and its influence on the interfacial aspects of blood compatibility. Biomaterials 2004;25:4659–73.

[30] Sanchis R, Fenollar O, Garcia D, Sanchez L, Balart R. Improved adhesion of LDPE films to polyolefin foams for automotive industry using low-pressure plasma. Int J Adhes Adhes 2008;28:445–51.

[31] Source of data: gasonics/IPC applications notes. San Jose, California: Gasonics International.

[32] Tetra-Etch® by WL Gore & Associates, Inc, <www.WLGore.com>.

[33] Holdsworth GS, Dalgleish A. Plasma advancement expands applications of fluoropolymer linings and coatings in the battle against corrosion. COMPOSITES 2001 Convention and Trade Show Composites Fabricators Association, Tampa, FL; 2001.

[34] Hansen GP, et al. Achieving optimum bond strength with plasma treatment, presented at the Soc. of Mfg. Eng. Adhesives Conference, paper No. AD89-537, Sep. 12–14, 1989.

[35] Occhiello E, Mora M, Morini G, Garbassi F, Johnson D. On oxygen plasma-treated polypropylene interfaces with air, water and epoxy resins: II epoxy resins. J Appl Polym Sci 1991;42(7):2045–52. doi:10.1002/app.1991.0704 20732.

[36] Kusano Y, et al. US Patent 5,425,832, assigned to Bridgestone Corp., Jun. 20, 1995.

[37] Da Youxian HJ, Griesser AW-H, Mau, Schmidt R. Surface modification of poly(tetrafluoroethylene) by gas plasma treatment. Polymer 1991;32(6):1126–30.

[38] Caro JC, Lappan U, Simon F, Pleul D, Lunkwitz K. On the low-pressure plasma treatment of PTFE (polytetrafluoroethylene) with SO_2 as process gas. Eur Polym J 1999;35:1149–52.

[39] Caro JC, Lappan U, Pleul D, Lunkwitz K. Insertion of sulfur-containing functional groups into polytetrafluoroethylene (PTFE) by low pressure plasma treatment. Surf Coat Technol 1999;116–119:792–5.

[40] Baumgartner K-M, Schneider J, Schulz A, Feichtinger J, Walker M. Short-time plasma pre-treatment of polytetrafluoroethylene for improved adhesion. Surf Coat Technol 2001;142–144:501–6.

[41] Coen MC, Lehmann R, Groening P, Schlapbach L. Modification of the micro- and nano-topography of several polymers by plasma treatments. Appl Surf Sci 2003;207:276–86.

[42] Xu H, Hu Z, Wu S, Chen Y. Surface modification of polytetrafluoroethylene by microwave plasma treatment of H_2O/Ar mixture at low pressure. Mater Chem Phys 2003;80:278–82.

[43] Chen Q. Investigation of corona charge stability mechanisms in polytetrafluoroethylene (PTFE) Teflon® films after plasma treatment. J Electrostat 2003;59:3–13.

[44] Tu C-Y, Wang Y-C, Li C-L, Lee K-R, Huang J, Lai J-Y. Expanded poly(tetrafluoroethylene) membrane surface modification using acetylene/nitrogen plasma treatment. Eur Polym J 2005;41:2343–53.

[45] Tua C-Y, Liua Y-L, Lee K-R, Lai J-Y. Surface grafting polymerization and modification on poly(tetrafluoroethylene) films by means of ozone treatment. Polymer 2005;46:6976–85.

[46] Sun H-X, Zhang L, Chai H, Chen H-L. Surface modification of poly(tetrafluoroethylene) films via plasma treatment and graft copolymerization of acrylic acid. Desalination 2006;192:271–9.

[47] Wang C, Chen J-R. Studies on surface graft polymerization of acrylic acid onto PTFE film by remote argon plasma initiation. Appl Surf Sci 2007;253:4599–606.

[48] Shi T, Shao M, Zhang H, Yang Q, Shen X. Surface modification of porous poly(tetrafluoroethylene) film via cold plasma treatment. Appl Surf Sci 2011;258:1474–9.

[49] Kolská Z, Rezníckova A, Hnatowicz V, Svorcík V. PTFE surface modification by Ar plasma and its characterization. Vacuum 2012;86:643–7.

[50] Murokh I. Atmospheric plasma surface treatment technique. El Segundo, CA: Tri-Star Technologies, <http://Tri-Star-Technologies/news/articles/atmosphericplasmasurfacetreatment.pdf>; 2005.

[51] Tendero C, Tixier C, Tristant P, Desmaison J, Leprince P. Atmospheric pressure plasmas: a review. Spectrochim Acta B 2006;61:2–30.

[52] Bujanda AA, Rodriguez-Santiago V, Ho CC, Stein BE, Jensen RE, Pappas DD. Atmospheric plasma treatment of polymer films and alumina ceramics for enhanced adhesive. United States Army Research Laboratory; 2008.

[53] Murokh IY, Kerner AA. US patent No.5,798,146, assigned to Tri-Star Technologies Corp., Aug 25, 1998.

[54] Wolf R. Atmospheric surface modification of polymers for biomedical device adhesion. Enercon Industries, <www.EnerconInd.com>; 2003.

[55] Kogoma M, Nakamura H, Jinno H, Okazakim S. Efect of atmospheric pressure glow plasma on adhesive bond strength of plastics. Faculty of Science and Technology, Sophia University; 1992.

[56] US Patent No. 5,124,173, H. Uchiyama, S. Okazaki, M. Kogoma, assigned to E. C. Chemical Co. et al., June 23, 1992.

[57] Tanaka K, Inomata T, Kogoma M. Improvement in adhesive strength of fluorinated polymer films by atmospheric pressure glow plasma. Thin Solid Films 2001;386:217–21.

[58] Villermeta A, Cocoliosa P, Rames-Langlade G, Coeuret F, Gelot JL, Prinz E, et al. ALDYNE®: surface treatment by atmospheric plasma for plastic films converting industry. Surf Coat Technol 2003;174–175:899–901.

[59] Sparavigna AC. Plasma treatment advantages for textiles, Cornell University Library, arXiv:0801.3727v1 [physics.pop-ph], pp. 1–16, <http://arxiv.org/ftp/arxiv/papers/0801/0801.3727.pdf>; 2008.

[60] Wolf R. Atmospheric plasma: the new functional treatment for nonwovens. Proceedings of TAPPI PLACE, Conference, Boston; 2002.

[61] Jinka S, Behrens R, Korzeniewski C, Singh V, Arunachalam AI, Parameswaran S, et al. Atmospheric pressure plasma treatment and breathability of polypropylene nonwoven fabric. J Ind Textil 2013;42(4):501–14.

[62] Kale KH, Desai AN. Atmospheric pressure plasma surface treatment of textiles using non-polymerizing gases. Indian J Fibre Text Res 2011;36:289–99.

[63] Okubo M, Tahara M, Abratani Y, Kurki T, Hibino T. Preparation of PTFE film with adhesive surface treated by atmospheric-pressure nonthermal plasma graft polymerization. In: Proceedings Electrostatics Joint Conference – Session 11: Non-thermal Plasma, Boston University, June 16–18, 2009.

[64] Murokh I. In-line plasma treatment of wire insulation materials. El Segundo, CA: Tri-Star Technologies; 2005.

[65] Wolf RA. High performance wire & cable jacketing materials. Enercon Industries Corporation, <www.EnerconInd.com>; 2013.

[66] Yasuda H, Gazicki M. Biomedical applications of plasma polymerization and plasma treatment of polymer surfaces. Biomaterials 1982;3:68–77.

[67] Chua PK, Chena JY, Wang LP, Huang N. Plasma-surface modification of biomaterials. Mater Sci Eng R 2002;36:143–206.

[68] Favia P, d'Agostino R. Plasma treatments and plasma deposition of polymers for biomedical applications. Surf Coat Technol 1998;98:1102–6.

[69] Yang J, Bei J, Wang S. Enhanced cell affinity of poly (D,L-lactide) by combining plasma treatment with collagen anchorage. Biomaterials 2002;23:2607–14.

[70] Wilson DJ, Rhodes NP, Williams RL. Surface modification of a segmented polyetherurethane using a low-powered gas plasma and its influence on the activation of the coagulation system. Biomaterials 2003;24:5069–81.

[71] Wan Y, Qu X, Lu J, Zhu C, Wan L, Yanga J, et al. Characterization of surface property of poly(lactide-co-glycolide) after oxygen plasma treatmen0t. Biomaterials 2004;25:4777–83.

[72] Wan Y, Tu C, Yang J, Bei J, Wang S. Influences of ammonia plasma treatment on modifying depth and degradation of poly(L-lactide) scaffolds. Biomaterials 2006;27:2699–704.

[73] Ozcan C, Hasirci N. Plasma modification of PMMA films: surface free energy and cell attachment studies. J Biomater Sci Polym Ed 2007;18(6):759–73.

[74] Khorasani MT, Mirzadeh H, Irani S. Plasma surface modification of poly (L-lactic acid) and poly (lactic-co-glycolic acid) films for improvement of nerve cells adhesion. Radiat Phys Chem 2008;77:280–7.

[75] Bax DV, Wang Y, Li Z, Maitz PKM, McKenzie DR, Bilek MMM, et al. Binding of the cell adhesive protein tropoelastin to PTFE through plasma immersion ion implantation treatment. Biomaterials 2011;32(22):5100–11.

[76] Schut JH. Plasma treatment: the better bond. Plast Technol 1992;64:64–9.

Chapter 10

Effects of Additives on Surface Treatment of Plastics

This chapter presents a brief discussion of the impact of additives on the surface treatment of materials. We will discuss important additives and the properties and characteristics that they impart to surfaces of materials (thermoplastics, thermosets, and elastomers) in which they are compounded. Generally speaking it is possible to argue that "all" materials can be compounded with additives. In practice plastics and elastomers are the only hosts to additives, with rare exceptions. An example is the incorporation of polytetrafluoroethylene particles into the Ni−P matrix to combine the different properties of Ni−P alloys and PTFE. The resultant yields unique properties when formed into coatings by electroless (www.electro-coatings.com/electroless-nickel-plating.php) processing, such as non-stick, higher dry lubricity, lower friction, good wear, and good corrosion resistance and resistance to fouling [1].

Chapter 12 describes adhesion promoters including the all-important silane class of additives.

10.1 INTRODUCTION

A typical dictionary defines an *additive* as "a substance added in small amounts to something else to improve, strengthen, or otherwise alter it." There are vast numbers of additives incorporated in materials to enhance and modify their processing, performance, or characteristics which render them more useful. Table 10.1 shows a lengthy, though probably incomplete, list of additive types. Additives are normally used in small quantities, and they enhance the value of materials such as plastics, paints, color prints, and lubricants, by improving their process ability, performance, and appearance during manufacture and in application [2].

Generally, the modifying effect of additives is either chemical or physical. For example, adding a few parts of a laser marking additive such as titanium dioxide to a plastic renders it laser-markable. A matting additive such as silica powder, when added to a plastic film, can drastically reduces its gloss via a physical action.

Surface Treatment of Materials for Adhesive Bonding. DOI: http://dx.doi.org/10.1016/B978-0-323-26435-8.00010-1

TABLE 10.1 List of Additive Types

Acid scavenger	Adhesion promoter
Antifoaming agent (foam suppressant)	Antifogging agent
Antioxidant	Antiozonant
Antislip agent	Antistatic agent
Antitack (antiblock) agent	Biocide/antibacterial agent
Blowing/foaming agent	BOPP film modifier
Carbon black	Chain extender
Chelating agent/complexing agent	Clarifying agent
Cling agent	Coupling agent/compatibilizer
Crosslinking catalyst/accelerator/initiator	Crosslinking or curing agent/hardener
Defoamer	Deodorant
Desiccant	Dispersing agent
Emulsifier	Fibers (extender reinforcer)
Filler/fiber conductive	Fillers (reinforcer/extender)
Flame retardant/smoke suppressant	Flatting (gloss control) agent
Flexibilizer	Fluorescent whitening agent
Fragrance	Fresh keeping agent
Gel inhibitor	Heat stabilizer
Hydrophilic agent	Impact modifier
Infrared filter	Inhibitor
Laser marking additive	Leveling agent
Light stabilizer/UV absorber	Lubricants and waxes
Matting/gloss agent	Melt strength enhancer
Metal deactivator	Nucleating agent
Organic pigment	Oxygen absorber
Peptizer	Photoinitiator
Photoselective agent (agriculture)	Plasticizer
Polymerization inhibitor	Polymerization initiator
Processing aid	Release agent
Slip agent	Solvent
Styrene suppressant	Tackifier
Thickening agent	Thixotropic agent
Titanium dioxide	Viscosity modifier
Vulcanizing agent	Wetting agent

From a surface treatment standpoint the key variable to consider is the surface energy. Any additive(s) which blooms to the surface affects the surface energy of the binder. This change would impact the outcome of any surface treatments. The most frequent effect is reduction of the effective of the treatment process, resulting in a smaller increase of surface energy than desired. The consequence could be poor wetting of the treated surface, which can easily result in weak adhesive bonding. Some of the surface-modifying additives are discussed here.

TABLE 10.2 Laser Marking Processes

Process
Black carbonization
Bleaching or changing the color of a colorant in the material
Physical modification of the surface finish
Scribing a shallow groove into the material by vaporization
Highly-controlled modification of the surface by melting

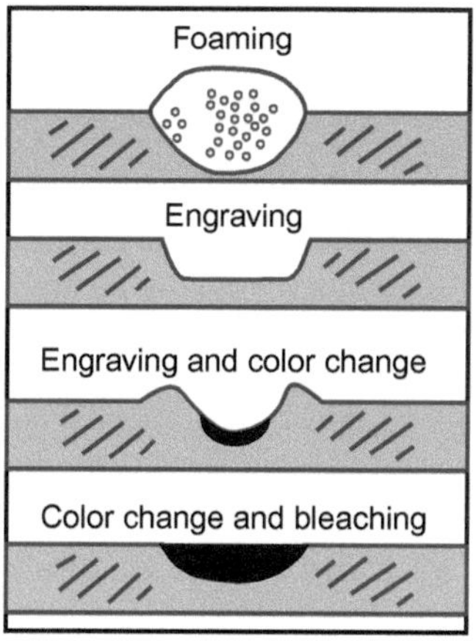

FIGURE 10.1 Examples of four types of laser marking on plastics [5].

10.2 LASER MARKING

Some additives have an unusual effect as opposed to an additive like silica which acts strictly physically. For example, laser marking additives [3] have chemical/physical impacts. Laser marking which is used to mark wires, cables, and other polymeric parts is performed by irradiating the part containing the additive with a laser beam. Letters are passed through letter or other templates to create letters and other shapes.

The marking process has a profound impact on a small part of a part surface. The marking processes include one of the following processes or a combination of two or more processes detailed in Table 10.2 [4]. Surface energy of a laser-marked surface (Figures 10.1 and 10.2) does not change markedly because of the small proportion of the involved surface.

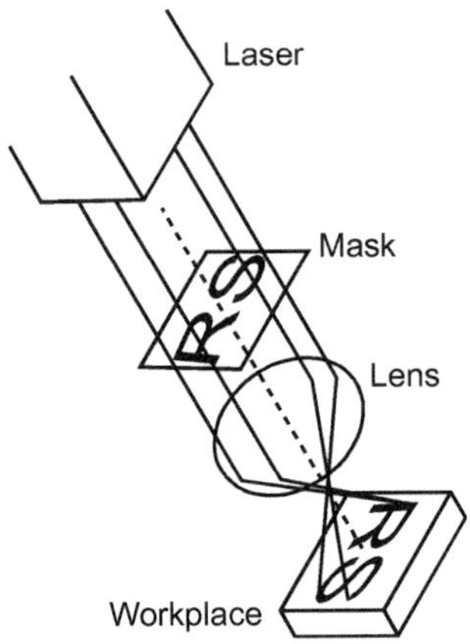

FIGURE 10.2 Laser marking using a letter mask [6].

10.3 ANTISTATIC ADDITIVES

Chemical composites of most plastics do not conduct electricity well because their chemical structures render them virtually neutral. This is a useful property for making insulation from plastics. Some polymers which contain benzene rings or double bonds can act as electrical conductors. The reason for this is the ability of the double bonds to disperse electrons along the polymer chain. The lack of electrical conductivity in the majority of plastics even to the point of dissipating static electricity poses a problem for many applications.

Antistatic chemicals are added to plastics during processing to facilitate static electricity dissipation. Antistats are usually compounds with much smaller molecules than polymers which are not miscible with the polymer (relatively low glass transition temperature). The chemical structures of the antistatic agents are designed such that along with moisture they can dissipate static electricity. To work, the antistats must migrate to the surface of the polymer. For high glass transition temperature (significantly above ambient temperature) antistatic agent must be incompatible with the polymer when cooling and solidification take place.

Clearly, the presence of a low molecular species material like antistat creates a problem for surface treatment and adhesion bonding. The intensity of the surface treatment must be sufficiently high to remove the antistatic compounds from the plastic's surface. Even then after a period of time migration of antistat to the surface reduces the level of the treatment on the surface.

10.4 ANTI-BLOCKING AGENTS

Plastic films exhibit self-adhesion properties; that is, when wound up on rolls, film layers stick to one another. This phenomenon is known as *blocking*. Film layers stick together during unwinding, which causes problem for the ensuing steps. A great many particulate additives such as kaolin and amorphous silicate are used to prevent these difficulties; these agents are known generally as anti-blocking agents. The anti-blockers are thought to work via a physical mechanism to prevent surface to surface adhesion with individual filler particles forming protrusions on the film surface. The friction and slip characteristics of the films are likewise affected. The overall particle size-distribution of the anti-blocking agents affect the micro surface smoothness of the film. Optical properties of the film should not be affected. Normally the anti-blocking agents are introduced as a dry additive (talcum powder) during compounding or processing of the polymer or, for specialized applications, as organic-solvent based dispersion prior to polymerization.

An example of an anti-blocking agent is erucamide pellets incorporated directly into the polymer during the extrusion process. They work by migrating to the surface as the polymer cools, forming a solid lubricating layer at the surface, lowering the friction, or reducing adhesion between contacting polymer surfaces and the polymer and other materials.

Surface treatment methods such as corona and flame can remove the anti-blocking agents from the surface because of their energetic nature. The removal of loose powders from the surfaces may require an extra processing step such as vacuum brushing.

10.5 INTERNAL LUBRICANTS

These lubricants are largely compatible with the polymers and impact the compounding and processing of polymers and fillers. They have the following effects:

a. Reduce friction heat.
b. Delay gelation.
c. Decrease melt viscosity.
d. Improve flow characteristics.

Some of the examples of external lubricants include natural paraffin, synthetic paraffin, micro paraffin, stearic acid, hydroxy-stearic-acid, and polyethylene waxes.

Silicones are a major class of organic materials which are used as both internal and external lubricants. Common silicone additives include polydimethyl siloxane and silicone resins. Their effects on plastic compounds vary depending on the amount of additive incorporated (Table 10.3).

Contrary to silicone, silicone polymers such as polysiloxanes and polyalkyl siloxanes such as polydimethyl siloxane have very low surface energy. They, therefore, wet the majority of solid surfaces and spread quite easily. Polydimethylsiloxanes have a surface energy of 20.4 dynes/cm and are capable of wetting most surfaces [8]. The methyl groups point to the outside in the

TABLE 10.3 Effect of Silicone Additives on Plastics [7]

Addition of 0.1–1% by Weight	Addition of 1–5% by Weight
Improved processing and flow properties	Better surface properties for the moldings
Less extruder torque	Improved surface smoothness and gloss
Significantly faster throughput	Lower coefficient of friction
Internal lubrication	Greater scratch and abrasion resistance
Better mold filling and mold release	Better mechanical properties
Reduced warpage of moldings	

molecule, yielding very hydrophobic films and a surface with good release properties, particularly if the film is cured after application [9]. Silicones have excellent thermal stability; polydimethyl siloxane is stable to 250°C. As such silicones do not simply degrade away thermally it is virtually impossible to treat or bond surfaces of silicon-containing plastics or elastomers by corona or flame treatment methods. Sections 8.3.8 and 8.7.12 describe solvent-based procedures for the treatment of silicone and fluorosilicone polymers. Section 8.7.3 describes treatment techniques for silicone rubbers including active activated gas plasma techniques.

10.6 ADDITIVES FOR WEAR AND FRICTION REDUCTION

Wear is a function of friction force and hardness of a surface. Friction is independent of the contact surface area and only dependent on the structure of the surface, material type, and the level of the force applied. A smoother surface acts in two ways: it reduces the friction force but it also increases the contact surface area. Reduction of wear for thermoplastics, elastomers, and thermosets has two requirements. Surface friction must be reduced and surface hardness increased.

Archard Eq. (10.1) describes the wear phenomenon mathematically [10]. This relation was originally derived for adhesive wear (see Ch. 9), but has proven useful in abrasive wear as well.

$$Q = \mathrm{K}\frac{WL}{H} \tag{10.1}$$

In this equation Q is the total volume of particles and debris produced due to wear. W is the total normal load, L is the sliding distance length, and H is the hardness of the softest contacting surfaces. K is a dimensionless constant. The role of friction is expressed by the term WL, which is proportional to the work of friction forces per Reye's hypothesis [11]. Consequently, lowering the work of friction (or coefficient of friction) and increasing the surface hardness reduce wear.

These two requirements are usually met by two sets of additives, although there are exceptions. For example, adding molybdenum disulfide (MoS_2) to coatings, lubricants, and plastics usually increases the surface hardness and reduces the friction force. Any hard filler, inorganic or organic, can raise the hardness of material surfaces to different extents, depending on the filler type. Molybdenum disulfide forms a film which, combined with its low coefficient of friction (0.03–0.06), reduces the work of friction and wear. Graphite can similarly reduce the friction between two moving solid surfaces [12].

Another class of material is fluorinated additives. The most common additive is low molecular weight polytetrafluoroethylene (PTFE), which is used as additive in polymeric materials, lubricants, inks and coatings [13]. There are several reasons for fluoroadditive addition but perhaps the most important is to reduce friction and wear. The extremely low coefficient of friction (0.04–0.07) of PTFE in combination with its good thermal stability makes it attractive as an additive for almost any material.

PTFE has an exclusively linear chain structure and very low intermolecular forces, which facilitate its polymer chains sliding over one another. Another consequence of weak intermolecular forces is softness of PTFE. Macroscopically speaking, PTFE can easily smear and form a thin film on just about any surface [14].

The most common method for adding low molecular weight PTFE and other fluorinated additives to plastics and other materials is the use of compounding. PTFE is added as a fine powder with extremely small particles (as small as $<1\ \mu m$) to the plastic pellets. After dry-mixing the blend is extruded, usually, using a twin-screw extruder, and repelletized. The pellets of the compound are then fabricated into parts by a standard process such as injection molding. Thermodynamic forces drive the PTFE particles towards the surface of the particles in order to minimize the surface free energy.

When a plastic part comes in contact with another surface the asperities of the two surfaces meet. There is PTFE either at the surface or near the surface. After a "break-in" period of time PTFE particles are engaged, which results in smearing and formation of a film on the surface. The presence of PTFE particles in the plastic allows the surface film to be renewed as contact and wear occur. The PTFE film has both low coefficient of friction and low surface energy. The latter property thus prevents the wetting of the surface by most liquids. Table 10.4 lists the surface energies of a select group of polymers including polytetrafluoroethylene, which has the lowest surface energy. Table 10.5 presents data on the surface energies of non-glass and metal substrates; these energies are all significantly larger than those of polymers.

The presence of additives such as molybdenum disulfide and PTFE has consequences relative to surface treatment and adhesive bonding (see Fig. 10.3). Both additives form a loose film ("boundary") that either prevents an adhesive bond from forming or which allows only a weak one to form. Surface treatment of a part containing MoS_2 requires removal of the MoS_2

TABLE 10.4 Surface Energies of Selected Polymers [15]

Material	Surface Energy (dynes/cm)
Polytetrafluoroethylene	19
Polydimethyl Siloxane	23
Natural Rubber	24
Polyethylene	30
Polypropylene	30
Acrylonitrile Butadiene Styrene	35
Polymethylmethacrylate	41
Polyethylene Terephthalate	42
Polycarbonate	46

TABLE 10.5 Surface Energies of Glass and Metal Substrates [15]

Materia	Surface Energy (dynes/cm)
Tin	526
Aluminum	840
Zinc	852
Silica Glass	~1000
Copper	1362
Stainless Steel	650–1200

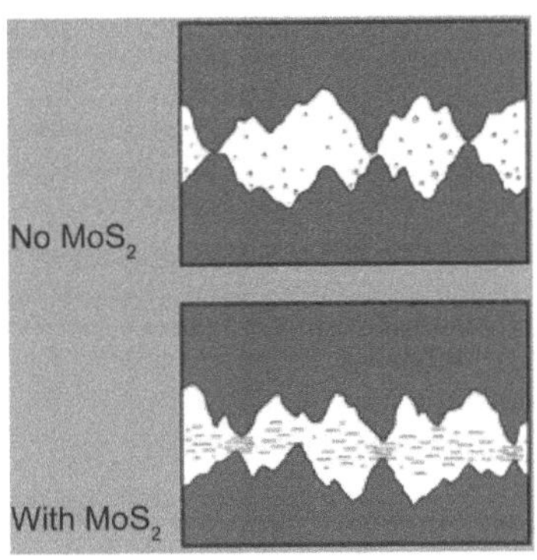

FIGURE 10.3 Lubricating action of MoS_2: molybdenum disulfide is dispersed between two metal surfaces; the asperities (surface irregularities on the metals) are prevented from coming into contact. (*Reproduced with permission of Acheson Industries*: www.achesonindustries.com/doc/guides/HowDoSolidLubricantsFunction.pdf)

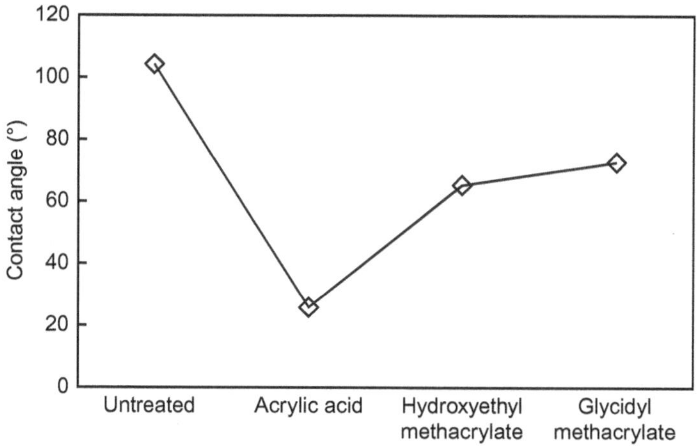

FIGURE 10.4 Contact angles on PTFE film treated using three kinds of monomers (Q_{Ar} = 100 l/min, 4 min treatment) [16].

from the surface. Aggressive wet chemical treatment techniques would enable treatment of a plastic or elastomer part made from compounds of molybdenum disulfide.

The presence of PTFE in a part is even more complicated because it is virtually impossible to remove from a surface completely. There are no inorganic or organic solvents that can dissolve PTFE. Even if the film was removed PTFE can only be treated by sodium etching (see Chapters 6, 8, and 9) or under special vacuum plasma. None of the atmospheric plasma treatment technologies (ATM) to date can provide effective treatment of PTFE for adhesive bonding. The only reliable successful results for the treatment of PTFE for adhesion have been obtained by grafting a chemical group to its surface.

For example, Okubo et al. [16] have demonstrated successful grafting of acrylic groups such as acrylic acid, hydroxyethyl methacrylate, and glycidyl methacrylate to the surface of a PTFE film using atmospheric plasma treatment (nonthermal plasma treatment, NPT). Acrylic acid grafted film surface exhibited the largest decrease in the contact angle, as seen in Fig. 10.4.

Figure 10.5 shows the results of the T-type peeling strength tests. In this figure, the horizontal axis shows the temperature of acrylic acid monomer and the vertical axis depicts the peel strength of 25 mm width PTFE film. Compared to the untreated samples which have peel strength is 0.5 N, the treated samples exhibit far better adhesive bond strengths.

Practically speaking, it is not possible to treat most plastics and elastomers which contain PTFE to impart adherability.

In summary, whether an additive affects the surface properties of a polymeric material (plastic, elastomer, or thermosetting) depends on a number of factors. Additive and polymeric material properties and processing methods

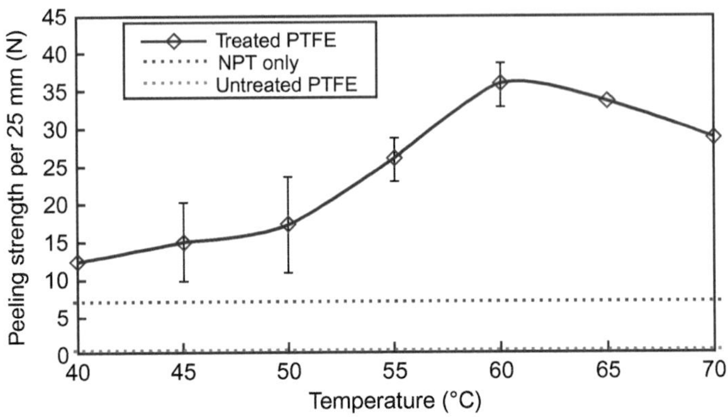

FIGURE 10.5 Peel strength vs. temperature of acrylic acid grafted PTFE film using NPT. (Q_{Ar} = 100 l/min, 4 min treatment) [16].

affect the surface properties of the parts made from them. Of prime importance is the surface energy of the material containing the additive(s). Fluoropolymer and silicone additives have the strongest impact on the surface energy of polymeric materials to which they are added.

REFERENCES

[1] Zhao Q, Liu Y, Abe EW. Effect of Cu content in electroless Ni–Cu–P–PTFE composite coatings on their anti-corrosion properties. Mater Chem Phys 2004;87:332–5.

[2] Murphy J. Additives for plastics handbook. Elsevier B.V.; 2001.

[3] Sabreen SR. “Smart additives” enhance plastics laser marking, by PennWell Corporation, Industrial Laser Solutions; January/February 2012.

[4] Han A. Analysis of the laser marking, nonconventional technologies review – no. 4, <www.revtn.ro/pdf4-2008/4-Han_A.pdf>; 2008.

[5] Laser marking on plastic materials. Rofin Baasel, Inc, <www.primross.com.mx/laser/marking/Laser_Marking_on_Plastics.pdf>; June 2013.

[6] Additives for surface and substrate. Munzing, <www.munzing.com/techinfo/181.pdf>; 2012.

[7] Silicones and silanes in the plastics industry, Pub No. 5778e, Wacker Chemie AG, <www.Wacker.com>; May 2006.

[8] Owen MJ. Chemtech. 1981 11, 288, an updated version of this article can be found in Chimie Nouvelle 2004; 85: 27.

[9] Andriot M, et al. Silicones in industrial applications. In: De Jaeger R, Gleria M, editors. Inorganic polymers. Nova Pub; 2007. pp. 61–161.

[10] Archard JF. Contact and rubbing of flat surfaces. J Appl Phys 1953;24:981–8.

[11] Reye T. Zur theorie der zapfengreibund. Der Civilingenieur 1860;4:235–55.

[12] Molybdenum-Sulfur Compounds in Lubrication. International Molybdenum Association, <www.imoa.info/moly_uses/moly_compounds/lubricants.php>; 2013.

[13] Ebnesajjad S, Morgan RA. Fluoropolymer additives, Plastics Design Library. Elsevier; 2012.

[14] Ebnesajjad S. Fluoroplastics, Non-melt processible fluoropolymers, the definitive user's guide and databook, Plastics Design Library, vol. 1. Elsevier; 2000.

[15] Bartholomew E. Acrylic pressure sensitive adhesives exhibiting enhanced adhesion to low surface energy substrates. Avery Dennison Performance Adhesive Center, pub. by Pressure Sensitive Tape Council, <www.pstc.org/files/public/BartholomewEric.pdf>; 2013.

[16] Okubo M, Tahara M, Abratani Y, Kuroki T, Hibino T. Preparation of PTFE film with adhesive surface treated by atmospheric-pressure nonthermal plasma graft polymerization. In: Proceedings Electrostatics Joint Conference, session 11, Nonthermal Plasma; 2009.

Chapter 11

Surface Treatment and Bonding of Ceramics

The significance of ceramics in today's world far exceeds their ancient position as a material of choice for earthenware and clay pots. Many new types of ceramics have been developed in the laboratory, thus expanding the choices beyond the naturally occurring ores. The natural ceramics consist of a few oxides and silicates ceramics. The new developments have added sulfides, nitrides, carbides, non-silicate glasses and many other classes of compounds to the ceramic class used as dielectric, magnetic, and optical materials [1]. Ceramics are often combined with other materials including metals, thermoplastics, thermosets, and elastomers either by compounding or by joining of distinct parts. In the latter case adhesives are required to bond ceramics to other materials. This chapter begins with a short description of industrially significant ceramics followed by important treatment methods to impart adhesive bondability to them.

11.1 INDUSTRIAL CERAMICS

The focus of this section is on the ceramics which are commonly used in various modern industries. Sometimes they are called "advanced ceramics" referring to ceramics used in specialized, recently developed, versus antiquity, applications. Advanced ceramics often have relatively simple chemical compositions but are difficult to manufacture. Advanced ceramics can be classified according to their chemical/crystal structure [2,3]. These include:

1. Oxide ceramics – oxides form the dominant class of ceramic compounds. They have a refractory character, and are chemically stable at high temperatures in oxygen-containing atmospheres. The subgroups include:
   - **(a)** Binary oxides
   - **(b)** Alumina ceramics
   - **(c)** Zirconia ceramics
   - **(d)** Spinels
   - **(e)** Perovskites
   - **(f)** Garnets

Surface Treatment of Materials for Adhesive Bonding. DOI: http://dx.doi.org/10.1016/B978-0-323-26435-8.00011-3

   (g) Silica and silicate ceramics
2. Chalcogenide ceramics
   (a) Binary sulfides, selenides, and tellurides
   (b) Binary sulfides, selenides, and tellurides
3. Nitride and oxynitride ceramics
   (a) Binary nitrides
   (b) Silicon and aluminum nitrides
4. Carbides and borides
   (a) Diamond and silicon carbide
   (b) Heavy metal carbides
   (c) Borides.

A subgroup within the advanced ceramics is called *technical ceramics.* Technical ceramics feature hardness, physical stability, extreme heat resistance, chemical inertness, biocompatibility, and superior electrical properties. As such, they are highly resistant to melting, bending, stretching, corrosion, and wear [4]. A list of these ceramics is given in Table 11.1.

Ceramic materials used in dentistry must have some important properties including [5,6] translucency, chemical stability, fluorescence, biocompatibility, a high resistance to compression, and a coefficient of thermal expansion similar to that for tooth structure. There are some clinical disadvantages and limitations, such as friability and susceptibility to fracture propagation [7].

Some of the ceramics are described in the following sections. The coverage of ceramic types in this chapter is neither comprehensive nor extensive. The reader may refer to ceramics reference books for in-depth reading [2,8,9].

TABLE 11.1 Examples of Technical Ceramics [4]

Technical Ceramics
Alumina
Aluminium nitride
Aluminium silicate
Zirconium silicate
Barium titanate
Fused silica
Magnesium oxide
Pyrolytic boron nitride
Silicon carbide (SiC)
Silicon nitride
Steatite
Titania
Zirconia
Lead zirconate titanate

11.1.1 Alumina Ceramics

Alumina is probably the most widely used advanced ceramic, and is made from aluminum oxide. Alumina possesses high ionic inter-atomic bonds leading to high chemical stability, thereby making it a good electrical insulator. It is extremely resistant to wear and corrosion and has a high mechanical strength.

Alumina ceramics are manufactured by different types of manufacturing processes including isotactic molding [10], injection molding, and extrusion. Precision grinding and lapping, laser machining, and other processes are among the finishing processes of this ceramics type. Alumina ceramics perform well when wear resistance, corrosion resistance, and strength are required at an economical price. They have high dielectric properties rendering them suitable for many electronic applications including electronic substrates and components of semiconductor processing equipment.

11.1.2 Steatite Ceramics

The composition of this advanced ceramic consists of magnesium silicate (Mg_2SiO_4). Steatite is a material of choice of for insulators for electrical components. Other properties of steatite include excellent dielectric strength, low dissipation factor, and high mechanical strength. Steatite's excellent insulating properties render it suitable for use in thermostats and many electrical household products.

11.1.3 Zirconia Ceramics

Zirconia is composed of zirconium oxide (ZrO_2). It has the highest strength and toughness at room temperature of all the advanced ceramic materials in addition to high resistance to corrosion, wear, and abrasion. The fine grain size allows for extremely smooth surfaces and sharp edges. Because of its high tolerance to degradation zirconia is the material of choice in the manufacture of bearings and grinding. It also has high resistance to crack formation or "fracture toughness." Zirconia is commonly used in dentistry in addition to applications such as automotive oxygen sensors or applications which are subjected to mechanical load (structural ceramics).

11.1.4 Silicon Carbide

Silicon carbide (SiC) particles are consolidated (or bonded together) by a process called *sintering*. Sintering is coalescence of a powder into a solid or a porous mass by heating it under pressure without melting of the powder. Silicon carbide forms an extremely hard ceramic making it useful for automotive clutches and brakes, and bulletproof vests. This ceramic exhibits the highest corrosion resistance among all the advanced ceramics in addition to retaining its strength at up to 1400°C. Silicon carbide also has excellent wear resistance and thermal shock performance. Other applications of ceramic carbide include pump parts, mechanical seals, nozzles, and plates for polishing silicon wafers.

11.1.5 Cordierite Ceramics

Cordierite consists of magnesium iron aluminum cyclosilicate. It usually occurs in argillaceous rocks. Argillaceous components include fine-grained (<2 μm) aluminosilicates, and specifically clay minerals such as kaolinite, montmorillonite-smectite, illite, and chlorite. Claystone and shales are thus predominantly argillaceous. Cordierite has a very high thermal shock resistance and thus is widely used in high temperature industrial applications such as heat exchangers for gas turbines.

11.1.6 Mullite

A rare grade of aluminum silicate ($3Al_2O_3 \cdot 2SiO_2$) in nature is mullite, which is colorless or white and has an orthorhombic crystal structure. It is commonly formed at high temperatures and under low pressure conditions in man-made ceramics. Mullite properties include low thermal conductivity and thermal expansion and outstanding resistance to creep. It exhibits high temperature strength and outstanding stability under harsh chemical environments. Applications of mullite are found in furnace muffles, thermocouple protection tubes, and rollers for kilns.

An example is aluminum silicate, with the chemical formula $3Al_2O_3 \cdot 2SiO_2$, which is formed upon firing aluminosilicate raw materials [11]. It is the most important constituent of ceramic whiteware, porcelains, and high-temperature insulating and refractory materials. It will not melt below 1810°C, whereas those with a lower ratio partially melt at temperatures as low as 1545°C [12].

11.1.7 Silicon Nitride

Silicon nitrides (Si_3N_4) feature an excellent combination of material properties. They are nearly as light as silicon carbide (SiC), but their microstructure gives them excellent thermal shock resistance and their high fracture toughness makes them resistant to impact and shock. They also offer an excellent combination of low density, high strength, low thermal expansion, and good corrosion resistance and fracture toughness. Applications include various aerospace and automotive engine components, papermaking machine wear surfaces, armor, burner nozzles, and molten metal processing parts.

11.1.8 Sapphire Ceramics

Sapphire is the single-crystal form of alumina oxide (Al_2O_3), a mineral known as *corundum.* It can be found naturally as gemstones or manufactured synthetically. Sapphire holds a unique combination of physical, chemical, and optical properties, making it the optimal choice for a wide range of applications. The outstanding features of the material include: high melting temperature (2053°C), extreme hardness (9 on Mohs scale), strong mechanical properties, thermal conductivity equivalent to that of metals, outstanding chemical resistance, low coefficient of friction, exceptional optical transparency, and superior radiation resistance.

Applications of sapphire include gallium arsenide carrier plates, point-of-sale scanner windows, microwave plasma tubes and windows, fixtures for high temperature equipment, and blue light emitting diodes.

11.2 TREATMENT OF CERAMICS FOR BONDING

This section will describe examples of surface treatment methods for important ceramics. There are many types of ceramics, if one includes the many commercial and experimental ceramics used in dentistry. Not surprisingly a great deal of surface treatment and bonding studies is conducted in the field of dentistry.

Ceramics have inherently high surface energy and are usually wetted at contact angles less than 90°, which is lower than that of most plastics. Yet some like zirconia are hard to bond, especially when strong durable adhesive bonds are required. Another difference from plastics is that most ceramics have a low coefficient of thermal expansion (CTE). Ceramics are different from plastics from the standpoint of surface treatment techniques. There are relatively few methods commonly just for imparting adherability to plastics surfaces. Conversely, other than a few methods for ceramics such as silane treatment, there are many variations in the manner by which those surfaces are prepared for bonding.

The techniques used for bonding ceramics are listed in Table 11.2. *Adhesive* bonding involves use of an adhesive compound (organic) which acts as an intermediary between the ceramic and the other adherend. Many adhesives including acrylics and methacrylics are used to bond ceramics.

Cementing uses an inorganic chemical or a mixture of inorganic and organic components as an adhesive layer with well-known applications in dentistry. Dental cements, by definition, are "luting agents" meant to fill the microscopic gap between restorative materials and tooth structure. Examples include zinc phosphate, polycarboxylate, and glass ionomer cements. Mechanical joining involves use of hardware such as screws and nuts.

Brazing is another technique that is often used to bond ceramics and metals. The brazing process creates a very thin, uniform joint. It is desirable

TABLE 11.2 Summary of Ceramic Joining Methods[a] [13]

Joining method	Shape Flexibility	Temperature/ env. cap.	Vac./ Hermiticity	Strength	Cost
Adhesive	M	L	L	L	L
Cementitious	M-H	M	L	L	L
Mechanical	L-M	L-M	L-H	L-M	L-M
Brazing	M-H	M	H	M-H	M
Diffusion welding	L	H	H	H	H
Fusion welding	M	H	H	M-H	M-H

[a] *L = low, M = medium, and H = high.*

to have the braze joint as thin as possible. In fact, the thinner the brazed joint, the stronger the final product will be.

Diffusion bonding or welding has been widely demonstrated for ceramics. It entails joining by placing a ceramic powder layer between two ceramic components and heating these components and the layer so they sinter together and are consolidated. Diffusion welding produces good properties but is limited by configurations to which it can be practically applied.

Fusion welding of ceramics, as with metals, is achieved by filling the joint between parts to be joined with molten material obtained by melting the edges of the parts making up the joint while they are in contact (i.e., in autogenous welding), or with additional molten material from a filler of a similar or compatible material. In fusion welding of dissimilar materials, their melts must also be compatible with one another. Besides chemical compatibility between substrates and any filler, the ceramics being fusion-welded must be compatible with the welding environment and must be physically compatible with one another. Commercial applications of fusion welding are limited [14].

Treatment methods for adhesive bonding, cementation, and brazing are covered in more detail in the rest of this chapter.

11.3 TREATMENTS FOR ADHESIVE BONDING

The bonding surface of ceramics must be completely cleaned to remove all oils and other organic contaminants. As is the case for any other bonding surface, oils, grease, and organic contaminants act as inhibitors for forming adhesive bonding. A good solvent such as methyl ethyl ketone, acetone, *iso*-propanol, toluene or others which do not leave a residue behind may be used to remove organic contamination and oils from the ceramic surfaces.

Generally ceramic surfaces can be treated in the following manner to impart bondability. Some of these techniques are more commonly applied than others.

1. Degreasing and cleaning.
2. Roughening of the surface by grit or powder blasting.
3. Acid etch usually using hydrofluoric (HF) or phosphoric (H_3PO_4) acids.
4. Silane treatment.
5. Laser treatment.
6. Flame treatment.

11.3.1 Degreasing

Degreasing and cleaning can be accomplished by methyl ethyl ketone, acetone, or *iso*-propanol wash. A metal brush may be used, if required. Other organic solvents could be used but care should be taken to avoid heavier solvents such as kerosene which could leave a residue behind. The cleaned surface may be dried in ambient air or by use of clean compressed air. A simple test to determine if the surface is clean and degreased is to drip a few drops of water on the cleaned area. If the water spreads to cover the area with a continuous film, the bond area is clean. If the water beads, that means the surface needs to be degreased further.

11.3.2 Abrasion or Surface Roughening

Abrasion or *surface roughening* is important to obtain the strongest and most durable bonds. Abrading a surface removes surface films, scale, and oxides and also gives a more suitable area for the adhesive to contact or grip. Abrasion should always be applied, if it is practical, to the adherend being bonded. In the case of ceramics grit blasting is superior to sand paper (120–200 grit), emery cloth, or steel wool. Just enough abrasion should take place to clean and roughen the surface. After completion of the abrasion treatment all loose particles should be removed from the surface before applying adhesive by either brushing it clean or, better yet, by blowing with compressed air. The next step is degreasing to clear away loose particles and residual oils.

Roughening by grit blasting accomplishes two purposes [15]. It removes loose debris from the bonding surface and increases the interfacial (contact) area of the ceramic and the other adherend. The increase in contact area adds to the strength of the bond between two adherends.

Typically, a controlled-particle size and shape grit (Table 11.3) such as alumina is used for surface roughening. It is then applied in the cleaning or preparing of a surface by forcibly propelling a stream of alumina grit against the surface.

TABLE 11.3 Grit Size Conversion [16]

Grit Size	Inches (Average)	Microns (Average)
16	0.043	1092
20	0.037	940
24	0.027	686
30	0.022	559
36	0.019	483
46	0.014	356
54	0.012	305
60	0.010	254
70	0.008	203
80	0.0065	165
90	0.0057	145
100	0.0048	122
120	0.0040	102
150	0.0035	89
180	0.0030	76
220	0.0025	63

Often the combination of abrasion and degreasing is sufficient surface preparation for most bonds except the most demanding industrial applications.

11.3.3 Acid Etch

Acid etching is a process where acid is applied to a porcelain or ceramic surface to create pores to enhance bond strength of an adhesive or a coating to the surface. Examples of etching acids include hydrofluoric, phosphoric, and hydrochloric acids. For example the bond surface is thoroughly etched (washed) with 3–10% hydrofluoric acid for a few minutes followed by thorough rinsing and drying. There are disadvantages to the use of acids, particularly hydrofluoric acid (HF), which is a toxic and corrosive acid. This acid has health risks and can also corrode the surrounding surfaces.

There are controversies surrounding the effectiveness of acid etching because of inconsistent results for its ability to improve long term adhesion. It is suggested that some acid residues from the etching process tend to remain in the surface pores of ceramic adherend. These residues can damage the adhesive layer and weaken or cause failure of the bond [17,18].

The effect of acids depends on the type of ceramics and the acid etching process. Some ceramics exhibit an increase in the surface roughness (porosity), i.e., an increase in the area of adhesive contact. Others, such as alumina,

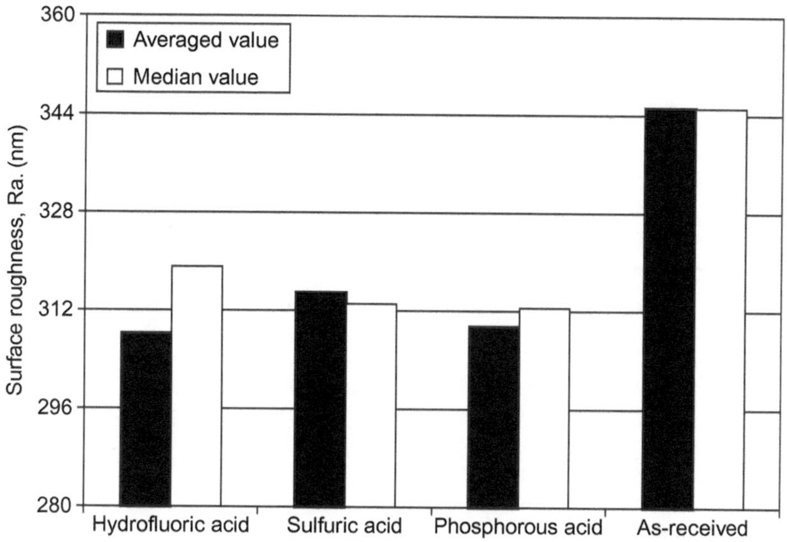

FIGURE 11.1 Surface roughness of substrates with various acid treatments [19].

actually undergo a *decrease* in surface roughness (Fig. 11.1), that is, a *decrease* in the area of adhesive contact.

In dentistry hydrofluoric acid (HF) is commonly used to etch the bonding surface of indirect porcelain restorations [20,21]. To avoid working with the hazardous HF, acidulated phosphate fluoride is used. Similarly their validity in achieving adequate bonding in dental restorations is still controversial. Figure 11.2 shows a comparison on (a) an unetched, (b) a phosphoric acid etched, and (c) a hydrofluoric acid etched specimen of feldspar. The unetched surface is smooth as opposed to phosphoric and hydrofluoric acid etched samples, which are porous. It is possible to observe that porosity actually reduces the contact surface of adhesive (and bond strength) unless the applied adhesive fills the pores and forms a continuous layer film. In that case the adhesive contact surface area increases, resulting in an increase in the bond strength.

11.3.4 Silane Coupling Agents

There are methods available for creating stronger interfacial bonds. Silane coupling agents or other adhesion promoters coupled with structural adhesive formulations tailored for specific applications have vastly improved the performance of bonded assemblies, including those involving ceramics. Chapter 12 offers an in-depth review of adhesion promoters including silane chemistry and its mechanism of adhesion promotion.

Silane groups usually have two different reactive groups, as shown in Figure 11.3(a). One group is reactive to the substrate and the other to the

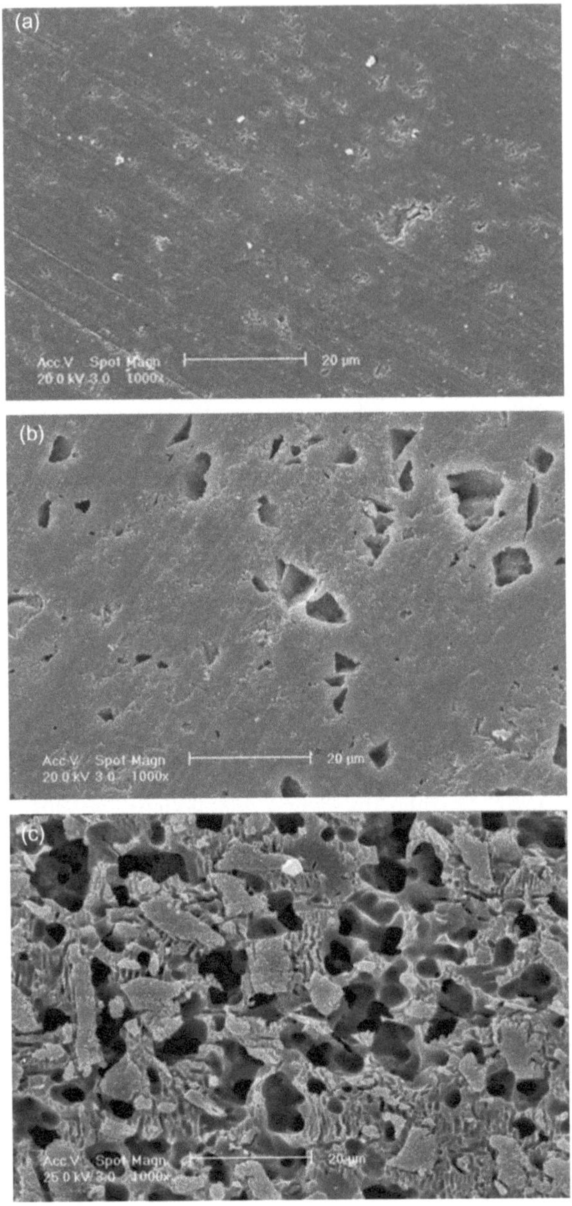

FIGURE 11.2 SEM micrographs of feldspar ceramic surfaces: (a) Specimen after 600 grit SiC wet grinding; (b) Specimen treated with 37% H_3PO_4; (c) Specimen treated with 8% HF [20,21].

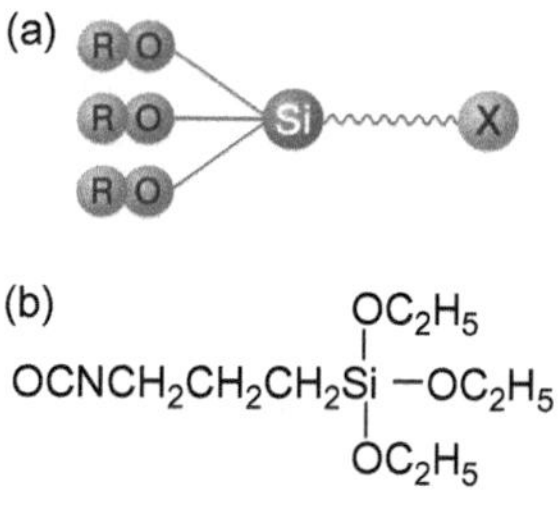

FIGURE 11.3 (a) General structure of silanes. Si = silicon; RO = interacts/reacts with inorganic materials (ceramic); X = Reactive groups form bond with organic materials (adhesive). (b) An example of a silane coupling agent is 3-isocyanatopropyl triethoxy silane.

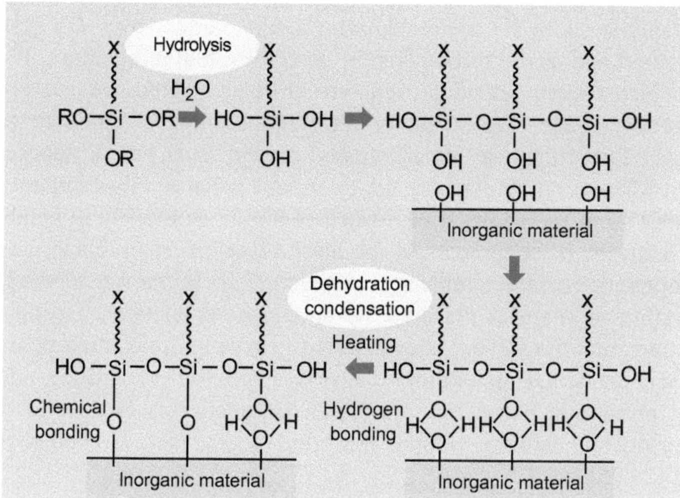

FIGURE 11.4 Reaction mechanism of an ideal silane [22].

adhesive. An example can be seen in Figure 11.3(b). Most metals and inorganic materials have hydroxyl groups in their structure which are capable of hydrogen bonding to other materials and giving tightly bound water on their surface. In the bonding process with silanes, silane coupling agents form silicon-hydroxyl (silanol) groups which then form bonds with these inorganic surface hydroxyl groups by covalent or hydrogen bonding. The usual inorganic-reactive group(s), —O-R, on the silanes which promote adhesion, are hydrolysable organic groups, like alkoxy or acetoxy silanes. The X, or organic, group on the silane is typically a reactive group with which the adhesive will react or interact. Figure 11.4 shows the reaction mechanism of an ideal silane.

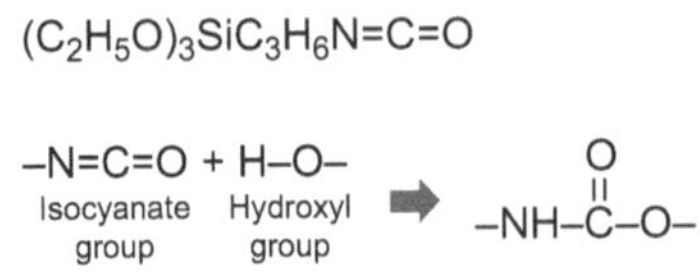

FIGURE 11.5 Reaction mechanism of silane with iscocyanate end group with a hydroxyl group.

Silanes come with a variety of organic-reactive end groups including epoxy, amine, isocyanate, and alkene-containing. The type of reaction and the required catalyst depend on the end groups. Figure 11.5 illustrates the reaction mechanism of the silane shown in Fig. 11.3(a).

Inorganic surfaces are pretreated with a silane either by a dry method or a wet method and at room temperature or at elevated temperature. If a filler surface which contains residual moisture is being treated, the moisture will hydrolyze the silane. Water and/or an alcohol may be added to enhance hydrolysis. For fillers a high-intensity mixer, such as a Henschel or Littleford Mixer, can be used for the treatment process. Filler treatment may also benefit from the addition of a trace of amine or titanate to catalyze the reaction with the filler surface. At the least a monolayer of silane must form on the inorganic surface to enhance adhesion [23]. In the wet method, a low concentration of silane is dissolved in water and/or alcohol, and applied to the substrate, and the solvent is evaporated to give a thin coating of silane on the surface. In dental restorations there is a problem with using silanes in adhesion promotion because of the bond degradation over time in the oral environment, but, without silanes, the bond degradation is even more pronounced [24,25].

11.3.5 Laser Treatment

Laser treatment has been under investigation as a tool for the treatment of ceramics for bonding [26]. The benefits of using excimer (or exciplex) lasers for surface pretreatment of ceramics include: ceramic materials will absorb ultraviolet radiation strongly but only at the very top surface layer (μm range), and such action not only cleans up surface contaminants such as fluorocarbons and silicones completely, but could also provide an extremely effective means for structuring ceramic surfaces via the action of ablation, hence improving the adhesion bond strength [27,28]. Previously, the ablation effects on the modification of surface structure of some engineering ceramics have been investigated [29].

Silicon nitride (Si_3N_4) based ceramic cutting material and a structural alloy steel (SAE 4340) were surface-treated using a 800 mJ KrF excimer laser with an goal of improving the ceramic-to-metal bond strength. Significant improvement in adhesion shear strength was obtained for the laser-treated samples as compared with those tested in as-received and mechanically-abraded conditions [26]. [More recent studies have shown that the effectiveness of laser treatment is dependent on the type of ceramic, and laser and other types of treatment combined with laser treatment [30].

11.3.6 Flame Treatment

Flame treatment is a well-known method of surface treatment to impart adherability to a number of plastics such as polyolefins and polyvinyl fluoride. Flame treatment oxidizes the surface of polymeric materials to introduce polar reactive groups such as hydroxyl and carboxyl which improves surface free energy and consequently the wettability and adherability of their surfaces.

The use of this method for the treatment of ceramics is rather recent [31,32]. The PyrosilPen flame treatment technology was investigated by surface-treating silicate, aluminum oxide, and zirconium oxide ceramics. It was found to yield good bonding to luting composites. However, sandblasting the ceramic surfaces prior to flame treatment resulted in a stronger bond.

The PyrosilPen-Technology is based on the Silicoater-Technology developed in 1984 [33]. For the first time, this system allowed a very strong and long lasting adhesive bond to be created between resins and precious as well as non-precious metals.

11.4 TREATMENT FOR CEMENTATION

Cementation requires similar surface treatment to that required for adhesive bonding including mechanical treatment and chemical treatment, namely silanation. Examples of cements and surface treatment methods used in dental restorations are listed in Table 11.4.

Mechanical treatment in laboratory-processed composites using sandblasting with aluminum oxide particles proved to be the best alternative to raise restoration surface energy because it promotes a nonselective degradation of the resin and results in a better adhesion to the composite cement protocols [7]. Acid etching is less effective and detrimental because of its ability to dissolve the inorganic constituents of the cement.

11.5 TREATMENTS FOR BRAZING

Brazing is one of the oldest processes [34] used for bonding metals to ceramics for structural purposes.

Brazing can join common industrial metals including aluminum, copper, stainless steel, and refractory metals to aluminum oxide, aluminum nitride, silicon carbide, and other oxides, nitrides, and carbides. The processing conditions depend on the materials and the joined assembly size and

TABLE 11.4 Laboratory-Processed Composite Composition and Surface Treatment Protocols [7]

Restorative Materials	Composition*	Surface Treatment Protocols
Solidex	61% UDMA and photostarters; 39% (vol) inorganic particles	Sandblasting with aluminum oxide for 10 s and silane application
Targis	33% BIS-GMA, DDMA, and UDMA; 67% (vol) inorganic particles	Sandblasting with aluminum oxide for 10 s and silane application
Artglass	30% methacrylates; 70% inorganic particles	Sandblasting with aluminum oxide for 10 s and silane application
belleGlass	26% UDMA and DDMA; 74% inorganic particles	Sandblasting with aluminum oxide for 10 s and silane application
Filtek Z250	40% UDMA, BIS-EMA, BIS-GMA; 60% inorganic particles	Sandblasting with aluminum oxide for 10 s and silane application

**BIS-GMA = bisphenol A glycidyl methacrylate; BIS-EMA = bisphenol-A polyethylene glycol diether dimethacrylate; DDMA = decane dimethacrylate; UDMA = urethane dimethacrylate.*

geometry. The biggest problem is the mismatch of the coefficient of thermal expansion (CTE) of the materials being joined. Materials expand at different rates depending on the composition (atomic elements), structure (atomic arrangement), and thermal properties.

Linear CTE of $23 \times 10^{-6}/°C$ of aluminum (Table 11.5) indicates it is one of the most expansive of metals when heated. In contrast, SiC, quartz, and tungsten have almost zero or little expansion when heated. The mismatch would be substantial if aluminum were to be welded to silicon carbide as an example. The differences would appear like the schematic shown in Fig. 11.6.

For example, for silicon carbide when the braze thickness is less than 10 μm, the strength of the braze joint is comparable to that of SiC material alone. Typically, the brazing process is conducted at temperatures exceeding 700°C and utilizes an intermediate silicon compound with a CTE closely matched to the base SiC material. The metal of choice is usually Kovar, a nickel/iron/cobalt alloy, which is a controlled expansion alloy most commonly used for high integrity glass and ceramic, together with metal seals, lids, lead frames, and electronic package bases. As specified in ASTM F15, the alloy (Kovar©) is available in rods, sheets, coils, and other forms. This is primarily because the CTE closely matches that of the alumina substrate.

Other common braze materials used are silver/copper, pure silver, copper, and gold. Molybdenum, manganese, or tungsten coatings ensure excellent interface bonding of conductor to substrate when brazing is desired.

TABLE 11.5 Coefficient of Thermal Expansion of Different Materials [35]

Material	Linear Coefficient, α, at 20°C (10^{-6}/°C)
Aluminum	23
Brass	19
Carbon steel	10.8
Copper	17
Diamond	1
Glass	8.5
Glass, borosilicate	3.3
Gold	14
Invar	1.2
Iran	11.1
Lead	29
MACOR	9.3
Magnesium	26
Molybdenum	4.8
Nickel	13
Platinum	9
Quartz (fused)	0.59
Sapphire	5.3
Silicon carbide	2.77
Silicon	3
Silver	18
Stainless steel	17.3
Tungsten	4.5

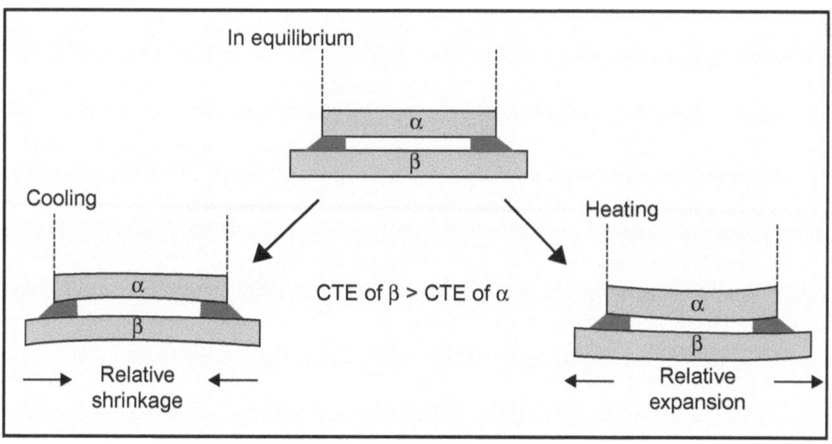

FIGURE 11.6 Schematic diagram of a brazing joint with mismatching CTE materials [35].

REFERENCES

[1] White WB. Basic science of advanced ceramics. In: Somiya S, et al., editors. Handbook of advanced ceramics. Elsevier; 2003.

[2] Somiya S, et al. Handbook of advanced ceramics. Elsevier; 2003.

[3] Taylor DA. Advanced ceramics. Mater Australia 2001;33(1):20–2.

[4] <www.morgantechnicalceramics.com/products-materials/about-ceramics/>; 2003.

[5] Borges GA, Sophr AM, De Goes MF, Sobrinho LC, Chan DCN. Effect of etching and airborne particle abrasion on the microstructure of different dental ceramics. J Prosthet Dent 2003;89:479–88.

[6] Mclean JW, Hughes TH. The reinforcement of dental porcelain with ceramic oxides. Br Dent J 1965;119:251–67.

[7] Soares CJ, Soares PV, Pereira JC, Fonseca RB. Surface treatment protocols in the cementation process of ceramic and laboratory-processed composite restorations: a literature review. J Esthet Restor Dent 2005;17:224–35.

[8] Carter CB, Norton MG. Ceramic materials: science and engineering. Springer; 2007.

[9] Wagh AS. Chemically bonded phosphate ceramics – twenty-first century materials with diverse applications. Elsevier; 2004.

[10] Ebnesajjad S. Fluoroplastics, vol. 1. Non-melt processible fluoroplastics. Plastics Design Library. Oxford, UK: Elsevier; 2000.

[11] Duval DJ, Risbud SH, Shackelford JF. Mullite. In: Shackelford JF, Doremus. RH, editors. Ceramic and glass materials:structure, properties and processing. Springer; 2008.

[12] Encyclopedia Britannica, <www.britannica.com/EBchecked/topic/396942/mullite>, June 2013.

[13] Roy W. Rice. Ceramic fabrication technology. 1st ed. CRC Press; 2002.

[14] <www.ukessays.co.uk/essays/environmental-studies/carbon-electrolyte-layer.php>; 2008.

[15] Thurmond JW, Barkmeier WW, Wilwerding TM. Effect of porcelain surface treatments on bond strengths of composite resin bonded to porcelain. J Prosthet Dent 1994;72:355–9.

[16] Technical Manual, Blastite® BT (virgin aluminum oxide), pub by Washington Mills, <www.washingtonmills.com/applications/blasting-media>; 2013.

[17] CTI REPORT 80-2-2 (R-85). Methods and equipment for the preparation of horizontal concrete slabs for bonding ceramic tile, CTI REPORT 80-2-2 (R-85). Ceramic Tile Institute of America, Inc; 2013.

[18] Kato H, Matsumura H, Atsuta M. Effect of etching and sandblasting on bond strength to sintered porcelain of unfilled resin. J Oral Rehabil 2000;27:103–10.

[19] Sun Z, Pan D, Wei J, Wong CK. Ceramics bonding using solder glass frit. J Electron Mater 2004;33(12):1516–23.

[20] El Zohairy AA, De Gee AJ, Hassan FM, Feilzer AJ. Dent Mater 2004;20:778–87.

[21] Powers JM, Farah JW, O'Keefe KL, Kolb B, Udrys G. Guide to all-ceramic bonding, sponsored by Kuraray America, Inc, <www.kuraraydental.com>; 2011.

[22] Silane Coupling Agents, pub. by Shin-Etsu Silicone, <www.Silicone.JP/e/>; 2012.

[23] Pape PG. Adhesion promoters. In: Ebnesajjad S, editor. Handbook of adhesives and surface preparation. Oxford, UK: Elsevier; 2011.

[24] Lung CY, Matinlinna JP. Aspects of silane coupling agents and surface conditioning in dentistry: an overview. Dent Mater, Elsevier; 2012;28:467–77.

[25] Doucet S, Picard B, Tavernier B. Effect of different ceramic surface treatments on one-step micro indentation bond strength. Eur Cells Mater 2005;10 (Suppl. 4), pub. by AO Research Institute Davos, <www.ecmjournal.org>.

[26] Zhang XM, Yue TM, Man HC. Enhancement of ceramic-to-metal adhesive bonding by excimer laser surface treatment. Mater Lett 1997;30:327–32.

[27] Miyamoto, MH. SPIE 1279, 66; 1990.

[28] Henari, BW. Appl Opt 1995;34:581.

[29] Man HC, Zhang XM, Yue TM. Proceedings of ICALE0' 95, San Diego, USA; 1995, p. 704.

[30] Yucel MT, Aykent F, Akman S, Yondem I. Effect of surface treatment methods on the shear bond strength between resin cement and all-ceramic core materials. J Non Cryst Solids 2012;358:925–30.

[31] Janda R, Roulet J-F, Wulf M, Tiller H-J. A new adhesive technology for all-ceramics. Dent Mater 2003;19:567–73.

[32] RLS Enterprises, New Brunswick, New Jersey, USA, <www.rlsenterprises.com/index.html>, July 2013.

[33] Nord RS. Kulzer's silicoater: a new technique for bonding methacrylates to metal. Trends Tech Contemp Dent Lab 1986;3:32–4. (See also pp. 36–4, 42–6.)

[34] Coortek, <www.coorstek.com/services/Brazing-Bonding-Joining.php>, June 2013.

[35] Joining dissimilar materials pub by S-Bond Technologies Corp, <www.s-bond.com>, April 5, 2011.

Chapter 12

Adhesion Promoters

12.1 GENERAL CONCEPTS

Adhesion promoters, or coupling agents, are chemicals that act at the interface between an organic polymer and an inorganic substrate to enhance adhesion between the two materials. Organic and inorganic materials are very different in many ways, for example, compatibility, chemical reactivity, surface properties, and coefficient of thermal expansion, such that forming a strong adhesive bond between these two dissimilar materials is difficult. An adhesion promoter, in its optimal sense, will act effectively at the organic–inorganic interface to chemically and physically wed these dissimilar materials into a strong cohesive bond structure.

Other approaches to improve adhesion involve chemical modification of the polymer surface via oxidation with ozone or plasma, and physical abrasion of the inorganic substrate to increase the bonding surface area by altering the surface roughness. Although these approaches can enhance adhesion, the use of adhesion promoters will not only alter physical and chemical forces at the interface, but also provide a "glue," or compatibility bridge, to give a much greater level of adhesion. Additionally, adhesion promoters can impart resistance to environmental and other destructive forces, such as heat and moisture, which often act on the bonded site to destroy adhesive strength.

Adhesion promoters are chemical materials that contain dual functionality in the molecular structure. A metallic central atom, such as silicon, zirconium, titanium, aluminum, or others, will give inorganic reactivity to the adhesion promoter, especially if methoxy, ethoxy, or hydroxyl groups are attached to the metal atom. An organofunctional group can also be attached to the metal atom through an alkylene, arylene, or other type of organic bridge, to give traditional organic reactivity to the adhesion promoter. The inorganic reactive groups can condense with themselves to give the adhesion promoter an oligomeric structure. An oligomeric adhesion promoter has dual- or multi-functionality and structural integrity, such that a stable chemical bond occurs between the dissimilar organic and inorganic surfaces to promote adhesion between the two dissimilar materials. This basic concept of the chemistry and action of adhesion promoters has allowed

Surface Treatment of Materials for Adhesive Bonding. DOI: http://dx.doi.org/10.1016/B978-0-323-26435-8.00012-5

great advances in reinforced plastics, adhesive bonding, and compatibilization of different materials in a wide variety of applications.

Other types of reactive adhesion promoters that are not based on *inorganic* reactivity are also useful as compatibilizers for dissimilar polymeric materials. These adhesion promoters include reactive organic oligomers or polymers, such as thermoplastics polymers (polyethylene, polypropylene, etc.) grafted with organofunctional groups. Nonreactive adhesion promoters include block copolymers that function by having polymeric segments with solubility parameters that are matched to the components in order to be adhered or compatibilized and function by atomic interactions of van der Waals, dipole interaction, and other atomic forces.

Organosilane coupling agents are the predominant chemical type of adhesion promoter and, therefore, are here the main focus of discussion.

12.2 SILANE ADHESION PROMOTERS

12.2.1 Typical Silane Coupling Agents

A silicon-based chemical that functions as an adhesion promoter, or as a coupling agent, has a general structure of four substituents attached to a single silicon atom. The most common structure has three inorganic–reactive alkoxy groups, methoxy or ethoxy, and one organic group (see Fig. 12.1), although sometimes it may contain only two alkoxy groups, one of the alkoxy groups being replaced by a methyl group. The organic group is either reactive (amino, epoxy, methacrylate, etc.) or unreactive (methyl, butyl, octyl, phenyl, etc.). Coupling agents, of which there are several different organofunctional types, are commercially available. Typical commercial coupling agents are listed in Tables 12.1 and 12.2.

General Structure: $R_xSi(OR')_y$ $x + y = 4$

- R = alkyl, aryl, or organofunctional group
- OR' = methoxy or ethoxy
- Examples:

$CH_3CH_2CH_2CH_2CH_2CH_2CH_2CH_2Si(OCH_2CH_3)_3$
(*n*-octyltriethoxysilane)

$H_2NCH_2CH_2NHCH_2CH_2CH_2Si(OCH_3)_3$
(aminoethylaminopropyltrimethoxysilane)

FIGURE 12.1 General structure of silane coupling agents.

TABLE 12.1 Silane Coupling Agents – Part 1: Matching Organic Group to Polymer Type

Chemical Type	Chemical Name	Polymer
Amine	Aminopropyltriethoxysilane	Acrylic, nylon, epoxy, phenolics, PVC, urethanes, melamines, nitrile rubber
Diamine	Diaminopropyltrimethoxysilane	Acrylic, nylon, epoxy, phenolics, PVC, melamines, urethanes, nitrile rubber
Methacrylate	3-Methacryloxypropyltrimethoxysilane	Unsaturated polyesters, acrylics, EVA, polyolefin
Epoxy	3-Glycidoxypropyltrimethoxysilane	Epoxy, PBT, urethanes, acrylics, polysulfides
Methyl	Methyltrimethoxysilane	Hydrophobing agent for mineral surfaces
Isobutyl	Isobutyltrimethoxysilane	Hydrophobing agent for mineral surfaces, masonry water repellent

TABLE 12.2 Silane Coupling Agents – Part 2: Matching Organic Group to Polymer Type

Chemical Type	Chemical Name	Polymer Type/Application
Phenyl	Phenyltrimethoxysilane	Hydrophobing , dispersing aid for mineral surfaces, blends, hydrophobe, thermal stability
Octyl	Octyltriethoxysilane	Hydrophobing for mineral surfaces; dispersion of minerals in polyolefins; masonry water repellent
Vinyl	Vinyltrimethoxysilane	Graft to polyethylene for moisture crosslinking, EPDM rubber, SBR, polyolefin
Chloroalkyl	3-Chloropropyltrimethoxysilane	Urethanes, epoxy, nylon, phenolics, polyolefins
Chloroalkyl	3-Chloropropyltriethoxysilane	Urethanes, epoxy, nylon, phenolics, polyolefins
Vinylbenzylamino	Vinylbenzylaminotrimethoxysilane	Epoxies for PCBs, polyolefins, all polymer types

12.2.2 Silane Chemistry

Silicon is in the same family of elements as carbon in the periodic table, but silicon compounds exhibit significant differences in chemical reactivity compared to analogous carbon compounds. Similarly to carbon, silicon will conveniently bond to four substituents in its most stable state. However, silicon is more electropositive than carbon, does not form stable double bonds, and is capable of very special and useful chemical reactions that enhance its effectiveness as an adhesion promoter component. A silicon molecule that has dual reactivity can be synthesized; that is, both organic and inorganic reactivities can be synthesized into a single molecule. The unique nature of silane chemistry, to have both inorganic and organic reactivities in one molecular structure together with the special chemistry of silicon, gives rise to the use of silane chemicals as coupling agents and adhesion promoters.

Monomeric silicon chemicals are known as silanes. A silane has the chemical formula **1** in which R_1, R_2, R_3, and R_4 can be the same substituent or four different substituents.

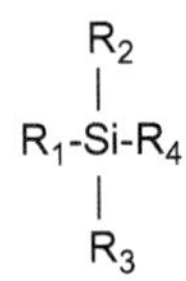

1

The electropositive nature of the silicon atom causes certain atoms or groups attached directly to silicon to be very reactive and exhibit inorganic rather than organic reactivity. Silane substituents, such as chlorine, amine, methoxy, acetoxy, or hydroxyl, are reactive with inorganic materials that contain hydroxyl groups, such as −Si−OH, −Al−OH, or metal−OH in glass, minerals, metals, and even water. A silanol, −Si−OH, or a silylamine, $-Si-NH_2$, is generally not stable and will condense rapidly to siloxane, −Si−O−Si−, or silazane, −Si−NH−Si−, respectively. A chlorosilane or a methoxysilane, $RSiX_3$, X = Cl or CH_3O, reacts with water (hydrolyzes) to give unstable silanetriol, $-Si(OH)_3$, condenses with a second silanetriol to form a siloxane, −Si−O−Si−, structure (Reaction (1)), or reacts with available metal hydroxyls on a metal or glass surface to form a silicon−oxygen−metal bond with the inorganic substrate (Reaction (2)).

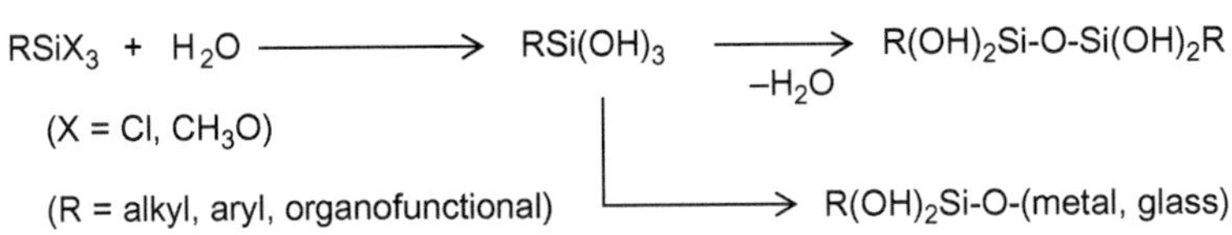

(1)

(2)

To be classified as an *organo*silane, one of the silane substituents must be an organic group that is directly attached to silicon via a Si−C bond.

Organic substituents attached to silicon are either nonreactive, such as methyl, ethyl, butyl, octyl, phenyl, and others, or organofunctional; that is, they contain an organic reactive group, such as amino, epoxy, methacrylate, sulfido, isocyanato, and so on. An organic spacer group usually separates the organofunctional group from the silicon atom. If the spacer group is at least three carbons in length, for example $-CH_2CH_2CH_2-$, then the organic reactivity of the organic group in the silane is similar to organic reactivity in carbon chemistry. If the spacer group is one or two carbons in length, then the silicon atom may influence the reactivity and change the chemistry of the organic group. Other reactive silanes, particularly vinyl silanes, $-Si-CH=CH_2$, and silicon hydrides, $-Si-H$, are useful reactive groups in silicon chemistry.

12.3 ADHESION PROMOTER MECHANISM WITH SILANES

12.3.1 The Interphase Region

Silane coupling agents will act in the interphase region, the area between an inorganic (such as glass, metal, or mineral) and an organic substrate (such as an organic polymer, coating, adhesive), and act as a bonding or bridging agent to improve the adhesion between the two dissimilar materials [1,2] (see Fig. 12.2). Modification of the interphase region produces other desirable changes: improves wet-out of the inorganic substrate by the polymer, improves dispersion of fillers in liquid polymers, reduces the viscosity of the uncured polymer–filler mixture, protects minerals against cleavage, heals flaws in the surface, and strengthens the polymer layer at the interface by interdiffusion with the coupling agent to form interpenetrating polymer networks (IPNs).

The interface (or interphase region) between a polymer and an inorganic substrate involves a complex interplay of physical and chemical factors related to adhesion, physical strength, and retention of properties of the product. Adhesive bonds are destroyed by migration of water into this interface to hydrolyze bonds and cause physical detachment. Silane coupling agents

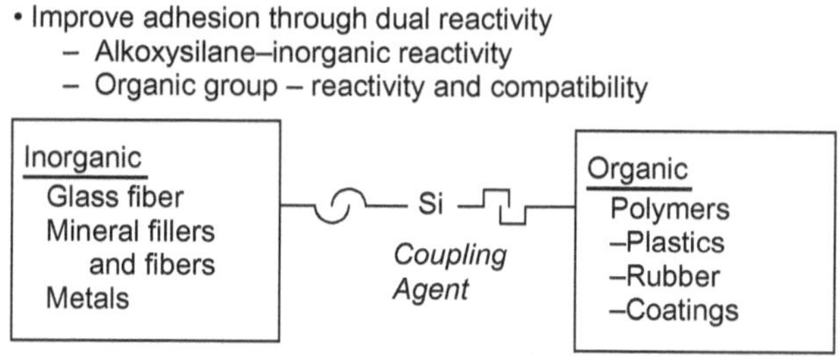

FIGURE 12.2 Silane coupling agents – dual reactivity.

have unique chemical and physical properties that not only enhance bond strength, but also prevent debonding at the interface. In composites, often a 40% increase in flexural strength is obtained using a silane coupling agent. In coatings and adhesives, silane coupling agents significantly increase bond strength and resistance to humidity and other adverse environmental conditions.

12.3.2 Bonding to the Inorganic Substrate

Silane coupling agents, with their inorganic reactive groups on silicon, bond well to most inorganic substrates, especially when the substrate contains silicon, aluminum, or most heavy metals. The bond with the inorganic substrate forms very rapidly and dynamically in most situations. The alkoxy groups on silicon hydrolyze to silanols, either with added water or from residual water on the inorganic surface (Reaction (1)). The silanols coordinate with metal hydroxyl groups on the inorganic surface to from an oxane bond, with elimination of water (Reaction (2)). Studies of the mechanism of bond formation of silanes with inorganic substrates show covalent oxane bonds with silica and glass reinforcements [3]. Acids and bases catalyze both hydrolysis and condensation. It is estimated that a silanetriol, $R'Si(OH)_3$ (the intermediate generated by a trialkoxysilane coupling agent), has an equilibrium constant 10^3 more favorable toward bonding to a silica substrate than a coupling agent from a corresponding monoalkoxy silane [4]. Trialkoxysilanes have approximately a 10-fold equilibrium constant relative to dialkoxysilanes toward bonding to the inorganic substrate.

The condensation of silanols with other silanols at the interface gives a multimolecular structure of crosslinked siloxane. The crosslinked structure exists as multiple layers of siloxane, with a very tight siloxane structure existing close to the inorganic surface and a more diffuse structure away from the surface. This "grid" type of structure allows organic coatings to interpenetrate, or diffuse, into the siloxane structure to allow electrostatic forces of interpenetration, very powerful adhesive forces, to become part of the bonding mechanism. This type of IPN structure is shown in Fig. 12.3 [5].

Once the coupling agent is attached to an inorganic surface, that surface takes on the surface chemistry and surface reactivity characteristics of the organic group attached to the coupling agent. The treated surface exhibits the surface energy of the organic group, and the surface becomes reactive, as determined by the reactivity of the organofunctional group in the coupling agent. A silane surface modifier can be selected to give the surface whatever property is desired to transform the nature of the surface into that which is needed to allow optimization of the inorganic material for the intended use.

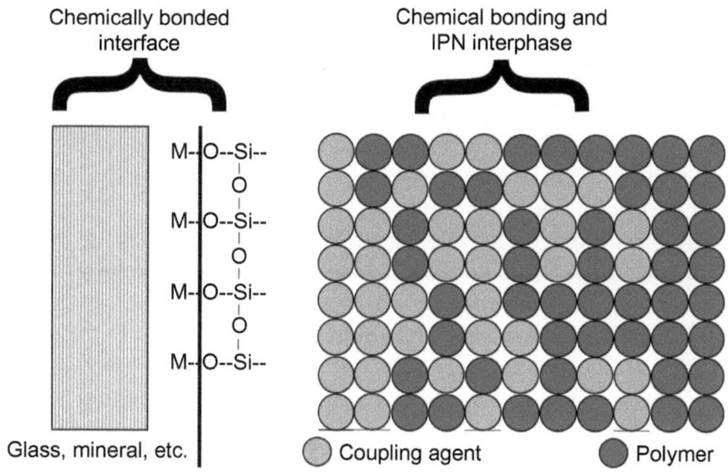

FIGURE 12.3 Interpenetrating network theory – bonding to polymers.

12.3.3 Bonding to the Organic Polymer

The interaction of the silane coupling agent with the organic polymer is complex. For a thermoset polymer, the silane bonds to the polymer through chemical reactivities of both. The most important consideration is to match the reactivity of the coupling agent with that of the thermoset polymer [6]. Coupling agents are optimum when the organofunctional group participates in the curing mechanism or has some way of reactively attaching to the polymer structure in the final composite. For example, either an epoxysilane or an aminosilane can participate in the curing mechanism of an epoxy resin; a methacrylate silane bonds through free radical crosslinking during the cure of an unsaturated polyester resin in the composite; and phenolic resins have several different reaction modes, with the result that several types of silanes (amino, epoxy, and chloroalkyl, for example) impart good properties to phenolic composites.

For thermoplastic polymers, especially polyethylene and polypropylene that lack reactivity on the polymer backbone, having covalent bonding to the polymer is difficult. However, good property improvements are obtained with silane coupling agents in thermoplastic systems. In these cases, bonding is best explained by interdiffusion of the polymer into the layered siloxane network at the inorganic surface and formation of an IPN in the interphase region. Interdiffusion and crosslinking of silane adhesion promoters, as depicted in Figure 12.3, were identified as the mechanism of adhesion of polyvinylchloride plastisols to silane-treated glass [5]. Such interdiffusion is an important factor in adhesion of thermoplastic elastomers to primers, comprising silane-modified tackifying resins [7], primers used for bonding polyolefins [8], silane-modified melamine resin primers for bonding engineering thermoplastics [9], and silane-modified epoxy primers for bonding crosslinkable ethylene–vinyl acetate copolymers to surfaces in solar cell modules [10]. The solubility parameter of the organic group on the silane, if matched with that of the polymer, produces the beneficial effect of enhancing the interpenetration of the polymer.

Thermoplastic polyester resins, such as PET (polyethylene terephthalate) and PBT (polybutylene terephthalate), contain residual phenolic or carboxylic reactive sites that make epoxysilanes effective. Octyl and phenyl silanes are also used in combination with organofunctional silanes, for thermoset applications, to improve dispersion of the resin by matching solubility parameters, increase hydrophobic character, and give greater resistance to attack of water in the interphase region. Aromatic structures in silane-coupling agents, such as phenylsilanes, impart increased heat resistance to the adhesive bond.

12.4 OPTIMIZING COUPLING AGENT PERFORMANCE

"Wet strength," the strength of the bond in a humid-aging environment, is often studied to evaluate the effectiveness of coupling agents. Physical properties, such as flexural strength of composites or a peel adhesion test for coatings and adhesives, are measured before and after a wet aging test. A test specimen is aged in warm or boiling water for 2, 4, or 24 h, for a week, or for longer. The length of time that a bond survives, or the coupling agent that gives the greatest retention of bond strength in the wet environment, gauges the effectiveness of a coupling agent.

The effect of the organic structure of the coupling agent in improving the flexural strength of a silica-reinforced unsaturated polyester composite is shown in Figure 12.4. A significant increase in dry and wet strength of the composite, after aging for 24 h in boiling water, manifests when a coupling agent is used. The choice of a coupling agent with optimum reactivity and chemical structure optimizes the retention of flexural strength. Two or more possible coupling agents can be effective in the same system, but often one gives the best combination of properties in the final bonded material.

Adhesion can also be improved by using blends of silanes. A hydrophobic silane in combination with a hydrophilic silane often improves bonding and retention of wet strength, compared to using the hydrophilic silane alone. The hydrophobic nature of some silanes is listed in Tables 12.1 and 12.2. Phenyltrimethoxysilane, a hydrophobic silane, impels a significant improvement in adhesion of a urethane polymer to glass when blended with a hydrophilic diaminosilane, even though the aminosilane adhesion promoter alone is a good coupling agent for urethane polymers [11]. The data in Table 12.3

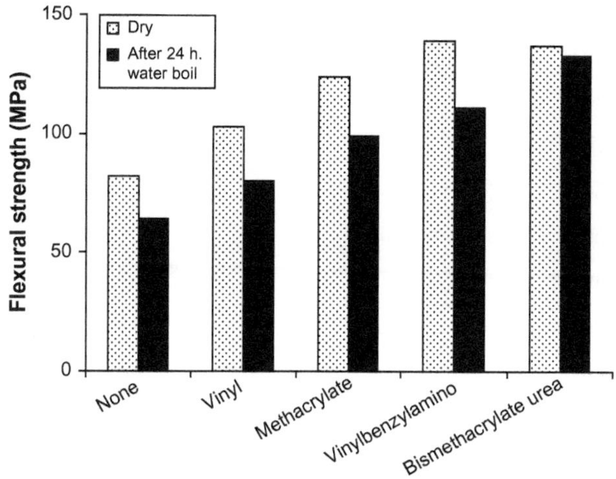

FIGURE 12.4 Silane organic group – effect on flexural strength of a silica–polyester composite.

TABLE 12.3 Silane Blends – Adhesion of Urethane to Glass

Ratio of Silanes	Adhesion to Glass (N/cm)		
I/F in Primer	Dry	2 h in Boiling Water	5 h in Boiling Water
Control	3.0	Nil	Nil
0/100	C	Nil	Nil
50/50	C	Nil	Nil
80/20	C	C	C
90/10	C	C	C
95/5	C	C	C
99/1	C	C	C

C = cohesive failure > 20 N/cm; I = phenyltrimethoxysilane; F = diaminopropyltrimethoxysilane.

show that the bond of urethane to glass fails in boiling water after 2 h if the aminosilane alone is used, but total bonding is retained even after 5 h in boiling water when 90% phenylsilane is blended with the aminosilane.

Silane blends also can be used to increase the thermal stability of a silane for use in high-temperature polymers. Many reinforced plastics, such as polyimides, are fabricated at temperatures of 400°C or higher, and adhesive

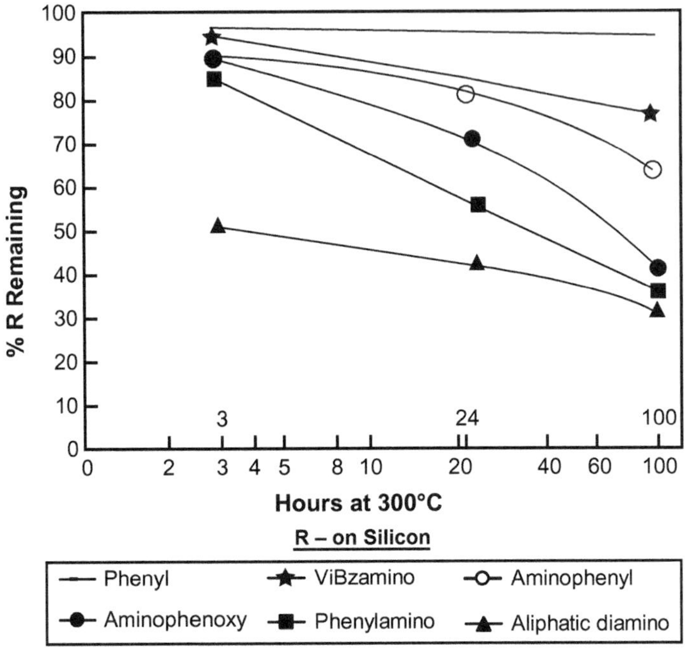

FIGURE 12.5 Stability of $RSiO_{3/2}$ in air at 300°C.

bonds often must survive high-temperature environments. Standard aliphatic silane coupling agents have adequate heat stability for fabrication temperatures up to about 250°C, but they may decompose at higher temperatures. Certain aromatic functional silanes, such as an aminophenylsilane imidized with an aromatic anhydride, have outstanding heat stability, but they are not commercially available.

The thermal stability of several commercial silanes was studied by determining isothermal weight loss of the hydrolyzed silsequioxane form of the silanes, $RSiO_{3/2}$, at 300°C [11]. Figure 12.5 shows that the phenylsilyl structure has very good thermal stability, but the diaminosilane has poorer thermal stability. The vinylbenzylaminosilane has surprising good thermal stability. Phenyltrimethoxysilane blends, or vinylbenzylaminosilane, have been recommended for high-temperature applications.

A blend of 90% diaminosilane and 10% phenylsilane in fiberglass-reinforced polyimide, processed at 400°C, was aged up to 2000 h at 260°C. Flexural strength data in Table 12.4 show the improved thermal performance of the phenylsilane blend compared to a monoaminosilane, the standard coupling agent that had been used in that application [12]. In an analogous system, proprietary sizes were formulated with the more

TABLE 12.4 Thermal Stability – S-Glass/Polyimide Laminates

Properties of Laminates (MPa)	Coupling Agents on Glass	
	Blend 9/1, Phenyl/ Diaminosilane	Monoaminosilane
Flex. Str., initial	544	476
1000 h at 260°C	409	258
2000 h at 260°C	306	134

Phenylsilane = $PhSi(OMe)_3$; monoaminosilane = $(EtO)_3Si(CH_2)_3NH_2$; diaminosilane = $(MeO)_3Si(CH_2)_3NH(CH_2)_3NH_2$.

TABLE 12.5 Si-C Fiber/Polyimide Laminates – Retention of Strength at 315°C

Silicon Carbide	Flexural Strength (MPa)			
Fiber size	Initial	100 h	500 h	1000 h
Epoxy organic	1480	1320	1000	610
DCC-1	2350	2320	1660	890
DCC-2	2120	2060	1670	980

DCC-1 and DCC-2 are proprietary – based on vinylbenzylaminosilane.

thermally stable vinylbenzylaminosilane and applied to a silicon carbide fiber–polyimide composite. The composite was aged at 315°C for up to 1000 h. The proprietary sizes improved flexural strength after 1000 h compared to an epoxy–organic size that was normally used in such systems (see Table 12.5).

An increase in siloxane crosslinking in the interphase region also improves performance. A multifunctional crosslinking additive can be added as an adhesion enhancer [13]. A hexafunctional methoxydisilane additive, $(CH_3O)_3SiCH_2CH_2Si(OCH_3)_3$, can be blended with conventional silanes. The data in Table 12.6 show that the crosslinking additive improves the adhesion of an epoxy film to glass compared to either a diaminosilane or an epoxysilane alone. Similarly, two silanes with organic functional groups that react with each other can be used to spur increased crosslinking. Pairings of aminosilane and epoxysilane, aminosilane and methacrylate silane, or chloroalkyl silane and an aminosilane, react jointly and in properly formulated

TABLE 12.6 Adhesion Enhancer/Crosslinker $(MeO)_3SiCH_2CH2Si(OMe)_3$ Adhesion of Epoxy Film on Glass

Primer on Glass	Curing Agent Type – Hours to Lose Adhesion		
	Polyamide	DMP-30	DEH-24
None	<1	<1	<1
Diaminosilane alone	1	1	1
Blend 9:1 diaminosilane:enhancer	C	C	C
Epoxysilane	C	8	C
Blend 9:1 epoxysilane:enhancer	C	36	C
Blend 8:2 epoxysilane:enhancer	C	C	C

C = No failure after 1 week in 70°C water.

amounts to give in situ formation of multifunctional alkoxy reactivity. Also, the use of tetraethoxysilane in a coupling agent formulation can increase crosslinking.

12.5 HOW TO CHOOSE A SILANE COUPLING AGENT

The nature of the organic group in the coupling agent is a key consideration when choosing a silane. With any silane coupling agent, the three alkoxy groups on silicon should bond equally well to an inorganic substrate, but matching the organofunctional group on silicon with the polymer type of the resin bonded dictates which silane will be used in a particular application.

Several silanes are recommended for one polymer type. For example, the cure system for an epoxy resin, amine versus anhydride, high-temperature versus low-temperature cure, and aliphatic versus aromatic amine, often affects the performance of the coupling agent.

A list of silane coupling agents and recommendations for their use in various applications is displayed in Table 12.7. A correlation exists between the chemical and physical characteristics of the coupling agent and the chemical and the physical characteristics of the polymer.

The same principles that are used in selecting a silane for reinforced plastics are used to select a silane for paints, inks, coatings, and adhesives (PICA). The chemical nature of the organic material used in the application is the major criterion that governs the effectiveness of a particular coupling agent as an adhesion promoter.

TABLE 12.7 Organosilanes – Applications

Chemical Type	Chemical Name	Coupling Agent for Polymer Type/ Application
Amine	Aminopropyltriethoxysilane	Acrylic, nylon, epoxy, phenolics, melamines, PVC, urethanes, nitrile rubber
Diamine	Diaminopropyltrimethoxysilane	Acrylic, nylon, epoxy, phenolics, melamines, PVC, urethanes, nitrile rubber
Methacrylate	3-Methacryloxypropyltrimethoxysilane	Unsaturated polyesters, acrylics, EVA, polyolefin
Epoxy	3-Glycidoxypropyltrimethoxysilane	Epoxy, PBT, urethanes, acrylics, polysulfides
Methyl	Methyltrimethoxysilane	Hydrophobing agent for mineral surfaces
Isobutyl	Isobutyltrimethoxysilane	Hydrophobing agent for mineral surfaces, masonry water repellent
Phenyl	Phenyltrimethoxysilane	Hydrophobing, dispersing aid for minerals, blends with silanes/thermal stability
Octyl	Octyltriethoxysilane	Hydrophobing/dispersion of minerals in polyolefins, masonry water repellent
Vinyl	Vinyltrimethoxysilane	Graft to polyethylene for moisture crosslinking, EPDM rubber, SBR, polyolefin
Chloroalkyl	3-Chloropropyltrimethoxysilane	Urethanes, epoxy, nylon, phenolics, polyolefins
Chloroalkyl	3-Chloropropyltriethoxysilane	Urethanes, epoxy, nylon, phenolics, polyolefins
Vinylbenzylamino types	Vinylbenzylaminotrimethoxysilane	Epoxies for PCBs, polyolefins, all polymers

12.6 GENERAL APPLICATIONS OF SILANE COUPLING AGENTS

Fiberglass-reinforced polyesters were developed in the 1940s and required an adhesion promoter to retain the physical strength of the composite during

TABLE 12.8 Silane Applications

- Fiberglass cloth, mat, chopped glass (surface treatment) for *adhesion* in reinforced plastics (PCBs, auto, boats, tanks, etc.).
- Mineral (surface treatment) for *adhesion and/or dispersion* in reinforced plastics (auto, W&C, electronics, rubber/tires, etc.).
- Paints, inks, coatings, adhesives (primer or additive) to give *moisture resistant adhesion* to surfaces.
- Moisture *crosslinking* of polyethylene (Sioplas).

ambient aging. Other types of coupling agents, such as chrome complexes, were good coupling agents. But silane coupling agents were developed to fill this need specifically, and they have proven to be the best adhesion promoters for these materials. Since then, other applications for adhesion promoters have been developed, including mineral-filled composites, laminates for printed circuit boards, insulation fiberglass, silica-filled rubber tires, rubber insulation for wire and cable, adhesion promoters in PICA, and moisture-cured crosslinking systems for polyethylene. Various applications of silane coupling agents are listed in Table 12.8.

12.7 INDUSTRY AND UTILITY

12.7.1 Surface Treatment of Fiberglass

Glass fiber is a very hydrophilic material. It requires a protective coating to eliminate the degradative effects of moisture in a variety of applications, particularly when it is used as a reinforcement in polymeric composites. The natural chemistry of silanes is similar to the silicon chemistry of glass, and the favorable equilibrium of bonding silanes to glass surfaces designates them as being ideal for treatment thereof. Virtually all glass fiber for most applications requires a coupling agent to render the glass fiber a viable material for intended applications. The fiberglass may be in the form of random glass matt, chopped strand, woven roving, continuous filament, or woven glass cloth; regardless of the form of the glass fiber, it must be treated with a coupling agent to render it useful as a reinforcement for polymers. The nature of the final application determines which type of silane is used and how the silane is applied in the process.

Other noncoupling benefits of silane adhesion promoters in composites are very important. These include the following: change in rheology of a filled system during processing, elimination of inhibition of cure caused by the reinforcement, better wet-out of the fiber, greater smoothness of the composite surface, increased clarity of a composite, and healing of flaws in glass fibers. A minuscule amount of coupling agent prompts a dramatic difference in the performance of the end product.

12.7.1.1 Fiberglass for Reinforced Plastics

The glass fiber that is used in reinforced plastics has a coupling agent on the surface. Fiberglass manufacturers apply a water-based "size" to it as the filaments are gathered into strands during the fiber-forming process. The formulation contains several ingredients, including a film-forming polymer, a glass lubricant, an antistatic agent, and other ingredients, but a critical component is 0.1–0.5% silane coupling agent. As the size is water-based (an important requirement of the fiberglass industry), any silane that is used must be dissolved in water. The addition of a silane to water generally gives a precipitated gel via hydrolysis and leads to the formation of insoluble hydrophobic siloxane structures. However, if the water is acidified to a pH of 4 with acetic or formic acid, a silane coupling agent hydrolyzes and dissolves in water without gelling, especially at concentrations less than 5%. The low pH stabilizes the silanol species against condensation to allow the silane to dissolve and remain soluble. A higher pH of 5–6 is used in acid-sensitive epoxy silanes. Hydrophilic silanes, such as aminosilanes, form aqueous solutions at all pHs, but pH 4 is preferred for glass treatment to give correct orientation on the glass surface [3].

The magnitude of the effect of a coupling agent regarding the improvement of strength properties of composites is usually significant (see Fig. 12.6). A fiberglass-reinforced phenolic composite with and without diaminopropyltrimethoxysilane on the glass surface was tested as prepared, and was also tested after immersion for 2 h in boiling water. Before aging, the flexural strength of the unaged composite showed a 38% increase with the silane versus without a silane; after aging, the sample without the silane lost 66% of its strength, whereas the silane-treated sample lost only 12% of its flexural strength. The wet strength properties represent an accelerated aging test comparable to long-term conditions in the environment.

The effect of the type of organic group in the silane coupling agent is shown in Table 12.9. Epoxysilane and vinylbenzylamino silane were used in

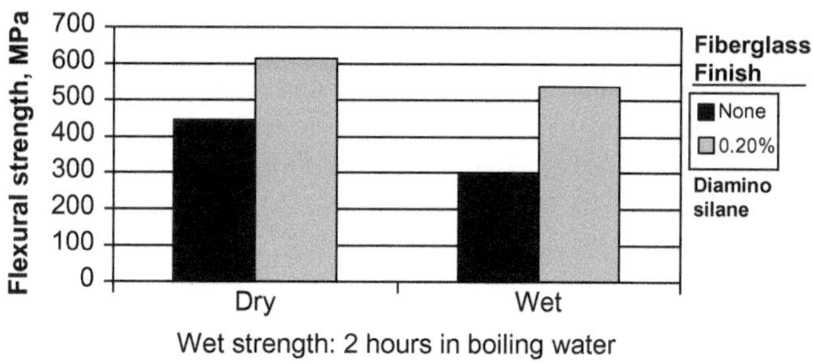

FIGURE 12.6 Fiberglass-reinforced phenolic laminate strength test.

TABLE 12.9 Coupling Agent – Organic Group Effect on Strength of Epoxy–Glass Laminate

Silane Coupling Agent Type	Flexural Strength (MPa)	
	Dry Strength	Wet, 72 h in Boiling Water
None	394	221
Epoxy	605	476
Vinylbenzylamino	670	518

fiberglass-reinforced epoxy composite formulations and the samples were aged in boiling water for 72 h. Both coupling agents gave a substantial increase in initial flexural strength and improved retention of strength after the aging test. That the performance of the vinylbenzylamino coupling agent improved may be because the coupling agent has a complex chemical structure and cationic character to coat the glass fiber more efficiently, and it forms an IPN network with the thermoset polymer. It is also more hydrophobic than the epoxy silane and has a very reactive amine functionality to participate in the curing process.

12.7.1.2 Printed Circuit Board Applications

The printed circuit board industry is based largely on high-quality fiberglass-reinforced epoxy laminates that are made from layers of electronic-grade woven glass cloth and on which a silane size becomes the critical chemical component between the glass and the resin. The industrial requirement for better bonding of the epoxy resin to the glass surface increases as the density of holes that are drilled in a laminate for conductive sites increases. It is very important to eliminate failure along the glass fiber in the laminate.

The fiberglass that goes into printed circuit boards is processed in a different way than the conventional fiberglass for reinforced plastics. The fiberglass is first woven into a glass fabric with a starch size and lubricant on the surface. The starch is burned off and the silane is applied to the heat-cleaned woven glass fabric as, usually, the only ingredient on the glass cloth. The silane is applied from an acidified aqueous treating bath at a pH of 4 and at a treatment level of 0.1–0.5%. The silane-treated woven glass cloth is dried, carefully inspected for flaws, and then supplied to laminators in the printed circuit board industry where lamination and fabrication of printed circuit boards are carried out.

The resins that are used in printed circuit board laminates determine the type of silanes that are used. High-performance epoxy resins are the most-used resins, so an epoxy compatible silane is required. Vinylbenzylaminosilane is used mostly for this application. Low ionic content (a low chloride level), ease of solubility in treatment baths, and minimization of agglomerates on the electronic glass surface to prevent flaws are important parameters in choosing a coupling agent.

For resin systems other than epoxy, or for other epoxy requirements, the glass weavers sometimes apply aminosilanes, epoxysilanes, and chloropropylsilanes.

12.7.1.3 Miscellaneous Fiberglass Applications

Other applications of fiberglass include insulation for construction applications, roofing shingles, and automotive insulation applications. These are not always composite applications, in the strictest sense, but still require a coupling agent to bond a resinous coating to the glass surface.

Insulation fiberglass for construction applications uses a phenolic binder to help retain the insulation value. An aminosilane, either a monoaminosilane or diaminosilane, can be mixed into an aqueous phenolic binder solution to bind the phenolic resin to the glass fiber surface. Without the silane present, the phenolic de-bonds during aging in the environment and causes the glass fiber to relax, so reducing the insulation value.

Fiberglass-based roofing shingles are similar to reinforced plastics' applications and require durable organic polymers to resist the sun and other degradative effects. Vinylbenzylaminosilane has been used in formulations to allow the products to meet the requirements of building codes and construction standards.

12.7.2 Surface Treatment of Minerals in Plastics

12.7.2.1 Nature of Bonding to Minerals

Mineral fillers require surface modification when used as reinforcements in plastics. Surface treatment improves mechanical strength and chemical resistance of composites by improving adhesion across the interface. Other benefits of surface treatment include improved dispersion of the mineral into polymeric media, lower viscosity of the mix, greater compatibility with the polymer, change of the mineral surface from hydrophilic to hydrophobic, improved electrical properties in a mineral-filled polymer system, and decreased cure inhibition of thermoset polymers. Some of these benefits are listed in Table 12.10.

In a mineral-filled composite, the interphase region between the polymer and filler involves the interplay of physical and chemical factors related to composite performance, including a tremendous mechanical stress of differential shrinkage during cure, wet-out, and dispersion of the mineral in the polymer, protection of the filler against abrasion and cleavage during mixing,

TABLE 12.10 Mineral Filler Surface Treatment Benefits

- Mineral Filler – 0.1%–10% on filler (particle size = 0.1–50 μm)
  - Plastics/rubber reinforcement
  - Better filler wet-out and dispersion
  - Lower viscosity of filled liquid resins
  - Improved moisture resistance of composite
  - Reduced cure inhibition of filler
  - Improved electrical properties of filled polymer

TABLE 12.11 Mineral Filler Applications

Fillers	Comments
Kaolin clay	Reinforced nylon, wire, and cable (EPDM)
Talc	Stiffness, abrasion – polypropylene (auto)
Mica	Stiffness, polypropylene (auto)
Silica	Reinforced rubber, epoxy PCBs
Wollastonite	Reinforced plastics, coatings
Glass fiber beads	Reinforced plastics
Aluminum trihydrate	Flame retardance
Magnesium hydroxide	Flame retardance
Crystobalite	Abrasion resistance/plastics
Titanium dioxide	Plastics, colorant, filler

optimum alignment of polymer segments at the interface, and adhesion and an interfacial structure to generate the desired improved properties.

Mineral fillers usually contain silicon, aluminum, and other metallic elements in chemical structure such that metal hydroxyl groups on the surface are very hydrophilic in nature. Some of the more commonly used mineral fillers include silica, wollastonite, talc, mica, glass beads, and kaolin clay (see Table 12.11). The reactive sites on the filler surface can function in a manner similar to the silanol groups on fiberglass reinforcement that is chemically treated in a similar way (see Fig. 12.7). Treatment of the filler can make the surface either chemically reactive via an organofunctional silane, or it can make it hydrophobic, organophilic, or hydrophilic, depending on the silane used. Some types of organosilanes and the characteristics of the organic group are listed in Table 12.12. Beyond these general characteristics of silanes, various silanes and polymer applications are catalogued in Table 12.7.

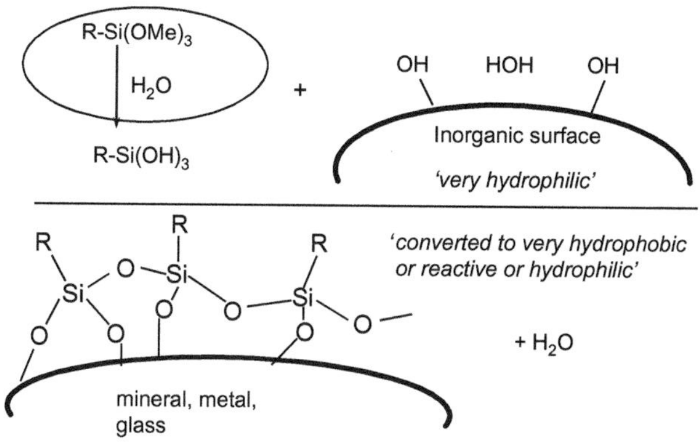

FIGURE 12.7 Silane bonding to inorganic surfaces.

TABLE 12.12 Organosilane Characteristics of R-Si(OMe)$_3$

R	Characteristics
• Me	Hydrophobic, organophilic
• Ph	Hydrophobic, organophilic
• *i*-Bu	Hydrophobic, organophilic
• Octyl	Hydrophobic, organophilic
• $NH(CH_2)3NH_2$	Hydrophilic, organoreactive
• Epoxy	Hydrophilic, organoreactive
• Methacryl	Hydrophobic, organoreactive

12.7.2.2 Methods of Applying Silanes to Minerals

Mineral fillers are often pretreated before compounding, but the silane is sometimes added to the filler–polymer mix during compounding (often referred to as in situ addition or integral blending) [14]. Sometimes, additional coupling agent is added during compounding to enhance adhesion or to provide performance benefits.

Particulate fillers are pretreated with a silane by dry blending at room temperature or at elevated temperature. The filler surface may contain residual moisture to hydrolyze the silane, but sometimes water and/or an alcohol is added. A high-intensity mixer, such as a Henschel or Littleford mixer, is used for the treatment process. Filler treatment may benefit from the addition of a trace of amine or titanate to catalyze the reaction with the filler surface.

At least a monolayer of silane is needed on the filler surface. In practice, about 1% silane is applied to the filler with a particle size of 1 μm, enough

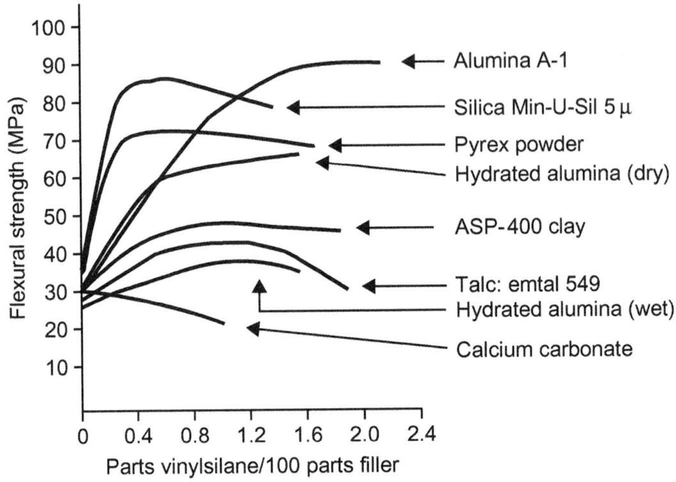

FIGURE 12.8 Silane treatment of fillers in a styrene–butadiene casting.

for several monolayers. Higher surface area fillers require higher levels of coupling agents. Figure 12.8 shows the effect of the level of silane treatment on several different fillers in relation to the flexural strength of a styrene–butadiene casting.

12.7.2.3 Effect of Silanes on Properties of Mineral-Filled Composites

A silane-coupling agent not only increases the flexural strength of a filler-resin composite; it also significantly lowers the viscosity of the uncured resin mix and allows for greater filler loadings and easier processing. An effect of silane on the viscosity of an uncured polymer–filler blend and on the flexural strength of the cured composite is exhibited in Figure 12.9. The use of different types of silanes in the treatment process can significantly alter the effect on viscosity of the mix, and can alter the resulting flexural strength of the composite.

Conveyed in Figure 12.10 are the benefits of silane filler treatment on the cure chemistry of a thermoset resin. The surface of the mineral filler can interfere with the curing chemistry and inhibit the cure. However, once the filler is treated with silane, the filler surface is changed chemically so that the inhibition of cure is minimized.

Electrical properties of filled resin systems are also improved by filler treatment. Filler particles are naturally hydrophilic via their metal hydroxide surfaces, and the particles naturally seek to agglomerate with each other, and so transport electrical charges through resin composite. Treatment with silane-coupling agent alters the chemistry of the filler surface, allows better

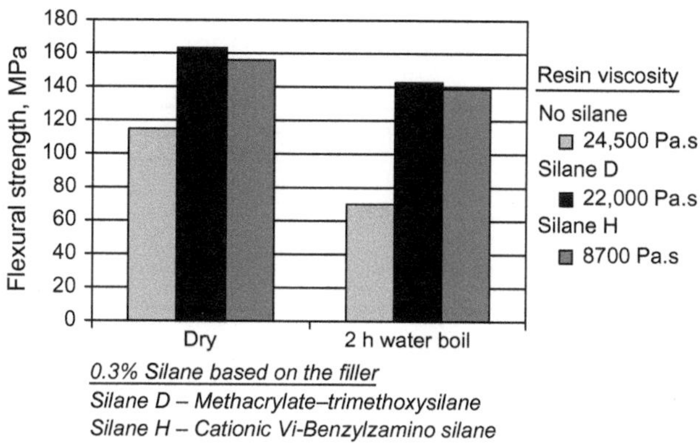

FIGURE 12.9 Viscosity and coupling effect – polyester castings with 50% silica.

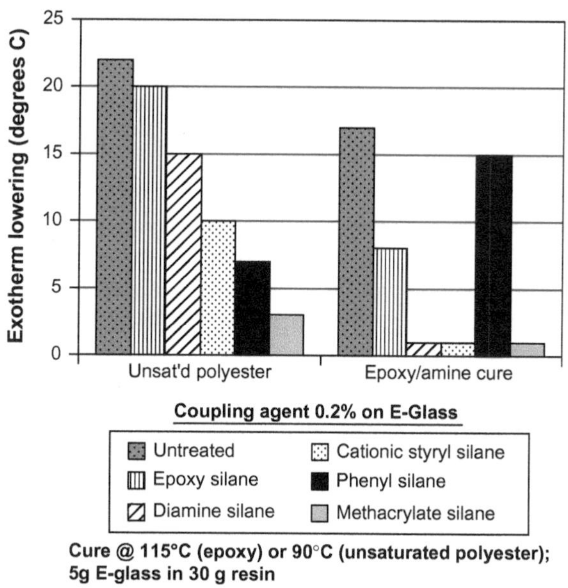

FIGURE 12.10 Resin exotherm lowering with silane-treated glass filler.

dispersion of the filler throughout the resin matrix, and imparts improved electrical properties to the composite. Table 12.13 indicates the improved electrical properties of a quartz-filled epoxy resin system with 0.3% silane admixed into the formulation. Improved insulation values, including reduced dielectric constant and reduced dissipation factor, are also denoted. Silica-filled epoxy molding compounds use an epoxysilane for such a reason.

TABLE 12.13 Epoxy Resin–Quartz Filler – Electrical Properties Versus Silane

Silane Added	Dielectric Constant		Dissipation Factor	
	Initial	H_2O Boil	Initial	H_2O Boil
Unfilled resin	3.44	3.43	0.007	0.005
Quartz, no silane	3.39	14.60	0.017	0.305
Quartz/epoxy-silane	3.40	3.44	0.01	0.024
Quartz/amino-silane	3.46	3.47	0.013	0.023

Epoxysilane = 3-glycidoxypropyltrimethoxysilane; aminosilane = aminopropyltriethoxysilane; water boil = 72 h.

In some cases, special epoxysilanes that have low chloride and low ionic content are used to further minimize negative effects on electrical properties.

Other filled resin systems that require a coupling agent include highly-filled sand cores with furan, urea–formaldehyde, urethane resins in the foundry industry, and highly-filled polymer concrete whereby polyester and epoxy binders are used to bond aggregate; and cultured marble, cultured onyx, and cultured granite are used, whereby a highly-filled thermoset resin is used to bond calcium carbonate and aluminum trihydrate.

Polypropylene and polyethylene applications require the use of fillers in many applications to give desired properties, such as stiffness, impact strength, physical strength, and scratch and mar resistance. Talc, mica, kaolin clay, and titanium dioxide are used in polyolefins in large volumes. The filler can be treated with an octylsilane or phenylsilane to render the filler in the polymer more dispersible. The low reactivity of polyolefins makes chemical bonding to the polymer very difficult. Maleated polyethylene or polypropylene adhesion promoters have been developed to yield improved bonding to fillers. The carboxy groups from the maleated polyolefin adhesion promoter can be used in combination with an aminosilane, offering good adhesion between the treated filler and the resin [15]. Combinations of aminosilanes and chlorinated paraffins have also shown promise experimentally for adhesion of fillers to polypropylene [16].

12.7.3 Mineral Fillers in Rubber Reinforcement

12.7.3.1 Rubber Tire Applications

Silica and kaolin clay can be treated with a mercaptosilane or polysulfidosilane to function adequately as a replacement for carbon-black reinforcement in sulfur-cured rubber applications. Carbon-black reinforced rubber is a huge

TABLE 12.14 Ethoxy Silane Additives Used by the Rubber Industry

$(EtO)_3Si(CH_2)_3{-}SS{-}(CH_2)_3Si(OEt)_3$	Disulfidosilane (1)
Silane (1) blended 50% with carbon black	
$(EtO)_3Si(CH_2)_3{-}SSSS{-}(CH_2)_3Si(OEt)_3$	Tetrasulfidosilane (2)
Silane (2) blended 50% with carbon black	
$(EtO)_3Si(CH_2)_3{-}SH$	Mercaptosilane

Note: Sulfidosilanes (1) and (2) are polysulfide mixtures.

industry, so the success of mineral-filled rubber has become a large application for silane coupling agents. Organic rubber normally uses carbon-black filler, a natural reinforcing agent that does not require a coupling agent. Kaolin clay and silica fillers require a coupling agent that bonds the "unnatural" filler to the rubber polymer and enters into the sulfur-based curing mechanism.

The first silane applications in rubber tires used mercaptosilane to modify silica and clay fillers [17]. Commercial applications were limited to specialty areas, such as white sidewall components, some tread stocks, and some off-road tires, but processing and physical property deficiencies limited commercial applications. However, Degussa Corporation, which has been a filler manufacturer and supplier of silica for a variety of applications, developed a tetrasulfidodisilane, Si-69, which eventually led to the largest single application for a silane-coupling agent [18]. The new silane fitted nicely into commercial rubber processing systems and allowed the new silica-reinforced rubber technology to become commercially successful. Silica-filled tires had several important advantages for autos compared to traditional carbon-black tires: reduced rolling resistance, improved fuel economy, longer tire life via reduced wear and abrasion resistance, less heat generation, and improved grip and skid resistance on wet surfaces.

The tetrasulfidosilane is largely used, but other variations have been developed for the rubber industry (consult Table 12.14).

12.7.3.2 Wire and Cable Applications

Mineral-filled rubber formulations were developed for wire and cable applications in the 1960s where exceptional electrical resistance was needed to prevent electrical breakdown under high humidity stress conditions. Carbon black is a conductive filler so the resulting rubber has poor insulation resistance properties. Silane-treated, mineral-reinforced rubber meets the strict electrical standards unattainable with a carbon-black rubber. Kaolin clay and silica filler are the most-used "white" fillers, and treatment with amino-, vinyl-, and mercaptosilane coupling agents gives good electrical properties.

Silane-treated kaolin clay has become the predominant filler for wire and cable insulation, with most applications using filler pre-treated with either a vinylsilane or a vinylsiloxane polymeric material.

Vinylsilane oligomers, such as partially hydrolyzed vinyltrimethoxysilane and vinyltrimethoxysilane and sometimes including dimethylsiloxy species in the structure, have been developed as premium treating liquids for kaolin clay reinforcements [19]. Silane additives, such as vinyltriethoxysilane or vinyl(tris-methoxyethoxy) silane, are often added during compounding to enhance properties.

12.7.4 Adhesion Promoters for PICA

Organofunctional silane coupling agents function as adhesion promoters for PICA in a manner similar to that for fiberglass and mineral-filled composites. The general conditions are similar; that is, an organic polymer is bonded to an inorganic substrate. Bond failure mechanisms in PICA adhesion systems are similar to those in other adhesion promoter applications. Moisture attack in the bonded interfacial region is a leading cause of loss of adhesion. A silane coupling agent, with its dual functionality, operates at the interface to induce improved adhesion when the bond is initially formed, and then to "protect" the bonded system from moisture-induced debonding and increase its useful life.

12.7.4.1 Choice of an Adhesion Promoter

Choice of the right coupling agent system for a PICA application requires similar thinking, as was discussed for reinforced plastics. If chemical reactivity is present in the polymeric component of the material bonded, then matching the reactivity of the silane coupling agent with the reactivity in the polymer is essential. Correctly choosing a silane system in order to maximize interpenetration of the coating into the silane network is also important. Silane recommendations for various polymers, as presented in Table 12.7, are useful for PICA applications as well. Blends of silanes can be used to give increased hydrophobicity, increased crosslinking of the silane structure, and increased thermal stability to the adhesion promoter system.

12.7.4.2 Application Methods

Silane adhesion promoters are applied to PICA in two possible ways – adding as a primer on the substrate before the coating is applied or directly to the coating system. When used as a primer, the silane is diluted in a solvent and sprayed, dipped, or wiped onto the inorganic substrate. Various solvents can be used: methanol, ethanol, isopropanol, an ether glycol, or sometimes a mixture of solvents. Water, at levels of 1–5%, is sometimes added to an alcohol solvent to promote hydrolysis of the silane during attachment to the surface.

Water can also be used as the solvent if the coupling agent is an aminosilane, because aminosilanes are soluble in water.

The concentration of silane in the solvent during application as a primer controls the thickness of the silane coating on the surface. Poor adhesion is obtained if the silane coating is too thick. A thick, crosslinked silane oligomer is a weak layer and, in such a case, the bond fails in the silane layer. Poor adhesion is also obtained if the silane coating is too thin because not enough silane is present to generate improved bond strength and water resistance. A concentration of 0.5–10% silane in solvent can be used in a primer, but usually levels of 2–5% are used. The solvent can be allowed to evaporate at room temperature or the coated substrate can be heated to drive off solvent; however, heating above 125°C crosslinks the adhesion promoter and reduces interpenetration of the coating into the primer matrix to lessen the primer's effectiveness. Silane primers are applied as a much thicker coating on a surface in comparison with comparable silane treatments on fiberglass or mineral surfaces for reinforcement applications.

12.7.4.3 Adhesion Promoter Benefits in Coatings

An example of silane benefits, when used as a primer, is shown in Table 12.15: urethane and epoxy paints were applied to aluminum surfaces using epoxysilane and aminosilane adhesion promoters to improve adhesion. The effects of degreasing and sandblasting the surface were compared to adhesion with the silanes [20].

Alternatively, the adhesion promoter can be mixed directly into a PICA system to give "unprimed" adhesion. Silane at levels of 0.5–5%, usually about 2%, is added to a coating system. Some techniques that can be used to promote migration of the silane to the interface are as follows: partially prehydrolyzing the silane with water before adding it to the coating system so that the crosslinked silane becomes less soluble in the coating; modifying the cure system of the coating; or choosing the functionality of the coupling

TABLE 12.15 Effect of Surface Primer on Adhesion

Paint	Bond Strength, Paint to Aluminum, Silane	Surface Preparation	Accelerated, Weathering, Adhesion
Urethane	None, none, Diamino (A)	Degreased, sandblasted, degreased	20, 95, 100
Epoxy	None, none, epoxy (D), diamino (A)	Degreased, sandblasted, degreased, degreased	70, 100, 100, 100

A: *N*-(2-aminoethyl)-3-aminopropyltrimethoxysilane; and D: 3-glycidoxypropyltrimethoxysilane.

agent to minimize interaction of the silane with the polymer system. Plueddemann [21] compares the effectiveness of a series of silanes as primers and additives in bonding thermoplastic and thermoset coatings to glass.

Silane adhesion promoters have been studied for the corrosion protection of metal surfaces. The use of a silane crosslinking system (Table 12.6) in combination with sulfur silanes (Table 12.14) provided promising coating systems to replace toxic chromates in steel [22]. Similar corrosion protection was realized on cold-rolled steel, iron, aluminum, and aluminum alloy surfaces by treatment with the same crosslinker system but in combination with aminosilanes in the admixture [23].

12.8 NONSILANE ADHESION PROMOTERS

Adhesion promoter chemistries, other than those of silanes, have been extensively evaluated for many years. These materials use the same concepts as silane coupling agents, but they are based on other metallic elements. They have inorganic reactivity on the metallic atom and contain organofunctional groups, such as methacrylate, carboxylate, and others, with dual- or multifunctionality built into the chemical structure to allot characteristics of dual reactivity, which silane-coupling agents possess.

Organo-titanates, organo-zirconates, and zircoaluminates have been advanced as adhesion promoters. Even so, they have not achieved the broad success that silane has achieved. In metal applications, the highly metallic nature of zircoaluminates promotes their reactivity with metal surfaces. Similarly, organo-titanates are well known as excellent wetting agents, they are able to function as adhesion promoters via chemical adhesion, and they provide more intimate contact between the adhesive or coating and the substrate.

Another family of adhesion promoters is based on the maleation of polypropylene, polyethylene, and other thermoplastic polymers. The polymer backbone in these coupling agents gives interpenetration into ethylene, propylene, or other thermoplastic polymers and adhesion to those polymers that occurs via van der Waals and other atomic forces. The carboxy group on the adhesion promoter allows for interaction and bonding to metals, mineral fillers, or glass, enabling adhesion to those materials.

Studies of maleated polypropylene in clay-based nanocomposities [24] suggest that the polar groups in the adhesion promoter (e.g., –OH, –COOH) intercalate between silicate layers, through hydrogen bonding or other chemical bonding, to the oxygen groups or negative charges in order to assist in exfoliation of the nanoclay reinforcement. In regular mineral-filled polyolefins, as was mentioned earlier [15], the use of a maleated polyolefin in combination with an aminosilane produces a beneficial adhesion promoter for mineral- or glass-filled polypropylene and polyethylene.

Chlorinated polyolefins are used as adhesion promoters for coatings and inks on polyolefin plastics. They can be used as primers on polyethylene, polypropylene, or thermoplastic polyolefins (polypropylene–elastomer blends). They are used for the adhesion of paint to plastic surfaces in the automotive industry.

12.9 SOURCES OF ADHESION PROMOTERS

Silanes are manufactured globally. They have traditionally been manufactured in the United States, Europe, and Japan, and new manufacturing facilities in China and Korea supplement the global supply with high-quality products. The emergence of large-volume applications for silanes, especially in silica-filled rubber tires, has substantially increased the number of suppliers.

Silane adhesion promoters are available globally from several manufacturers including Dow Corning Corporation [25], Degussa (now Evonik Degussa) [26], OSi Specialties (now Momentive Performance Materials) [27], Chisso Corporation [28], Shin-Etsu Chemical [29], KCC Silicones [30], and several other Chinese suppliers such as Nanjing Capatue [31] and Wuhan University Silicone New Material Co., Ltd [32].

Major suppliers of organo-titanates, organozirconates, and zircoaluminates are Kenrich Petrochemicals [33], Manchem/FedChem [34], and Wuhan University Silicone New Materials Co., Ltd [32].

Maleated polyolefin adhesion promoters are available from Eastman Chemical [35] (Epolene®) and DuPont [36] (Fusabond®).

Chlorinated polyolefin adhesion promoters are available from Eastman Chemical [35] (CPO), Nippon Paper Chemicals [37] (Superclon®), and DuPont [38] (Automotive Chemicals).

Other suppliers of adhesion promoters exist for the various types of adhesion promoters. All suppliers have websites on which product line lists, data sheets, Material Safety Data Sheets, and technical information are available. With regard to the various suppliers, the author's intent is not to favor or recommend one supplier over another.

REFERENCES

[1] Plueddemann EP. Silane coupling agents. 2nd ed. New York, NY: Plenum Press; 1991.

[2] Pape PG, Plueddemann EP. History of silane coupling agents in polymer composites. In: Seymour RB, Deanin RB, editors. History of polymeric composites. Utrecht, the Netherlands: VSP; 1987. pp. 105–39.

[3] Ishida H, Koenig JL. Molecular orientation of the coupling agent interphase of fiberglass reinforced plastics. J Polym Sci–Phys 1807;17(1979).

[4] Osterholtz FD, Pohl ER. Kinetics of the hydrolysis and condensation of organofunctional alkoxysilanes: a review. In: Mittal KL, editor. Silanes and other coupling agents. Utrecht, the Netherlands: VSP; 1992. pp. 119–41.

[5] Chaudhaury MK, Gentle TM, Plueddemann EP. Study of adhesion through interdiffusion and IPN formation in the interphase region of composites. J Adhes Sci Techol 1987;1:29–38.

[6] Matching reactivity of coupling agents and resins. In: Plueddemann EP, editor. Silane coupling agents. 2nd ed. New York, NY: Plenum Press; 1991. p. 3.

[7] Plueddemann EP. Silane primers from tackifying resins. In: Proceedings of 21st Annual Conference Composites Institute, paper 3-D. Washington, DC: Society of the Plastics Industry; 1966.

[8] Plueddemann EP, Stark GL. Silane primers for bonding to polyolefins. In: Proceedings of 35th Annual Conference Composites Institute, paper 20-B. Washington, DC: Society of Plastics Industry; 1980.

[9] Plueddemann EP. Dow Corning. Primer composition. US patent 4,231,910; November 4, 1980.

[10] Coulter DR, Cuddihy EF, Plueddemann EP. Chemical bonding technology for terrestrial photovoltaic modules, DOW/JPC-1022-91 Publication. Pasadena, CA: NASA Jet Propulsion Lab; 1983. pp. 83–6.

[11] Pape PG, Plueddemann EP. The use of mixed silane coupling agents. In: Proceedings of 40th Annual Conference Composites Institute, paper 17-F. Washington, DC: Society of the Plastics Industry; 1985.

[12] Pape PG, Plueddemann EP. Improvements in silane coupling agents for more durable bonding at the polymer-reinforcement interface. Eng Plas 1993;6:196–207.

[13] Plueddemann EP, Pape PG. Adhesion enhancing additives for silane coupling agents. In: Proceedings of 42nd Annual Conference Composites Institute, paper 21-E. Washington, DC: Society of Plastics Industry; 1987.

[14] Canova LA, Stafford SC, Ulrich AD. Influence of silane addition technique on composite preparation and properties. In: Proceedings of ANTEC, Society of Plastics Engineers, Los Angeles, CA, May 3–7, 1987:1250. Newtown, CT: Society of Plastics Engineers.

[15] Smith G. Evaluation of silane coupling agents in mineral-filled polypropylene. In: Proceedings of ANTEC, Society of Plastics Engineers, Dallas, TX, May 6–10, 1990:1946–1948. Newtown, CT: Society of Plastics Engineers.

[16] Hartlein RC. Dow Corning. Silane and chlorinated organic compounds. US patent 3,630,827; December 28, 1971.

[17] Speier JL. Dow corning. Preparation of alkyl polysulfides. US patent 4,125,552; November 14, 1978.

[18] Pletka HD, Michel R. Degussa. Process for the production of sulfur containing organosilicon compounds. US patent 4,072,701; February 7, 1978.

[19] Pickwell RJ, Osterholtz FD. Union carbide. Organofunctional silane–siloxane oligomer coupling compositions. US patent 4,526,922; July 2, 1985.

[20] Walker P. Organosilanes as adhesion promoters. In: Mittal K, editor. Silanes and other coupling agents. Utrecht, the Netherlands: VSP; 1992. pp. 21–47.

[21] Plueddemann EP. State of the art of silane coupling agents. In: Proceedings of American Society for Composites, First Technical Conference, Dayton, OH. Boca Raton, FL: CRC Press; 1985, pp. 264–79.

[22] van Ooij WJ, Subramanian V, Zhang C, Univ. of Cincinnati. Method of preventing corrosion of metals using silanes. US patent 6,261,638; July 17, 2001.

[23] Brown K, Bines EB, Song J, Tang N, Brent International PLC. Method of treating metals using amino silanes and multi-silyl-functional silanes in admixture. US patent 6,132,808; October 17, 2000.

[24] Dong Y, Bhattacharyya D, Hunter PJ. Experimental characterisation and object-oriented finite element modeling of polypropylene/organoclay nanocomposites. Compos Sci Technol 2008;68:2864–75.

[25] Dow Corning Corporation, Midland, MI. <www.dowcorning.com>; 2010.

[26] Evonik Degussa, Essen, Germany. <http://corporate.evonik.com/en/chemicals/products/Pages/default.aspx>.

[27] Momentive Performance Materials, North Greenbush, NY. <www.momentive.com>.

[28] Chisso Corporation, Tokyo, Japan. <http://www.chisso.co.jp/fine/en/>.

[29] Shin-Etsu Chemical, Tokyo, Japan. <www.silicone.jp/e/products/type/silanecup/index.shtml>.

[30] KCC Silicones, Seoul, Korea. <www.kccsilicone.com>.

[31] Nanjing Capatue, Nanjing, China. <http://www.capatue.com/english/index.asp>.

[32] Wuhan University Silicone New Material Co. Wuhan, China. <http://www.wdsilicone.cn/en/Info/index.asp?id = 6>.

[33] Kenrich Petrochemicals, Bayonne, NJ. <www.westnet.com/~kpiemm/>.

[34] Manchem/FedChem, Cleveland, OH. <www.manchem.com/>.

[35] Eastman Chemical Company, Kingsport, TN. <www.eastman.com>.

[36] DuPont, Wilmington, DE. <http://www2.dupont.com/Fusabond/en_US/>.

[37] Nippon Paper Chemicals, Tokyo, Japan. <www.npchem.co.jp/english/product/superchlon/>.

[38] DuPont, Wilmington, DE. <http://www2.dupont.com/Automotive/en_US/products_services/paintCoatings/plasticExterior.html>.

Index

Note: Page numbers followed by "*f*" and "*t*" refer to figures and tables, respectively.

D

E

F

G

R

S

FSC
www.fsc.org
MIX
Papier | Fördert gute Waldnutzung
FSC® C083411

Kolibri 360 GmbH
Ferdinand-Jühlke-Straße 7
99095 Erfurt, Deutschland
produktsicherheit@kolibri360.de